Environmental Science and Engineering
Subseries: Environmental Science

Series Editors: R. Allan • U. Förstner • W. Salomons

Jaime Klapp, Jorge L. Cervantes-Cota
José Federico Chávez Alcalá (Eds.)

Towards
a Cleaner Planet

Energy for the Future

With 116 Figures

 Springer

EDITORS:

DR. JAIME KLAPP
DR. JORGE L. CERVANTES-COTA
INSTITUTO NACIONAL DE INVESTIGACIONES
NUCLEARES (ININ)
DEPARTAMENTO DE FÍSICA
CARRETERA MÉXICO-TOLUCA KM 36.5
S/N
LA MARQUESA, OCOYOACAC, EDOMEX
52750 MÉXICO

E-mail: klapp@nuclear.inin.mx
       jorge@nuclear.inin.mx

DR. JOSÉ FEDERICO CHÁVEZ ALCALÁ
LABORATORIO DE METALURGIA
ESIQIE- INSTITUTO POLITÉCNICO
NACIONAL
APARTADO POSTAL 118-392
C.P. 07051
MÉXICO, D.F.

E-mail: jfchaveza@ipn.mx

Sponsoring Organizations: Deutscher Akademischer Austauschdienst (DAAD) Consejo Nacional de Ciencia y Tecnología (CONACYT) Instituto Nacional de Investigaciones Nucleares (ININ) Universitaet Karlsruhe (TH) Instituto Nacional de Astrofísica, Óptica y Electrónica (INAOE)

ISSN 1863-5520
ISBN 10   3-540-71344-1 **Springer Berlin Heidelberg New York**
ISBN 13   978-3-540-71344-9 **Springer Berlin Heidelberg** New York

Library of Congress Control Number: 2007923955

Springer is a part of Springer Science+Business Media
springeronline.com
© Springer-Verlag Berlin Heidelberg 2007

Cover design: deblik, Berlin
Production: A. Oelschläger
Typesetting: Camera-ready by the Editors
Printed on acid-free paper   30/2132/AO 543210

Preface

The world has entered a period of significant changes regarding the future of energy generation, mainly caused by the apparent exhaustion of hydrocarbons in the near future, and the Greenhouse gases (GHG) effect of altering the climate worldwide.

Mexico, as a developing country, and as the eleventh-most populated nation with the thirteenth-largest territory, and as the owner of important oil production resources, is a good example of a State needing to improve and increase its energy generation. A concerted action amongst the economic sectors involved – both governmental and industrial – could drastically improve the present situation regarding the development of its alternative energy resources.

The main motivation for organizing the German-Mexican Symposium 2006 *Energy for the future: towards a cleaner planet*, was to get a global perspective for changing the present energy-mix in Mexico, currently based on fossil fuels, towards cleaner energy resources. To achieve this, a wide range of relevant topics must be analysed, such as the state of the art of each energy type, their potential use and benefits, social, economic, political and environmental aspects, and their inclusion as a real alternative to energy generation programs. It will take time for every country to reduce its dependence on hydrocarbons and to increase its alternative energies share.

There are, and will long continue to be, opportunities to improve the energy efficiency of specific devices used in industrial applications. It has long been recognized that the very substantial gains in energy efficiency, result from changes in systems rather than from individual devices. The development of advanced controlled technology – and their application to industrial systems – has the potential to dramatically reduce industrial energy use.

The clean and alternative energy resources have potential benefits for Mexico and many other countries; additionally to their use for energy production, they are environmentally sound and can help to solve hard recurring local problems. For instance, urban solid waste can be used as fuel, thus reducing the problem of final disposal and reclaiming already deforested land for the production of energy crops. On the social dimension, renewables are often the only reasonable possibility for providing electricity-based services to remote communities, for improving quality of life and for launching productive projects for local economic development. In the urban and industrial sectors, renewables can constitute a "democratic force" to move away from centralised forms of energy supply: homeowners and entrepreneurs can generate their own electricity and hence financially contribute to the development of energy infrastructure. Green power technologies are well within the existing capabilities of the Mexican industry and represent a good opportunity for investors to participate in this new power technology market. This will mean new jobs, new ways of energy marketing and reactivation of stagnant industries. For instance, domestic use of photovoltaic systems, oil-producing plants as feedstock for biodiesel production, and a number of such possibilities, which could represent an important driving force for utilizing renewables in an economy in bad need of jobs.

This book points out the need of actions and programs concerning the reduction of local environmental pollutants and GHG emissions in the near and long term. The approach has been the understanding of the underlying principles and the discussion of what can be done to guarantee the availability of energy for the future.

In the first part of the book, the background of the present situation regarding energy generation in the world is analyzed. The fact that in the near future the energy supply will continue to rely widely on systems and processes involving thermal conversion or chemical fuels is also discussed. One of the major challenges researchers will face, concern with providing energy based on novel fuels and new technologies on a worldwide scale, against a background of vanishing resources, climatic changes and economic impact.

In a second part, several studies concerning traditional methods of producing electrical energy and the trends in the development of new technologies are presented. Energy efficiency, residuals produced, social impact and related implications are discussed. Energy efficiency could be improved through the incorporation of new technologies in a systematic approach, so that in the longer term, non-combustion technologies are likely to have a significant impact, such as fuel cells and gasification of biomass from in-plant residues. Both environmental concerns and efficiency improvement will lead the main energy consumption industries to reduce GHG emissions in the long term. The understanding of fundamentals in physics, chemistry, metallurgy, and biotechnology will allow the development of novel manufacturing processes; this knowledge, along with advanced modeling and simulation software, improved industrial materials, and measurements (sensors) and intelligent control systems, will result in major improvements and lead to fundamental breakthroughs.

The final part of the book involves contributions concerning alternative energies, its potential applications and how the future will be like. In the longer term, the very existence of a clean energy planet will depend critically on technologies that either do not presently exist in the marketplace or are in early stages of commercial trial. The commercial success of these technologies – which range from those producing energy with low or zero pollutant emissions (renewable, hydrogen, and nuclear power systems), to those that dramatically reduce energy use per activity or output – will make the difference between energy futures with high or low economic, social and environmental impacts. If the process of designing, constructing, starting up, controlling and maintaining intelligent building systems is done properly, the final product will deliver comfort, safety and a healthy environment, operating efficiently at reasonable cost. If part of this process breaks down, the product will fail to deliver these benefits.

The editors hope that the present book helps in going deeply in all the clean and alternative energy generation possibilities, and that a collaboration between Mexico and Germany can contribute to reach this goal.

The editors are very grateful to the Institutions that made possible the realization of the German-Mexican Symposium 2006 *Energy for the future: towards a cleaner planet*, and also this book, especially the German Academic Exchange Service (DAAD) and the University of Karlsruhe for generously supporting the organization, and the Instituto Nacional de Investigaciones Nucleares (ININ), for supporting and even involucrate its directive and staff members, such as MSc José Raúl Ortiz Magaña (General Director) and Dr. Luis Carlos Longoria Gándara (Research Director), in different stages of the organization. We also thank the support of the Consejo Nacional de Ciencia y Tecnología of México (CONACYT), the Comisión Federal de Electricidad (CFE), the Instituto Nacional de Astrofísica, Óptica y Electrónica (INAOE), and the Science Faculty of UNAM (FCUNAM).

We thank the whole organizing committee, in particular Manuel Corona, Angelica Gelover, José Manuel Huerta, Rosa María Salcedo and Luis Gil from the Mexican Alumni Society (SEMEXDAAD), for taking several tough responsibilities. Also, to the directives and personnel of the MUTEC Museum, for their invaluable help and their excellent installations for the Symposium, and to the Karlsruhe team, in particular Horst Hippler and Victor Martínez.

We acknowledge the help of the Edition Committee: Salvador Galindo Uribarri, Guillermo Covarrubias Maldonado, Guillermo Arreaga Garcia,

David Bahena Bustos, Guadalupe Rivera Loy (and her INAOE team), Maria Escribano Rodríguez, Itnuit Janovitz, Rosa Hilda Chavez, Jose Guillermo Cedeño, Gonzalo Mendoza Guerrero, and specially Ruslan Gabbasov for the format revision and correction of many of the articles.

Last but not least, we are also grateful to Arnold Spitta, Director of the DAAD Regional Office for Mexico, for his decisive support to the meeting and the book.

Mexico City,   *Jaime Klapp*
January 2007   *Jorge L. Cervantes-Cota*
                                                              José Federico Chávez Alcalá

Contents

List of contributors

Alonso, G., Dr.
Instituto Nacional de Investigaciones Nucleares, Apartado Postal 18-1027, México DF 11801, México. E-mail: galonso@nuclear.inin.mx

Araiza-Martínez, E., Eng.
Instituto Nacional de Investigaciones Nucleares, Apartado Postal 18-1027, México DF 11801, México. Email: earaiza @nuclear.inin.mx

Arancibia-Bulnes, C. A., Dr.
Departamento de Sistemas Energéticos y Departamento de Materiales Solares, Centro de Investigación en Energía, Universidad Nacional Autónoma de México, Privada Xochicalco s/n, Colonia Centro, A. P. 34, Temixco, Morelos 62580, México.

Balcázar-García, M., Dr.
Instituto Nacional de Investigaciones Nucleares, Apartado Postal 18-1027, México DF 11801, México. Email: mbg@nuclear.inin.mx

Bauer, M., Dr.
Instituto de Física, Universidad Nacional Autónoma de México, Ciudad Universitaria, 01000 México, DF, México. E-mail: bauer@fisica.unam.mx

Bazán-Perkins, S. D., Dr.
División de Estudios de Posgrado, Facultad de Ingeniería, Universidad Nacional Autónoma de México, Coyoacán 04510, México DF. E-mail: bazanperkins@hotmail.com

Best-Brown, R., Dr.
Centro de Investigación en Energía, Universidad Nacional Autónoma de México (UNAM), Privada Xochicalco s/n, 62580, Temixco, Morelos, México. E-mail: rbb@cie.unam.mx

Birkle, P., Dr.
Instituto de Investigaciones Eléctricas, Gerencia de Geotermia, Calle Reforma 113, Col. Palmira, Cuernavaca, Morelos, 62490 México. E-mail: birkle@iie.org.mx

Borja, M. A., Dr.
Instituto de Investigaciones Eléctricas, Calle Reforma No. 113, Col. Palmira, Cuernavaca, Morelos, C.P. 62490, México. E-mail: maborja@iie.org.mx

Castillo-Durán, R., M. Sc.
   Instituto Nacional de Investigaciones Nucleares, Apartado Postal 18-1027, México DF 11801, México. Email: rcd@nuclear.inin.mx
Cervantes-Cota, J. L., Dr.
   Instituto Nacional de Investigaciones Nucleares, Apartado Postal 18-1027, México DF 11801, México. E-mail: jorge@nuclear.inin.mx
Chávez, R.H., Dr.
   Instituto Nacional de Investigaciones Nucleares, Apartado Postal 18-1027, México DF 11801, México. Email: rhch@nuclear.inin.mx
de Buen-Rodríguez, O. D., Dr.
   Energía, Tecnología y Educación, Calle Puente Xoco 39, Col. Xoco. C.P. 03330 México DF. E-mail: demofilo@prodigy.net.mx
Dolores-Duran, M. D., Dr.
   Departamento de Ingeniería Mecánica, Facultad de Ingeniería, UAEM, Cerro de Coatepec s/n, Toluca, Estado de México, México. E-mail: mddg_2210@hotmail.com
Durán-de-Bazúa, C., Dr.
   UNAM, PECEC, Paseo de la Investigación Científica s/n, 04510 México DF, México. E-mail: mcduran@servidor.unam.mx
Eibenschutz, J., Eng.
   Comision Nacional de Seguridad Nuclear y Salvaguardias, Doctor Barragan 779, Colonia Narvarte, Mexico DF 03020, Mexico. E-mail: je@energia.gob.mx
Enríquez-Poy, M., Dr.
   Cámara Nacional de las Industrias Azucarera y Alcoholera, Calle Río Niágara 11, 06500 México DF, México.
Estrada, C. A., Dr.
   Departamento de Sistemas Energéticos y Departamento de Materiales Solares, Centro de Investigación en Energía, Universidad Nacional Autónoma de México, Privada Xochicalco s/n, Colonia Centro, A. P. 34, Temixco, Morelos 62580, México. E-mail: cestrada@cie.unam.mx
Fernandez, A., Dr.
   Departamento de Materiales Solares, Centro de Investigación en Energía. Universidad Nacional Autónoma de México, Temixco, Morelos, México. E-mail: afm@cie.unam.mx
Fernández-Valverde, S. M., Dr.
   Instituto Nacional de Investigaciones Nucleares, Apartado Postal 18-1027, México DF 11801, México. E-mail: smfv@nuclear.inin.mx
Flores-Ruiz, J. H., Dr.
   Instituto Nacional de Investigaciones Nucleares, Apartado Postal 18-1027, México DF 11801, México. Email: jhfr@nuclear.inin.mx
Gabbasov, R., Dr.

Instituto Nacional de Investigaciones Nucleares, Apartado Postal 18-1027, México DF 11801, México. E-mail: ruslan@nuclear.inin.mx

Galindo, S., Dr.
Instituto Nacional de Investigaciones Nucleares, Apartado Postal 18-1027, México DF 11801, México. E-mail: sgu@nuclear.inin.mx

García-Ramírez, E., M. Sc.
Instituto Nacional de Investigaciones Nucleares, Apartado Postal 18-1027, México DF 11801, México. E-mail: jenrique@nuclear.inin.mx

Guadarrama, J. J., M. A.
Instituto Tecnológico de Toluca, Av. Tecnológico S/N, Metepec, 52140 México. Email: jguad4@aol.com

Hernández-López, H., M. Sc.
Instituto Nacional de Investigaciones Nucleares, Apartado Postal 18-1027, México DF 11801, México. Email: hhl@nuclear.inin.mx

Herrera-Velázquez, J. E., Dr.
Instituto de Ciencias Nucleares, Universidad Nacional Autónoma de México, A.P. 70-543, Ciudad Universitaria, Del. Coyoacán, 04511 México DF, México. E-mail: herrera@nucleares.unam.mx

Hippler, H., Prof. Dr.
Universität Karlsruhe (TH), Kaiserstraße 12 - 76131 Karlsruhe. E-mail: Rektor@verwaltung.uni-karlsruhe.de

Huacuz, J. M., Dr.
Non-Conventional Energy Unit, Electrical Research Institute (IIE), Av. Reforma 113, Cuernavaca, México, C.P. 62490, E-mail: jhuacuz@iie.org.mx

Jaramillo, O. A., Dr.
Departamento de Sistemas Energéticos y Departamento de Materiales Solares, Centro de Investigación en Energía, Universidad Nacional Autónoma de México, Privada Xochicalco s/n, Colonia Centro, A. P. 34, Temixco, Morelos 62580, México. ojs@cie.unam.mx

Jiménez-Gonzalez, A., Dr.
Departamento de Materiales Solares, Centro de Investigación en Energía. Universidad Nacional Autónoma de México, Temixco, Morelos, México. E-mail: ajg@cie.unam.mx

Klapp, J., Dr.
Instituto Nacional de Investigaciones Nucleares, Apartado Postal 18-1027, México DF 11801, México. E-mail: klapp@nuclear.inin.mx

Longoria-Gandara, L. C., Dr.
Instituto Nacional de Investigaciones Nucleares, Apartado Postal 18-1027, México DF 11801, México. E-mail: longoria@nuclear.inin.mx

XXII

López, A., M. Sc.
Instituto Nacional de Investigaciones Nucleares, Apartado Postal 18-1027, México DF 11801, México. alopezg@nuclear.inin.mx

Manahan, S. E., Dr.
ChemChar Research, Inc. 123 Chem. Bldg. Univ. of Missouri-Columbia, Columbia, Missouri 65211, USA.

Mathew, J., Dr.
Departamento de Materiales Solares, Centro de Investigación en Energía. Universidad Nacional Autónoma de México, Temixco, Morelos, México.

Ortiz-Magaña, J. R., M. Sc.
Instituto Nacional de Investigaciones Nucleares, Apartado Postal 18-1027, México DF 11801, México. E-mail: rortizm@nuclear.inin.mx

Ortiz-Villafuerte, J., Dr.
Instituto Nacional de Investigaciones Nucleares, Apartado Postal 18-1027, México DF 11801, México. Email: jov@nuclear.inin.mx

Palacios, J. C., Dr.
Instituto Nacional de Investigaciones Nucleares, Apartado Postal 18-1027, México DF 11801, México. Email: palacios@nuclear.inin.mx

Peña, P., M. E.
Instituto Nacional de Investigaciones Nucleares, Apartado Postal 18-1027, México DF 11801, México. Email: ppg@nuclear.inin.mx

Ramirez, J. R., M. E.
Instituto Nacional de Investigaciones Nucleares, Apartado Postal 18-1027, México DF 11801, México. Email: jrrs@nuclear.inin.mx

Rojas-Avellaneda, D., Dr.
Centro de Investigación en Geografía y Geomatica "Ing. Jorge L. Tamayo". Contoy No. 137, Lomas de Padierna, Tlalpan, México 14740 DF, México. E-mail: dariorojas@centrogeo.org.mx

Sanchez-Juarez, A., Dr.
Departamento de Materiales Solares, Centro de Investigación en Energía. Universidad Nacional Autónoma de México, Temixco, Morelos, México. E-mail: asj@cie.unam.mx

Sebastián, P. J., Dr.
Departamento de Materiales Solares, Centro de Investigación en Energía. Universidad Nacional Autónoma de México, Temixco, Morelos, México. E-mail: sip@cie.unam.mx

Tan-Molina, L., Dr.
Molina Center for Energy and the Environment, 3262 Holiday Court, Suite 201, La Jolla, California, 92037, USA.

Verfondern, K., Dr.

XXIII

Research Center Juelich. Institute for Safety Research and Reactor Technology (ISR), D-52425, Juelich, Germany. E-mail: k.verfondern@fz-juelich.de

Part I

General Overview and Energy Efficiency

Energy for the Present and Future: A World Energy Overview

Jaime Klapp, Jorge L Cervantes-Cota, Luis C Longoria-Gandara, Ruslan Gabbasov

Instituto Nacional de Investigaciones Nucleares, Apartado Postal 18-1027, México DF 11801, México. E-mail: klapp@nuclear.inin.mx

Abstract

For many years coal and oil have been used as energy sources. Currently oil is the dominant source of energy, but experts predict that in a few decades it will no longer be profitable. Burning fossil fuels generates atmospheric contaminants that give rise to the greenhouse effect that artificially warms the earth, damages the earth ozone layer, and produces acid rain, all of which are very dangerous for living beings. As a consequence, abnormal phenomena such as the melting of glaciers, changes in the Gulf Stream, unprecedented heat waves, floods, hurricanes, and damage to marine organisms are now occurring. Although there are many skeptics, there is a general consensus that the earth is warming up. In order to reduce CO_2 and other greenhouse gas emissions, the extensive use of alternative and cleaner sources of energy have been proposed, including CO_2 emission-free nuclear energy and other alternative sources, among which are hydraulic, hydrogen, solar, eolic, biomass and geothermal. Another future alternative is to use methane hydrates, which have clean combustion and are believed to have considerable, still-unexplored reserves. It is also important to implement measures to improve energy efficiency for reducing greenhouse gas emissions. In this paper we review present and future energy sources, including both primary non-renewable and alternative sources of energy. There is still time to take corrective measures by replacing some polluting fuels with clean sources of energy that can contribute to inherit a clean and sustainable world for future generations.

1 Introduction

We are living worldwide challenging times in regard to energy consumption and production, when industrialized nations consume lots of energy and poorer countries are facing an energy lack or misadministration that would be needed to boost their development. Energy is one of the main driven issues for local and world economies.

The world consumes as much as two-and-a-half the energy consumed in the sixties. Quantitatively, the energy consumption in 1965 was nearly 4 billion tons of oil equivalent (toe= 10 Giga cal$_{th}$= 41.868 Giga Joule) and in 2005 was of about 10.5 billions of toe. For the year 2020 it is expected an additional demand of about one third of the energy demanded at present. If consumption continues as foretold in about 40 years, the oil resources will run out; gas will last only for 60 years and coal for 250 years. Also, in a few decades no Uranium will be left. Eventually, we will have to face the lack of these fuels for future energy needs. Of course, other alternatives will come into play. The question arises on what types of fuels are convenient for the development and wellness of our societies. This is to be considered in the present book.

The energy consumption varies from wood to the high technical nuclear energy, passing through intermediate technologies. Since not all countries have the same resources and technologies, not all of them achieve the energy production that they would need for their development. In fact, as much as one third of the world population has problems to get access to energy consumption. This inequality may in turn imply local and/or global instabilities, including war scenarios, escalating geopolitical tensions in some parts of the world and increased speculation in the Futures market. More industrialized countries, belonging to the Organization for Economic Cooperation and Development (OECD), have less energy needs than Non-OECD countries. In the recent International Energy Outlook (IEO 2006), made by the American Energy Information Administration (EIA), future projections indicate continued growth in world energy consumption. Countries belonging to the OECD are expected to increase their energy demand by 1 % in average within the period 2003–2030, whereas non-OECD countries will demand 3 %; this figure is based on assumptions about economic and population growth, as well as in energy intensity developments. In the later group, particularly important are China and India. These two nations are expected to leader the energy demand of the non-OECD Asia region, which will need an energy growth of about 3.7 % in average per year within the projected period, see Fig. 1.

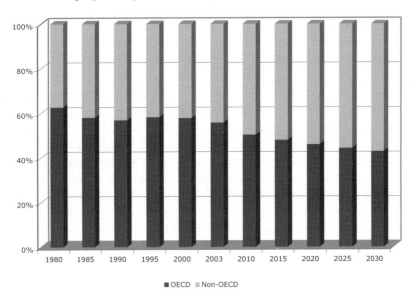

■ OECD ■ Non-OECD

Fig. 1. World energy consumption by region, 1980–2030

The energy production in 2004 was obtained from the following energy sources: 38 % petroleum (crude oil and natural gas plant liquids), 26 % coal, 23 % dry natural gas, hydro and nuclear 6 %, and others 1 %, see Fig. 2 (data from http://www.eia.doe.gov/iea/).

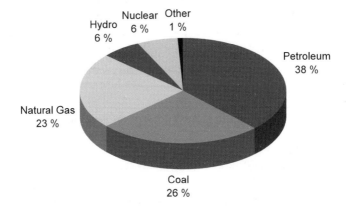

Fig. 2. World production of primary energy by energy type in 2004

Fossil fuels (oil, coal, and natural gas) continue to supply much of the energy use worldwide throughout the projections. Oil remains the dominant energy source, but its share of total world energy consumption declines from 38 % in 2003 to 33 % in 2030. Worldwide, transportation and industry are the major growth sectors for oil demand. On a global basis, the transportation sector – where there are currently no alternative fuels that compete widely with oil – accounts for about one-half of the total projected increase in oil use in the next two-and-a-half decades, with the industrial sector accounting for another 39 % of the incremental demand. Natural gas demand grows faster, driven mainly by power generation. It might overtake coal as the world's second-largest primary energy source by the year 2015. In this scenario, the share of coal in world primary demand slightly declines, with demand growth concentrated in China and India. Nuclear power's market share declines marginally, while that of hydropower remains broadly constant. On the other hand, unconventional resources (including biofuels, coal-to-liquids, gas-to-liquids, and solar, wind, tidal and wave energy) are expected to become more competitive. In 2003, world production of unconventional resources totaled only 1.8 million barrels per day, about 2.3 % of total world petroleum supply; in the IEO (2006) reference case, unconventional resource supplies rise to 11.5 million barrels per day and account for nearly 10 % of total world petroleum supply in 2030, see Fig. 3.

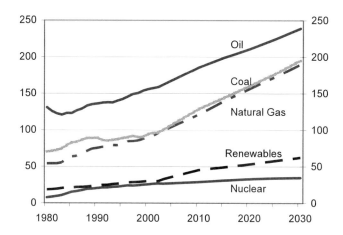

Fig. 3. Projection of world energy use by energy type (data in quadrillon Btu)

Although energy resources are thought to be adequate to support the growth expected in the next few decades, the massive consumption of fossil fuels inevitably provokes the global warming of our planet. Residual products of fossil fuels such as carbon dioxide (CO_2), methane (CH_4), or nitrous oxide (N_2O) –and other such as hydrofluorocarbons (HFCs), perfluorocarbons (PFCs), sulphur hexafluoride (SF_6) – act as "greenhouse gases". These gases allow sunrays to enter the atmosphere freely. When sunlight strikes the Earth's surface, some of it is reflected back towards space as infrared radiation, i.e., heat. Greenhouse gases absorb this infrared radiation and trap the heat in the atmosphere. Over time, the amount of energy sent from the sun to the Earth's surface should be about the same as the amount of energy radiated back into space, leaving the temperature of the Earth's surface roughly constant. An excessive amount of greenhouse gases provokes the over warming of the planet, changing its natural cycles, see illustration 1.

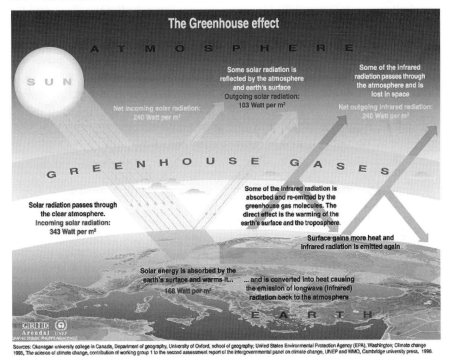

Illustration 1. The greenhouse effect, taken from the web page of the United Nations Framework Convention on Climate Change (UNFCCC): http://unfccc.int/essential_background/feeling_the_heat/items/3157.php

This has been happening during the last decades. Because of this reason nations of the world met in Kyoto, Japan, in 1997 to sign an intentional document called the "Kyoto protocol" (this protocol can be found is http://unfccc.int/resource/docs/convkp/kpeng.html). In this document it is stated that signed nations should achieve a quantified emission limitation and reduction of greenhouse gases, in order to promote sustainable development, in light of climate change considerations. However, some important energy consumers of the world have not signed the protocol or pulled out of it later on. For example, the United States (US) uses fossil fuels that currently provide more than 85 % of all the energy consumed in this country, nearly two-thirds of their electricity, and virtually all of their transportation fuels. The US holds less than 5 % of the world's population but produces nearly 25 % of carbon emissions. Moreover, it is likely that the nation's reliance on fossil fuels to power an expanding economy will actually increase over at least the next two decades even with aggressive development and deployment of new renewable and nuclear technologies, see web page of the US Department of Energy (DOE) for details: http://www.energy.gov/energysources/fossilfuels.htm.

Table 1 shows the world carbon dioxide emissions by regions from 1990 to 2003 and the expected for 2003–2030.

Table 1. World carbon dioxide emissions by region, 1990–2030 (data in million of metric tons, taken from http://www.eia.doe.gov/iea/)

	History		Projections		Average annual per-cent change	Average annual per-cent change
Region	1990	2003	2010	2030	1990–2003	2003–2030
OECD	11,378	13,150	14,249	17,496	1.1	1.1
North America	5,753	6,797	7,505	9,735	1.3	1.3
Europe	4,089	4,264	4,474	5,123	0.3	0.7
Asia	1,536	2,090	2,269	2,638	2.4	0.9
Non-OECD	9,846	11,878	16,113	26,180	1.5	3.0
Europe and Eurasia	4,193	2,725	3,113	4,352	-3.3	1.7
Asia	3,626	6,072	9,079	15,984	4.0	3.6
Middle East	704	1,182	1,463	2,177	4.1	2.3
Africa	649	893	1,188	1,733	2.5	2.5
Central and South America	673	1,006	1,270	1,933	3.1	2.4
Total World	21,223	25,028	30,362	43,676	1.3	2.1

In Fig. 4 we present the world carbon dioxide emissions by fuel type. Fossil fuels account for 80 % of the world's energy usage, with the respective greenhouse emissions. After the Kyoto protocol, the Rio and Toronto protocols were signed with stronger CO_2 emission requirements. In Fig. 5 energy related CO_2 emissions for some countries and groups of countries are shown (EIA 2006) from which it is clear that for the countries in the figure, only few countries are controlling their CO_2 emissions.

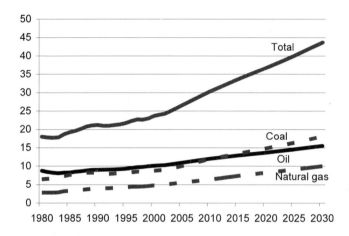

Fig. 4. World carbon dioxide emissions by fuel type, 1980–2030 (data in billion metric tons)

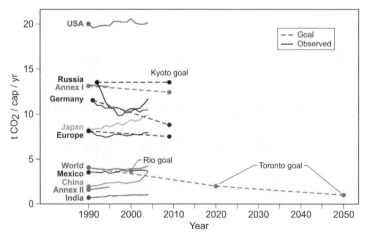

Fig. 5. Energy related CO_2 emissions per capita for some countries and groups of countries from the year 1990 (EIA 2006). The requirements of the Kyoto, Rio and Toronto protocols up to year 2050 are shown

Climate change compels a global restructuring of the world's energy economy. Worries over fossil fuels supplies reach crisis proportions only when safeguarding the climate is taken into account. It seems that this is happening now and, in consequence, we have to take the appropriate actions promptly. The year 2006 registered two events that shook the scientific community and the world. As a result of the global warming, presumably caused by greenhouse gases, great ice blocks came off the Arctic Ocean and Greenland, threatening the ecological balance and navigation in the North Atlantic Ocean. These are the types of events that we are expecting to happen by artificially warming our planet. Nations of the world must deal with this problem before it is too late. In fact, current levels of carbon dioxide are at its highest level ever –in Homo sapiens' modern times–. Because these gases have long –decades to centuries– atmospheric lifetimes, the result is an accumulation. In this way, carbon dioxide has been gradually gathering in the atmosphere and increased 31 % since preindustrial times (mid-19th century), from 280 parts per million by volume (ppmv) to more than 370 ppmv today, and half of the increase has been since 1965. Although natural events such as volcanic eruptions and changes in the sun's irradiance affect the earth's temperature, such influences are minor. By far the greatest changes are anthropogenic, resulting from the release of gases that exacerbate the greenhouse effect. This has yielded a warning in the atmosphere, as it is shown in Fig. 6 (Karl and Trenberth 2003), where it can be seen how the global average surface heating approximates that of carbon dioxide increases. In the past 25 years the global surface temperature has increased by 0.4°C. This may seem not to be very much, but it is equivalent to the temperature increase for the preceding 100 years. Each 1°C increase allows the atmosphere to hold 6 % more water; most experts are convinced that the far-reaching implications of this are already evident in floods and droughts, widening geographical distribution of vectors of infectious disease, increased air pollution and rates of respiratory illness, and so on (Klotz 2004). Taking into account the expected emissions, see Table 1, estimates of additional temperature increase by the end of this century range from 1.7°C to 4.9°C (Karl and Trenberth 2003).

At the expected pace of greenhouse emission production, the CO_2 will surely surpass 500 ppmv in the atmosphere in the next decades, unless qualitatively changes are made in the present way to consume energy, i.e., changing the current share of Fig. 2 to a cleaner energy consumption. Climatologists think that above 500 ppmv CO_2 in the atmosphere, there will be no return to catastrophic consequences on the earth.

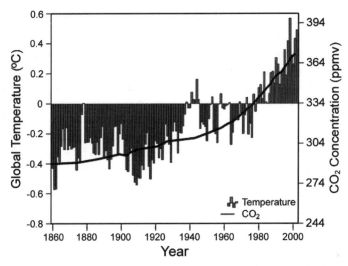

Fig. 6. Earth's temperature relative change due to global CO_2 emissions. Figure taken from Karl and Trenberth (2003)

There exist proposals by which greenhouse emissions can be gradually diminished. For instance, professors Pacala and Socolow (2004) put forward two 50-years futures for our planet. One in which the emissions rate continues to grow, from the present 7 billions tons of carbon a year to 14 billions tons in 2054, that will then produce above 560 ppmv of carbon dioxide in the atmosphere, with unknown consequences for our planet. In the other future, emissions are stabilized in the present rate value for the next 50 years and then reduced by about half over the following 50 years. However, to hold global emissions constant without stopping the world economy is a daunting task. In the last decades the gross world product of goods and services have grown at about 3 % a year on average, and carbon emissions rose half as fast. In spite of this, the authors argued that with present technology one can substitute carbon emissions by, for instance, massive use of alternative energy sources, such as ethanol and hydrogen as fuel for cars; and eolic, solar, and nuclear to displace coal. Being realistic, they propose to continue with some coal-fired plants, but with a higher efficiency and replace some of them by gas plants. They also propose to capture carbon dioxide and storage it in deep saline formations under the earth, among other portfolio of technologies that now exists to meet the world's energy needs over the next 50 years and to limit atmospheric CO_2 to a trajectory that avoids a doubling of the preindustrial concentration. This and other proposals should be discussed by scientists, industrials, governmental representatives, and society in general.

2 Traditional energy resources

We now review the present status and future of the main energy sources used worldwide. The projections are mainly based on IEO (2006).

2.1 Oil

Economic growth is among the most important factors to be considered in projecting changes in the world's energy consumption. This can be measured in terms of Gross Domestic Product (GDP) to underlie the projections of regional energy demand. During the past years, the GDP has increased in average 2–3 % yearly in OECD countries and 4–7 % in non-OECD countries. The projected growth for OECD countries is of about 2.6 % in the next twenty five years and of about 5 % for non-OECD countries. Thus, we expect a world oil demand in consistency with this growth rates. In Fig. 7 is shown the world oil consumption by region in 2003 and 2030, where it can be seen that North America –mainly the US– and Non-OECD Asia –mainly China and India– will be the most oil demanders of the world.

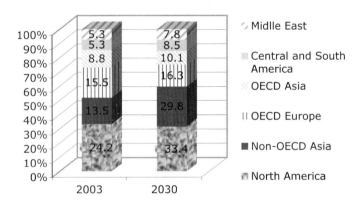

Fig. 7. World oil consumption by region in 2003 and 2030 (data in million barrels per day)

On the other hand, much of the world's incremental oil demand is projected for use in the transportation sector –as it is today the case–, where there are almost no competitive alternatives to petroleum; however, several of the technologies associated with unconventional liquids (gas-to-liquids, coal-to-liquids, and ethanol and biodiesel produced from energy crops) are expected to meet a growing share of demand for petroleum liquids during the projection period, see Fig. 8. Of the projected increase in oil use in the reference case over the 2003 to 2030 period, one-half occurs in the transportation sector. The industrial sector accounts for a 39 % share of the projected increase in world oil consumption, mostly for chemical and petrochemical processes.

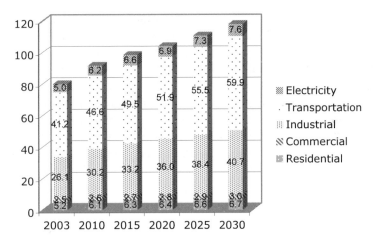

Fig. 8. World oil consumption by sector 2003–2030 (data in million barrels per day)

Global oil reserves today exceed the cumulative projected production between now and 2030, but more reserves will need to be proved up in order to avoid a peak in production before the end of the projection period. The exact cost of finding and exploiting energy resources over the coming decades is uncertain, but will certainly be substantial. Cumulative energy-sector investment needs are estimated at about USD 16 trillion (in year-2004 dollars) over 2004–2030, about half in developing countries. The bulk of this, almost USD 10 trillion, will have to be spent on the electricity industry, more than USD 2 trillion of the total in China. Financing the required investments in non-OECD countries is one of the biggest challenges facing the energy industry. Also, Germany must replace half its total power plant.

2.2 Coal

In the IEO (2006) reference case, world coal consumption nearly doubles from 5.4 billion short tons in 2003 to 10.6 billion short tons in 2030. From this amount, non-OECD countries accounts for 81 % of the increase. This is led by the strong economic growth and rising demand for energy in China and India; together they shall demand 3.6 billion tons, 3.9 % per year from 2003 to 2030, representing 86 % of the increase for the non-OECD region. Coal consumption in the other non-OECD countries grows by an average of 1.7 % per year, expanding by 0.6 billion short tons from 2003 to 2030, see Fig. 9.

Of the coal produced worldwide in 2003, 67 % was shipped to electricity producers, 30 % to industrial consumers, and most of the remaining 3 % to coal consumers in the residential and commercial sectors, see Fig. 10. Coal's share of total world energy consumption increases from 24 % in 2003 to 27 % in 2030, as can be seen in Fig. 11. It is also shown the coal share of world energy consumption by sector. The share by sector will remain roughly constant.

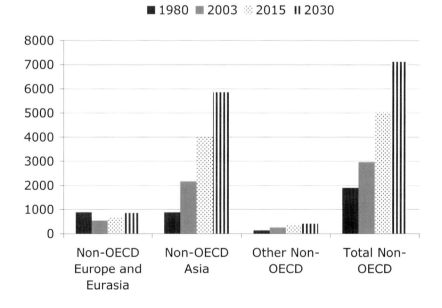

Fig. 9. Non-OECD regions coal consumption, 1980–2030 (data in million short tons)

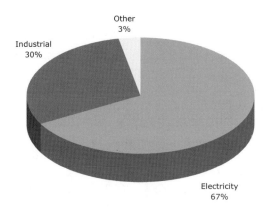

Fig. 10. World coal consumption by sector in 2003

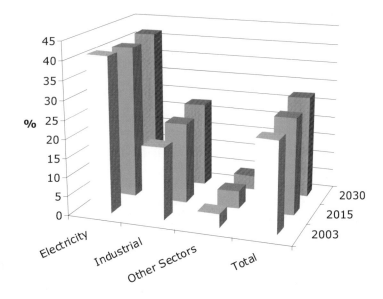

Fig. 11. Coal share of world energy consumption by sector, 2003–2030

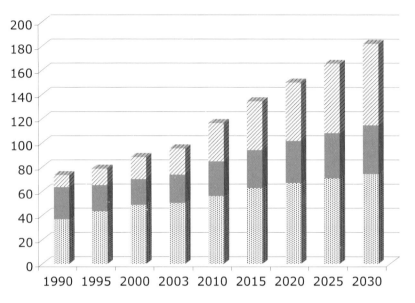

Fig. 12. World natural gas consumption by region, 1990–2030 (data given in trillion cubic feet)

2.3 Natural Gas

Natural gas is behind coal as the fastest growing primary energy source in the next two-and –a-half decades. The natural gas share of total world energy consumption increases from 24 % in 2003 to an expected 26 % in 2030.Consumption of natural gas worldwide is expected to increase from 95 trillion cubic feet in 2003 to 182 trillion cubic feet in 2030. Although natural gas is expected to be an important fuel source in the electric power and industrial sectors, the annual growth rate for natural gas consumption in the projections is slightly lower than the growth rate for coal consumption, see Fig. 3. This is because higher world oil prices increase the demand for natural gas, but its price increases too, making coal a more economical fuel source. Nevertheless, natural gas remains a more environmentally attractive energy source and burns more efficiently than coal; therefore, it still is expected to be the fuel of choice in many regions of the world. Figure 12 shows the world natural gas consumption by region for the years 1990 to 2030. Especially, non-OECD countries shall increase their demand more than linearly.

On the other hand, the industrial and electric power sectors are the largest consumers of natural gas. The industrial sector is the most demanding, and future projections indicate that this will continue so. Natural gas use grows by 2.8 % per year in the industrial sector and 2.9 % in the electric power sector from 2003 to 2030, see Fig. 13. In both sectors, the share of total energy demand met by natural gas grows over the projection period. In the industrial sector, natural gas overtakes oil as the dominant fuel by 2030. In the electric power sector, however, despite its rapid growth, natural gas remains a distant second to coal in terms of share of total energy use for electricity generation.

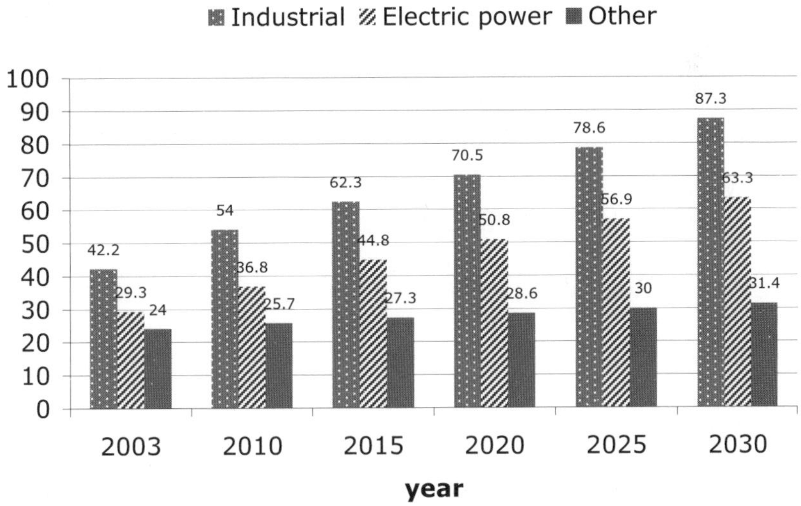

Fig. 13. World natural gas consumption by end-use sector, 2003–2030 (units in trillion cubic feet)

2.4 Nuclear

Nuclear generation began 50 years ago and now generates as much global electricity as was produced then by all sources. As of December 2005, there were 443 nuclear power reactors which produced electricity around the world. More than 15 countries rely on nuclear power for 25 % or more of their electricity. In some European countries and Japan, the nuclear share of electricity is over 30 %, see Fig. 14.

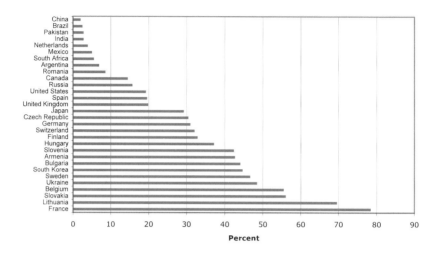

Fig. 14. Nuclear shares of national electricity generation as of 2005

The world's nuclear-powered generating capacity increases in the IEO (2006) reference case from 361 GW in 2003 to 438 GW in 2030. Few new builds are expected in the OECD economies outside of Finland, France, Japan, South Korea, and the US In this later country, nuclear capacity is expected to increase by 3 GW as a result of upgrades at existing plants and by 6 GW as a result of new construction. On the other hand, rapid growth in nuclear power capacity is projected for the non-OECD economies that are expected to add 33 GW of nuclear capacity between 2003 and 2015 and another 42 GW between 2015 and 2030. The largest additions are expected in China, India, and Russia.

Around the world, scientists in more than 50 countries use nearly 300 research reactors to investigate nuclear technologies and to produce radioisotopes for medical diagnosis and cancer therapy. Meanwhile, on the world's oceans, nuclear reactors have powered over 400 ships without harm to crews or the environment.

The big issue of nuclear energy is its waste, that lives long and it causes non acceptance in the societies of the world. Technically, there are safe ways to keep the waste in proper repositories under the earth, but this topic will be treated in another contribution to this book.

As far we have only discussed man-made fission reactors. For a very interesting *natural fission reactor* that operated long time ago on earth, see the paper of Miguel Balcazar in this volume.

A promising alternative although still far into the future is nuclear fusion that for over half a century has promised a clean and safe energy generation alternative. It is well known that the sun is supported from the nuclear fusion of hydrogen into helium in the central and hottest regions of the star, but down here on earth, after several decades of hard work and various devices developed in several countries it has not been possible to get more energy than is inverted. But the failures of the past may come to an end. The ministers of seven countries or groups or countries, namely China, the European Union, India, Japan, the Russian Federation, South Korea, and the United States of America, have signed on November 12, 2006 the International Thermonuclear Experimental Reactor agreement (ITER, www.iter.org) for the development of a fusion device that hopefully will produce more energy than in inverted. We hope that the ITER device will be successful and that fusion will become a reality as a clean energy source (see also the contribution of Julio Herrera, this volume).

3 Alternative energy resources

The world's growing energy demand presents a formidable challenge for the future. Apart from environmental considerations, the fact is that sooner or later the non-renewable sources of energy will run out, although this might be further into the future than is presently estimated. However, as Sheik Yamani pointed out, the Stone Age did not come to an end because of lack of stones. In the same way, the fossil fuels era will come to an end because of economic and environmental constraints as more competitive and cleaner sources becomes available.

The potential of renewable energies are shown in Fig. 15, from which it is clear that the solar energy potential is enormous, followed by wind, biomass, geothermal and hydropower and it is interesting to note that each of these renewable sources of energy have a potential larger than the world's energy need. Hydrogen is also a clean and essentially unlimited source of energy, although some of the mechanisms of hydrogen production involve the use of non-renewable sources of energy. The various governments and organizations of the world have to take decisions now in order to expand the use of clean and alternative energy sources. Some countries have started to introduced the required legislation, for example, the 25 EU countries have a goal for the renewable energies for the year 2010 of 25 % in electricity and 12 % for the total energy consumption, while for the year 2020 the goals of Spain and China are 20 % and 15 %, respectively. China aims to increase the renewable energy generation to 60,000 MWe. In the US the budget increased significantly: biomass +65 %, solar +79 % and eolic +13 %.

The use of hydrogen, solar, geothermal, biomass, wind, as well as other renewable energy sources, will provide the diversification of energy resources that will help the world to reduce the dependence on fossil fuels and decrease CO_2 and other greenhouse gas emissions. In Europe and specifically in Germany important decision has been taken that have reduced greenhouse gas emissions. The average increase in the percentage of installed capacity/consumption has been increased during the 1993–2003 period; the highest rates are for wind and PV with 29.7 % and 21.6 %, respectively, with much lower values for natural Gas (2.2 %), Coal (1.7 %), Oil (1.5 %) and Nuclear (0.6 %). In Fig. 16 we present a summary of the renewable power capacities in 2005 (in GW) for developing countries, EU and top six individual or groups of countries.

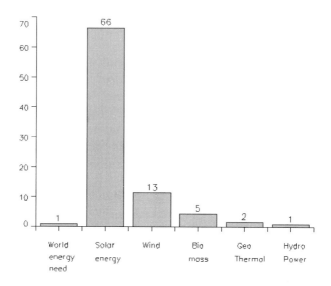

Fig. 15. The potential of renewable energies (data taken from Wuppertal Institut/DLR, Enquete commission of Deutscher Bundestag and Bucher 2006). The energy is in units of 140,000 bin kWh=140,000 TWh=5×10^{20} J

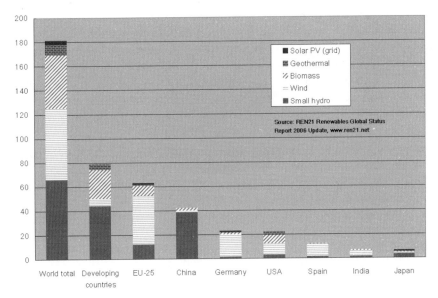

Fig. 16. Renewable Power Capacities in 2005 (GW). Figure taken from REN21 - Renewables Global Status Report 2006: http://www.ren21.net

It is interesting that a large proportion comes from developing countries where small hydro and biomass have a large proportion followed by wind and solar PV. In EU, Germany, US, Spain and India wind have the large proportion while in India small hydro takes the larger proportion. Despite the enormous potential of renewable energy sources, new technologies have to be developed that will allow its large scale exploitation. We now give a summary of the main renewable energy sources.

3.1 Hydrogen

Hydrogen and fuel cells have the potential to solve several major challenges facing most countries, e.g., dependence on energy (petroleum) imports, poor air quality, and greenhouse gas emissions. But this is expected to happen in a few decades, not today. Approximately 600 billion Nm^3 of hydrogen are produced every year through well established commercial processes for the hydro-cracking of oil, the production of ammonia and hydrogenation of edible fats, 96 % of which is captive (consumed on-site). Merchant users consume hydrogen at sites other than where it is produced, which accounted for 4 % in 1999, see Fig. 17.

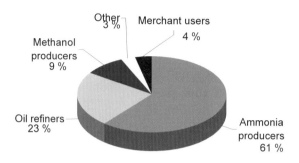

Fig. 17. World hydrogen consumption by end-use category, 1999. Data taken from Center for transportation analysis, Oak Ridge National laboratory: http://cta.ornl.gov/data/chapter6.shtml.

Significant progress is needed for hydrogen to become a widely available "consumer fuel" which meets energetic, environmental and production cost constraints, that is, to develop a "hydrogen economy". A key attraction of hydrogen as an energy vector is that it can be produced using diverse, domestic resources including fossil fuels, such as natural gas and coal (with carbon sequestration); nuclear; and biomass and other renewable energy technologies, such as wind, solar, geothermal, and hydroelectric power. The overall challenge to hydrogen production is cost reduction. For transportation, a key driver for energy independence, hydrogen must be cost-competitive with conventional fuels and technologies on a per-mile basis in order to succeed in the commercial marketplace.

Fuel cells offer a significant advantage over traditional combustion-based thermal energy conversion, in that they provide efficiencies of electrical power supply in the range of 35 to 55 %, while causing very low levels of pollutant emission. Fuel cells can in principle be built in a wide range of power ratings, from a few mW to several MW, and can be used in a wide variety of applications, from miniaturized portable power (effectively substituting the battery in portable electronic devices) through transport (as a zero-emission propulsion system) to power generation in a variety of sizes (from domestic combined heat and power systems, through to full size power stations and quad- generation). They offer advantages of weight compared with batteries, and instantaneous refueling, similar to combustion engines. However, fuel cells today have to further evolve from laboratory prototypes into rugged, robust units that can cope with mechanical as well as electrochemical 'stress' and be operated at will with as little restrictions as possible. No matter what fuel cell type is considered, the issue of ageing and ruggedness in everyday operation is of major concern. If cost reduction is one main driver in bringing fuel cells to the market, long lifetime and robust and reliable operation are the issues to be addressed to ensure quality and suitability (see European Full Cell and Hydrogen projects, report of the European Commission, Eur 22398, http://ec.europa.eu/research/energy.

The production of hydrogen today is mainly performed by steam reforming, partial oxidation of gaseous or liquid fuels or the gasification of coal. But the hydrogen economy is still under development. It is widely viewed that, during the transition period, hydrogen will predominantly be produced from fossil fuels. Therefore, the development of CO_2 capture and storage (sequestration) technologies is very important for the reduction of CO_2 and other greenhouse gas emissions. These technologies are critical to the process of removing CO_2 from the gas stream during the production of hydrogen from fossil fuel-based gasification. Since the technologies require more research before they can be commercialized, many countries are aiming to reduce the emissions through efficiency improvements in the short run. Capture and storage technologies however will play a critical role in emissions reduction in the long run. In 2050, 13 % of electricity generation shall stem from plants equipped with carbon capture and storage, which will be really developed only after 2040 (see World Energy Technology Outlook-2050 (WETO) H_2. Report of the European Commission, Eur 22038: http://ec.europa.eu/research/energy/gp/gp_pu/article _1100_en.htm).

Research in the next decades is likely to focus on developing and improving CO_2 capture and storage technologies – reducing costs, increasing capture rates and evaluating the different storage options. Beyond 2030, research will focus on the transition towards the hydrogen economy and the development of zero-emission power plants, see State and prospects of European Energy research, report of the European Commission, Eur 22397, http://ec.europa.eu/research/energy.

3.2 Solar

Energy production from sunlight is based on solar thermal energy systems and on the so called solar cells, also known as photovoltaic (PV) cells that use semiconductor materials to generate electric energy. After the 1973/74 winter energy crisis, solar energy systems became commercially available.

By that time in some countries like Israel and Cyprus, solar cells were used to produce warm water. At the present time solar energy only represents a very small proportion of the total world's energy consumption. It has been estimated that from solar energy our planet could generate only about 0.15 % of what can be generated from all sources of energy. The world's total solar energy generation for the year 2005 was 1,727 MW, with 833 MW in Japan, 353 in Germany, 153 in the USA and very little in the rest of the world. The cost of photovoltaic cells varies strongly depending upon the materials used to build them; more expensive materials give higher efficiencies. The more expensive crystalline cells give laboratory efficiencies of 30 % and even more, while the commercially available photovoltaic cells give 15–20 %. The first solar cell was built at Bell Laboratories by Chapin, Fuller and Pearson in 1954 and had an efficiency of 5.4 %, so important improvements have been achieved. In recent years, the US, Germany and Japan are promoting the use of solar energy. In California the "Million Solar Roof" project aim to produce 3,000 MW by the year 2018 and with similar projects the US is proposing to increase in about 20 years its solar energy production to about 10,000 MW.

Despite the enormous potential of solar energy as indicated by Fig. 15, its main problem is that the cost of photovoltaic cells is still very high. The current price of generating electricity with photovoltaic crystalline cells is 20 to 25 USD cents per kilowatt/hour, while the kilowatt/hour cost is 4–6 cents from coal, 5–7 cents from natural gas, 6–9 cents from biomass, and 2–12 from nuclear plants. The cost of photovoltaic cells have been decreasing during the past decade, in Japan in a single year decreased 8 % and in California has decreased by 5 % per year during the past few years. For the crystalline PV cells, the price in USD/Watt-p was ~100 in 1958, ~200 for the 1965–70 period, ~20 in the 1980's, ~4–6 in the 1990's and is expected to reach ~1–2 by the year 2010. As the price of photovoltaic cells is reduced, solar energy generation becomes more competitive.

The main problem of present day photovoltaic cells is its low efficiency. Research groups at the National Renewable Energy Laboratory in Colorado and the Los Alamos National Laboratory in New Mexico are developing new solar cells that use nanoparticles to produce nano solar cells with much higher efficiencies that will allow producing electricity from sunlight at a price about a factor of 10 lower than current prices with photovoltaic cells.

Looking into the more distant future, we can imagine placing giant solar collectors in a geosynchronous orbit, absorbing solar energy during the day and night without concerning any kind of weather conditions down in the earth. The energy is then focused and a microwave beam delivers the energy to an antenna on the ground. Important technological developments have to be carried out before this could be an option for the future.

3.3 Hydropower

The world electricity demand will increase at an average rate of 2 % yearly in the next three decades. Within this scenario, renewables increase their share of total world energy consumption slightly in the projections, and the renewable share rises from 8 % in 2003 to 9 % in 2030 (IEO 2006). Much of the growth in renewable energy sources results from large-scale hydroelectric power projects in non-OECD regions, particularly among the nations of Asia. China, India, and Laos, among others, are already constructing or planning to construct large hydroelectric projects in the coming years.

In this way, the use of hydroelectric power shall continue to expand over the projection period 2003–2030, increasing by about 2 % per year (IEO 2006). The generation of hydroelectricity was of 2,809 TWh in 2004 and it is estimated to be of 3,682 TWh in 2030. Then, its share to the world electricity generation shall be as much as 14 % (World Energy Outlook 2006, OCDE/IAE). Although hydroelectricity is an important source, its contribution could be, also, inferior to biomass and other renewable sources. However, hydroelectricity will have certain expansion in the next decades.

In the past, hydroelectricity has contributed appreciably as power source, being second after petroleum. But, also, it has been affected for environmental reasons and social effects. At present, the contribution of hydroelectricity has descended in rank as electricity source, but one expects that its potential could be important in future. Until now, most of the reserves have been used for the water provision, mainly, for irrigation. Only 25 % of the reservoirs of the world are associated to energy generation (World Energy Outlook 2004, OCDE/IAE). That is to say, at present, it is considered that only approximately a third of the economic potential of this source has been operated. One hopes that the new hydroelectric capacity is added in developing countries where the potential is high. As mentioned, some non-OECD Asian countries have plans to increase their hydroelectric capacity. In Central and South America, many nations are planning to expand their already well-established hydroelectric resources. Brazil generates more of 80 % of its electricity from hydroelectric sources. In the OECD, however, environmental limitations for new exploration sites as well as restrictions exist. Thus, hydroelectric capacity in OECD economies is not expected to grow substantially, and only Canada is expected to complete any sizable hydroelectric projects over the projection to the year 2030.

The use of the hydroelectricity as power source could help the reduction of atmospheric polluting emissions due to the fossil fuel combustion. Regions with strong hydroelectric economic potential are found in the south of Africa, China, Russia, Brazil and South America. Table 2 shows world consumption of hydroelectricity and other renewable energy by region, reference case, 1990–2030.

Table 2. World consumption of hydroelectricity and other renewable energy by region, reference case, 1990–2030, data in Quadrillion Btu taken from IEO (2006)

Region/Country	History			Projections					Average Annual Percent Change, 2003-2030
	1990	2002	2003	2010	2015	2020	2025	2030	
OECD									
OECD North America	9.5	9.8	9.5	11.5	12.0	12.8	13.9	14.7	1.6
United States[a]...............	6.1	5.9	5.7	7.2	7.5	8.1	8.7	9.1	1.7
Canada.....................	3.1	3.6	3.5	3.8	3.8	4.0	4.3	4.6	1.0
Mexico......................	0.3	0.4	0.3	0.6	0.7	0.8	0.9	1.1	4.5
OECD Europe	4.8	6.1	6.0	7.9	8.0	8.3	8.4	8.5	1.3
OECD Asia...................	1.6	1.7	1.9	2.2	2.3	2.4	2.4	2.5	0.9
Japan	1.1	1.1	1.4	1.4	1.5	1.5	1.5	1.6	0.4
South Korea	0.0	0.0	0.1	0.2	0.3	0.3	0.3	0.3	6.7
Australia/New Zealand	0.4	0.5	0.5	0.6	0.6	0.6	0.6	0.6	0.8
Total OECD	15.9	17.6	17.4	21.7	22.3	23.5	24.7	25.7	1.5
Non-OECD									
Non-OECD Europe and Eurasia...	2.8	3.0	3.0	4.0	4.6	4.7	4.9	5.0	1.9
Russia......................	1.8	1.9	1.8	2.4	2.9	2.9	3.0	3.1	2.0
Other.......................	1.0	1.1	1.2	1.6	1.7	1.9	1.9	1.9	1.8
Non-OECD Asia...............	3.0	4.9	5.1	9.8	10.8	11.9	13.2	14.8	4.0
China.......................	1.3	2.8	2.9	6.1	6.8	7.1	7.3	7.6	3.7
India	0.7	0.7	0.7	1.3	1.3	1.5	1.8	2.2	4.2
Other Non-OECD Asia	0.9	1.4	1.5	2.4	2.6	3.3	4.1	4.9	4.5
Middle East	0.1	0.2	0.2	0.6	0.6	0.7	0.7	0.8	4.5
Africa	0.6	0.9	0.9	1.2	1.3	1.3	1.3	1.5	1.9
Central and South America	3.9	5.7	6.0	8.1	9.5	11.1	12.9	14.7	3.3
Brazil.......................	2.2	3.0	3.3	4.2	5.0	5.8	6.7	7.6	3.2
Other Central and South America..	1.7	2.7	2.8	3.9	4.5	5.3	6.3	7.1	3.6
Total Non-OECD	10.3	14.6	15.2	23.6	26.8	29.6	33.1	36.7	3.3
Total World	26.3	32.2	32.7	45.2	49.1	53.1	57.8	62.4	2.4

3.4 Eolic

Of the renewable energies, the technology for transforming wind energy into electricity is one of the most mature. This is reflected by the fact that as seen from Fig. 16, in the year 2005 about one third of the world's renewable power capacity comes from wind, in the US represents about 50 %, and, in the 25 countries of the UE (largely in Germany and Spain), and in India represent the largest share. In Mexico, wind power is just starting, but the goal is to reach 6 % at national level by the year 2030. In Fig. 18 is shown the La Venta facility in the state of Oaxaca, where the most favorable wind conditions of the country are found. It has been estimated that

Mexico's main potential wind energy is about 5,000 MW, and 2,000 MW of these are in the Isthmus of Tehuantepec.

The theoretical wind potential is enormous, see Fig. 15. Wind energy is actually solar energy; a small proportion (less than 1 %) of the sunlight energy that gets into the atmosphere is transformed into the wind's kinetic energy. The present wind devices have been placed at low altitude where the wind current and power is in general not very strong. It has been estimated that about two thirds of the earth's total wind energy is located in the upper troposphere, which is at a much higher altitude than current wind generators. In several high altitude places around the world wind power surges to 5,000–10,000 watts per square meter, there is still intermittence but the wind never stops. In the northern Hemisphere between 20 and 40 degrees latitude and at about 10,000 meters is located the mother lode which is an intensive jet stream. However, the technology for using the high altitude winds is still not available. At the present time there are some projects for placing wind generators in the atmosphere that range in altitude from about 100 to 10,000 meters. The autogiro designed by Sky WindPower is proposed to be placed at 3,000 meters and computers adjust the pitch of the four blades in order to maintain the position and altitude of the autogiro. For the expansion of the wind power in Mexico, one of the main problems is the legislation, by constitutional mandate, electricity generation for the public service can only by done by the State, and until this kind of legislation is not changed, the expansion of wind power and other alternatives energies will be somehow limited.

3.5 Geothermal

Geothermal energy is energy available as heat yielded from within the earth's crust. It comes usually in the form of hot water or steam. It is exploited at suitable sites for electricity generation, or directly as heat with its many applications. Geothermal energy is commonly regarded as benign, but geothermal fluids usually contain small quantities of CO_2 and other toxic chemicals; consequently they are naturally emitted from thermal areas, see Renewables in Global Energy Supply, International Energy Agency fact sheet report, 2007: http://www.iea.org/Textbase/publications/free_new_Desc.asp?PUBS_ID=1596. However, alone in 2005 the geothermal generation reduced CO_2 emissions by 46.4 Mt (Geothermal Energy Annual Report, 2005, International Energy Agency).

Fig. 18. Wind power plant in La Venta, Juchitán, Oaxaca, where Mexico's largest wind energy resource is found in the Isthmus of Tehuantepec (picture courtesy of Marco A. Borja, IIE, Mexico, see also this volume)

An important property of geothermal power plants is that they can operate the whole day without interruption, providing base-load capacity. In 1980 about 4,000 MWe capacity of electricity generation from geothermal energy was operating in the world and currently there is about 8,000 MWe. In fact, the growth is linear with time; therefore we expect to have about 10,000 MWe by the year 2010. Figure 19 shows the historic evolution of the world geothermal capacity and its electricity generation (data taken from Bertani 2005). Figure 20 shows the energy produced from geothermal sources by several countries (data taken from Bertani 2005). The US is by far the main producer.

There are also prospects in some parts of the world, where the hot mantle is close to the surface, for injecting water underground and for recovering steam to produce electricity. Over the next 30 years, the world potential capacity for geothermal power generation is estimated at 85 GW. The costs of geothermal energy have dropped substantially during the past few decades. However, geothermal power is accessible only in limited areas of the world, the largest being the US, Mexico, Central America, Indonesia, East Africa and the Philippines. Challenges to expanding geothermal energy include very long project development times, and the risk and cost of exploratory drilling.

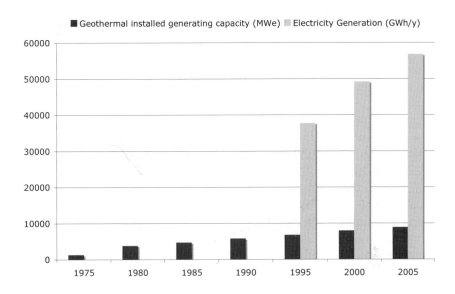

Fig. 19. The geothermal installed generating capacity (MWe) from 1975 to 2005 and the electricity generation (GWh/y) from 1995 to 2005

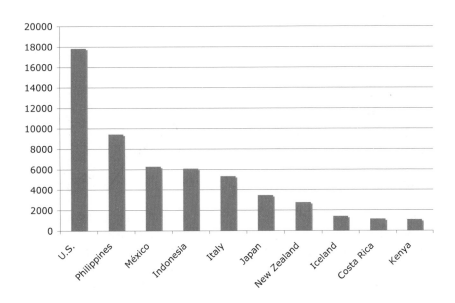

Fig. 20. Annual energy produced (GWh/y)

Geothermal heat generation can be competitive in many countries producing geothermal electricity, or in other regions where the resource is of a lower temperature, see Renewables in Global Energy Supply, International Energy Agency fact sheet report, 2007: http://www.iea.org/Textbase/pub–lications/free_new_Desc.asp?PUBS_ID=1596.

Considering that 70 % of the world's energy needs can be met with water at temperatures less than 200°C and that power plants are becoming increasingly efficient, the geothermal energy may represent a major renewable and sustainable source of energy for some parts of the world, see Enhanced Geothermal Innovative Network for Europe (ENGINE), Annex I – "Description of Work", contract No. 019760, 2006. However, a lot of research is needed before geothermal energy can be enhanced. There are big differences in the state of development of the key technical elements required for integration of geothermal electricity generation or heat/power cogeneration into an overall system. While, for example, drilling technology – as a key technology in the petroleum and natural gas industry – is technologically mature, stimulation technology is still in the pilot stage. Further development of stimulation technology to enhance the production of geothermal reservoirs is of utmost importance, as this will make the exploitation of a vast energy potential possible, particularly in crystalline rocks. Stimulation technology is also important in reducing the prospecting risk when drilling aquifers and fault zones. There is also still major potential for optimization and further development of plant engineering and power plant technology, see The State and Prospects of European Energy Research, report of the European commission, EUR 22397, 2006.

In fact, there is an effort to develop new technologies, called Enhanced Geothermal systems (EGS), which is thought to extract the natural heat in high temperature, water-poor rocks in formations that are either too dry or too impermeable to transmit available water at useful rates, see Geothermal Energy Annual Report, 2005, International Energy Agency.

3.6 Biomass

Biomass is a broad term that refers to all plant and animal matter existing on earth, both living and recently living. However, the concept of biomass in the last few years has acquired special relevance within the energy production industry. In this context biomass mostly refers to the growing of plants with the purpose of producing biofuel as a source of energy, heat, production of fibers, and chemicals. Some examples of these biofuels are biobutanol, biodisel, bioethanol and biogas coming from a variety of plants such as hemp, corn, willow or sugarcane. Petroleum or coal are not considered biomass, because they have suffered a transformation by means of natural geological processes. The aim of biomass development projects is to identify whether it is possible to utilize biomass in the energy production sector, by substituting a portion of conventional fuel by biomass, to perform combined combustion.

Fig. 21. Switchgrass (left) and rice chaff (right)

The importance of biomass is growing due to the environmental politics searching for sustainable fuel sources. Biofuels, in fact, represent a possible sustainable fuel source. Since plants normally grow at a certain rate, they are considered to be a renewable source. In order for them to grow, they convert carbon present in the atmosfere into biological matter through photosynthesis. When biofuels combust, they deliver to the atmosphere the carbon that was previosly taken from it. As new plants grow, they absorbe again the carbon, thus closing the cycle and leading to a stable level of atmospheric carbon.

Biomass production has also the advantage of using those organic products that are otherwise considered as waste to generate electric power. Wood residuals, tree branches and other scraps are transported from farms and factories to the biomass power plant. Then, the biomass is burned inside of a furnace. The heat generated translates into steam, and then used to propel turbines and generators.

Another use of biomass to generate energy is at the landfill with burning waster products. As garbage decomposes, methane gas is generated and conducted through pipelines to collectors. This landfill gas can be used in many applications, just like normal methane gas. Another approach can be done at animal feed lots. Animals in farms produce manure. When manure decomposes, it also gives off methane gas in a similar way to garbage.

However, when the product of biofuel burned is higher than the amount of new plant growth, the balance is lost. This leads to deforestation and an increase in greenhouse gases in the atmosphere. In this way, fossil fuels are not part of the carbon cycle since the carbon they contain corresponds to the long past and when burned, additional amounts of carbon are delivered. For example, 60 million tones of bone dry biomass are produced in California each year. Approximately only eight percent of this total is used to make electricity. However, according to some calculations, if the entire biomass production in California was used, the power output could be close to 2,000 MW. That amount of energy could provide electricity for two million homes.

Today, new ways of using biomass are still being discovered. One way is to produce ethanol, a liquid alcohol fuel. Ethanol can be used in special types of cars that are made for using alcohol fuel instead of gasoline. The alcohol can also be combined with gasoline. This reduces our dependence on oil – a non-renewable fossil fuel (Wald 2007).

Hoogwijk et al. (2001) have analyzed 17 such scenarios of biomass as a potential major source of energy. A classification into two groups was done: i) Research Focus, and ii) Demand Driven. The estimated potential of the Research Focus varies from 67 EJ (ExaJoules) to 450 EJ for the period 2025 – 2050, and that of the Demand Driven from 28 EJ to 220 EJ during the same period. The share of biomass in the total final energy demand lies between 7 % and 27 %. For comparison, current use of biomass energy is about 55 EJ. With about 4,500 EJ available, biomass resources are potentially the world's largest and most sustainable and renewable energy source (Hall and Rao 1999). Although the annual bio-energy potential is about 2,900 EJ, only 270 EJ can be considered to be available at competitive prices and sustainable.

Crops especially dedicated to produce biomass are not very popular right now, making residues currently the main sources of bio-energy and this will continue to be the case in the short to medium term, with dedicated energy forestry/crops playing an increasing role in the longer term. There is going to be an increase in biomass energy usage, particularly in its modern forms, thus having impact in the energy sector, in agriculture, and rural development.

4 Conclusions

A key message of the diverse world energy outlooks is that by 2030 the world's energy consumption will have doubled and, most important, that fossil fuels will continue to dominate the energy consumption. In absolute terms, some 87 % of total primary energy demand in 2030 will be fossil-fuel based. This implies that carbon dioxide emissions will be nearly twice those recorded in 1990. Developing countries are expected to have a serious influence on the global energy picture, representing more than 50% of the world's energy demand, as well as a corresponding level of CO_2 emissions. In addition, in relation to 1990 figures, the US's contribution to CO_2 emissions is expected to increase by 50 %, compared to an 18 % European Union increase. This will cause the atmosphere to have more than 500 ppmv of CO_2 and other greenhouse gases, with unknown catastrophic consequences to our planet.

The development and use of renewable energy technologies are important components for the future of a balanced global energy economy. Renewables can make major contributions to the diversity and security of energy supply, to economic development, and to addressing local environmental pollution. Furthermore, much attention has been attracted to their potential to avoid global warming through zero or near zero net greenhouse gas emissions, see Renewable energies: Research, Development and Demonstration Priorities, report of the International Energy Agency, 2006.

If we are realistic, the future does not look very promising with growing energy demand based on fossil fuels and CO_2 and other greenhouse gases possibly increasing above 500 ppmv. There are contradictions between the government's postulates and the projections; humanity is not really prepared for the energetic transition. At the present time the available alternative energy sources are limited, or still in the developing stage. The scenarios that we have analyzed (IEO 2006) assume that there are sufficient oil reserves for the next decades. However, other groups, for example in Princeton University and the University of Upssala in Sweden, claims that we have reached the peek in oil production and that are not significant reserves to be found and that the world is entering a possible severe energy crisis. A restriction to the IEO (2006) analysis is an explicit OCDE vision, especially in the United States. The OCDE assumes that there will be oil and that the remaining needs can be satisfied with nuclear energy. In 2003, a group at MIT (Ansolabehere et al. 2003) proposed very ambitious but uncertain nuclear plans.

The burning of fossil fuels will continue for some time into the future and something that could help to limit the growth in CO_2 and other greenhouse gases is to take certain energy efficient measures. In the International Energy Agency 2006 report "Energy Technology perspectives – Scenarios and Strategies to 2050", below the heading "Energy Efficiency is Top Priority" we find the sentence "Improving energy efficiency is often the cheapest, fastest and most environmentally friendly way to meet the world's energy needs".

There are skeptics who doubt about the extent and pace of global warming. But the consequences to now neglect the situation could be worse than the feared economic damage that has bred over caution. We have to take measures, not to wait to start seeing ice caps to vanish. Then, it will be too late.

Acknowledgements

This work has been partially supported by the Mexican Consejo Nacional de Ciencia y Tecnología (CONACyT) under contract numbers U43534-R and 44917.

References

Ansolabehere et al. (2003) The Future of Nuclear Power. MIT
Bertani R (2005) World Geothermal Generation 2001–2005: State-of-the-Art, Proc World Geothermal Congress 2005, Antalya, Turkey, 24–29 April 2005
Bucher E (2006) Solar Energy: Status Report 2006 and its Potential. Preprint
Hall DO, Rao KK (1999) Photosynthesis 6th Edition. Studies in Biology, Cambridge University Press
Hoogwijk M, den Broek R, Berndes G, Faaij A (2001) A Review of Assessments on the Future of Global Contribution of Biomass Energy. In: 1st World Conference on Biomass Energy and Industry. Sevilla, James & James, London (in press)
IEO (2006) International Energy Outlook 2006, Report #:DOE/EIA-0484. http://www.eia.doe.gov/oiaf/ieo/pdf/0484(2006).pdf
Karl TR, Trenberth KE (2003) Modern Global Climate Change. Science 302:1719–1723
Klotz L (2004) Kyoto commons-A tragedy? Can Med Assoc J 170:165
Pacala S, Socolow R (2004) Stabilization wedges solving the climate change problem for the next 50 years with current technologies. Science 305:968–972
Wald M (2007) Is Ethanol for the long Haul? Sci Am 296:28–35

Energy, Present and Future

Juan Eibenschutz

Comision Nacional de Seguridad Nuclear y Salvaguardias (National Commission for Nuclear Safety and Safeguard). Doctor Barragan 779, Colonia Narvarte, Mexico D.F. 03020, Mexico. E-mail: je@energia.gob.mx

1 Introduction

It is useful to look back into the past when trying to understand the present, and when attempting to analyze the future, particularly in the case of energy, whose role in development has been and will continue to be fundamental.

Energy use is strongly linked with both, the industrial activities and the well being of mankind. When one examines the growth in energy consumption over a long time period, several facts are worth noting.

Between 1850, and say 1900, the growth in energy consumption is important, going from something like 10 exajoules per year, to about 35 exajoules in 1900, which is about 3.5 times in fifty years. During these 50 years wood, the main source of energy, yielded to coal, with oil beginning to come into the picture.

During the next five decades, the growth was considerably more important, since it reached some 400 exajoules in year 1950, with coal being the most important source. Compared with the previous fifty years, it is more than a tenfold increase.

In 1965, the composition of primary commercial energy of the world was: oil 39.6 %; coal 38.4 %; gas 16.4 %; hydro 5.5 %, and nuclear 0.1 %. Forty years latter, in 2005, the composition was as follows: oil 36.4 %; coal 27.8 %; gas 23.5 %; hydro 6.3 %, and nuclear 6.0 %.

Total energy consumption went from 3,863 million tons of oil equivalent (TOE) in 1965, to 10,537 million TOE in 2005, that is 2.7 times more energy consumed by the world per year, over a period of forty years.

Per capita energy consumption was 1.15 TOE in 1965, and 1.63 TOE in 2005. During that time period, population increased 1.9 times, and per capita energy consumption 1.4 times, indicating improvements in energy efficiency.

Although presently oil continues to be the dominant energy source, coal, mainly due to environmental causes, and in spite of its growth in absolute terms, has a much lower participation in the energy balance, a loss of ten percentage points, whereas gas went up from 16.4 to 23.5 %, hydro increased slightly in participation and nuclear energy experienced the highest growth in relative participation, going from 0.1 to 6 %.

Obviously the energy scene is dominated by hydrocarbons, with a logical trend of oil displacement by natural gas, as a result of environmental concerns.

The world has moved from being driven by supply-demand considerations as the ruling factors in energy, to a more complex scheme, where the environment, geopolitics, public preference, "nobility" of sources, etc. have become the dominant factors. However, socio-economic inequalities are also present in the energy field, since about one third of mankind has no access to commercial energy sources.

In the case of technology, the fact is that except for nuclear, the basic technologies of the mid 20th century are the ones presently in use, albeit with major advances, particularly in materials and electronics that have resulted in impressive increases in conversion efficiencies, as well as in control and reliability of supply.

The challenges for the world's future are formidable, particularly in the case of energy. The market is enormous when considering the need to incorporate into the modern world those segments of the population presently deprived of the benefits of modern life.

It is interesting to note that many of the energy poor regions of the world experience more benign climate than the affluent regions, and this is particularly the case of low rather than high temperatures. Life is more difficult in cold weather, where survival without heat sources and adequate buildings is simply not possible.

Paradoxically, the more developed societies are subject to more environmental hardship than the less developed ones; it seems that mankind reacts to threats better than to the natural protection provided by a benign environment.

Furthermore, environmental, societal, economic, and political constraints add to the challenges of growing markets and diminishing resources, regardless of the failure of past predictions about the overall decline in oil production, oil as well as gas are non-renewable resources, and eventually they will run out. Moreover, as Sheik Yamani pointed out, the Stone Age did not come to an end because of lack of stones. Most likely, the oil era will end because of environmental and economic constraints, and perhaps because more competitive sources will become available.

Most analysts are forecasting a 50 % increase in energy consumption for the next 25 years. One of the many possible scenarios would have hydrocarbons shifting in relative participation in the market, from 60 % to 35 %, coal, increasing from 28 % to 35 %, nuclear going from the present 6 % to 10 % and hydro from 6 to 5 %. This implies that 25 % of the energy forecast to be needed in 2030 shall have to come from renewable or other sources.

As many have said, the future is very uncertain, but the fact is that demand will almost certainly be supply-constrained, and therefore solutions must be found to keep the world running.

For example, coal incremental participation in the energy balance, is only conceivable if carbon sequestration materializes, and this is not a trivial question. Nuclear electricity offers a real solution, but the political hurdles are serious; nuclear can physically take care of deficits, the question is, will the politics of energy succeed in deciding on facts, rather than on subjectivity?

2 Scenarios of the future

Scenarios for the next half century are open. Assuming no major disruptions, like a new world war, three possible futures can be postulated:

- **Soft transition:** Conservation and energy efficiency, more nuclear and gas, renewable energy, overall economic growth, and world population stabilizing.
- **Readjustment:** Economic reaction to subsidization, declining reserves, materialization of climate change, supply constraints.
- **Breakthrough:** Fusion, direct energy conversion, world wide shift from destruction to world wide well being, values.

Half a century is a long time. The three generic future scenarios presented above are all possible. However, with respect to the previous half century, there are important differences: climate change is showing signs of being real, non-renewable resources are declining, society is really participating in the decision making processes, and subjectivity is challenging objectivity. The question is: where do we go from here?

2.1 Energy conservation and efficiency enhancement

Firstly, common to every energy future, are climate change challenges, energy efficiency, and technology, some features of these are presented in what follows.

Greenhouse gas reduction

The consequences of severe climate change are so serious, that they may make our globe inhabitable. Fossil fuel combustion is a major contributor to greenhouse gas emissions, therefore future energy solutions will have to favor "clean" energy systems, as well as conservation.

Opportunities to save energy are significantly more important in the developed economies, where relative consumption is more than one order of magnitude higher than the world average consumption.

Carbon sequestration provides a theoretical solution that may allow coal to regain its preeminence in the world energy balance; unfortunately, the amounts of "sequestered" carbon to be disposed of are enormous and the technologies have yet to become commercially competitive.

Perhaps the non-emitting nuclear power will have a really crucial role in the future, since it is the only energy source that can provide large amounts without producing greenhouse gases.

Energy efficiency

As evidenced by the differences in energy efficiency indexes among different countries, energy pricing is one of the most effective efficiency enhancers. Future policies may promote pricing mechanisms that discourage waste. Solutions may include incentives associated with consumption indexes for industrial plants and large consumers, such as shopping centers, large hotels, etc.

Indexes would be determined for each type of industry or facility, and a special tax or credit would be applied to the bill, depending on the measured energy vs. production, or service, compared with the official indexes.

Cogeneration (combined heat and power)

One of the most effective energy efficient technologies involve the simultaneous production of electricity and heat, or other energy vectors, such as hydrogen, process steam, cold water, desalination, etc.

Fuel savings can be as high as 100 % considering that with the same amount of fuel consumed by the power plant, it is possible to produce the other energy vectors without additional fuel.

The regulatory framework plays a crucial role in cogeneration, because plants have to be optimized in order to obtain the maximum efficiencies, and this requires operating rules, and rates that allow for the necessary flexibility of operation.

Transport

Transportation of merchandise and people, accounts for more than one third of world energy consumption. Sea or waterways are the most fuel-efficient means of transport for merchandises, followed by railroad, with trucks being the least efficient.

Overall economics dictate the solutions. In most countries, the transport industry operates in a highly competitive market, and energy efficiency is not necessarily the main objective, since opportunity of delivery, or security of handling, or even refrigeration, can be the overriding factors.

Nevertheless, the per kilometer-ton energy consumption varies over a very wide range, and this is why the transportation sector offers a very important area of opportunity for energy savings.

People transport is another example of enormous opportunity for energy savings. On the one hand, the differences in energy consumption between individual transport and mass transport systems are more than one order of magnitude. Urban mass transit solutions not only save very large amounts of fuel, they solve pollution problems and facilitate the functioning of cities. On the other hand, private vehicle consumption can be as much as five or more times lower for the same service.

Recycling

Technologies for substituting fabrication with recycling are having more and more impact on environmental solutions, and they have the added benefit of lowering energy demand, for example, the energy requirement for producing aluminum from its ore is more than ten times the energy required to remanufacture from aluminum scrap.

Something similar happens with glass, paper and some plastics, where environmentally friendly solutions are also energy savers.

Information and communications

The present state of the art allows for intellectual work to be performed out of the office, in home computers, lap-tops, and even mobile phones. The more this practice becomes the rule, the less physical displacement requirements; because of this, huge energy savings are foreseeable all over the world, since personal transport needs are likely to decrease considerably.

In more and more countries, electronic banking, internet shopping, electronic payments of taxes, utilities, and even fines, are already saving energy since more and more people carry out their activities from home, without the need to go to banks, government offices, etc.

2.2 Energy production options

The previous section describes energy efficiency enhancers, as well as some energy conservation systems, in the following part some supply options are briefly discussed (see Scientific American Vol. 295, number 3, Sept. 2006), bearing in mind that these are technologically feasible, although they are not yet ready for deployment, since the economic feasibility is yet to be demonstrated.

High altitude wind generators

The concept is to tap the high altitude currents, characterized by much stronger wind velocities than available at the relative low altitudes of conventional wind generators, and therefore having the potential of much better power to weight ratios, that would allow more competitive energy production devices.

Satellite solar collectors

As a result of the space programs, the concept of very large satellites equipped with solar converters was developed, more than twenty years ago. The idea is to generate power with solar cells, in the TW range, and beam it to earth via microwave transmission to antennas located in suitable places.

Studies indicated that most of the technology was available, including outer-space manufacture of solar cells and the flimsy structural material required for structural purposes.

The main drawback is that for the scheme to work the power range is one terawatt, and even if the estimated per KW cost was of the order of

500 USD per KW, the investment would be of the order of half a trillion dollars, with a non-negligible risk due to the operation of spacecraft and outer-space manufacturing facilities.

Nuclear power

This form of energy presents itself as the most technically viable solution for future, non-polluting energy supply. Its re-birth comes as a result of climate change considerations, and the different initiatives for improved technical nuclear power plants will almost certainly contribute to solve, with environmentally friendly solutions the future energy needs.

In this case, technology is not the problem; presently nuclear power provides as much energy as hydro-power. The real problem is political and societal approval. It must be recognized that the bomb, the first large scale demonstration of nuclear energy, continues to exert its toll.

Regardless of the fact that nuclear power has demonstrated its safety, a biased public perception of the risks truly exists, and society has to come to grips with the facts and change to objective reasoning, instead of opposition based on subjective considerations.

Other energy supply solutions

Among the many possible sources, energy crops, photo-mechanical systems, deep geothermal, methane hydrates, massive wave energy generators, etc. may have increasing roles in future energy supply.

2.3 Breakthrough possibilities

Since this presentation attempts to discuss the future of energy, the possibilities of scientific breakthroughs may be worth some thoughts, because it is possible that unlimited amounts of energy could become available as a result of these breakthroughs.

A good example of the possible breakthroughs is cold fusion; it didn't work, but if it becomes real, maybe all energy needs of the future would be satisfied, without affecting the environment.

Direct conversion

To be usable, energy must be transformed, converted from primary energy to secondary energy. This means that the energy contained in the original resource is not the useful energy, since the transformation processes consume part of the original energy, with varying degrees of efficiency.

The fuel cell is almost a direct energy conversion device; unfortunately it requires secondary energy as fuel, but the conversion of the potential chemical energy to electricity has the highest efficiency of all power producing systems.

Nuclear energy comes from the potential energy obtained when nucleons are changed from certain elements to others. So far this potential energy must be used as heat, and this is the case even with fusion energy systems, that will produce heat, to raise steam and drive a turbine, which in turn rotates the generator that produces the electricity.

It is not unthinkable that advances in knowledge about matter could make it possible to use potential nuclear energy to directly produce mechanical or electrical energy.

Energy bacteria

Living organisms can transform products. Some processes of transformation occur naturally, like in fermentation. Research and genetic engineering may perhaps develop systems to generate organic energy vectors, or even hydrogen from vegetal matter. Such a breakthrough could provide a final solution to the energy needs of mankind.

3 Summary

Energy is a basic input to sustainability of mankind. No energy means no society. This presentation has attempted to show that although demand will grow there are ways to manage it, it has also tried to discuss different supply-side solutions, some of them dependant on scientific breakthroughs, that may or may not happen within the forthcoming years.

The present world is characterized by inequality. Eventually mankind will solve its problems, it always has. Transitions are the hardest, at least for some.

What is important is that globalization is real. Nowadays the world is really small; thanks to instant communications, practically whatever happens anywhere is known everywhere.

Energy is only one of the factors that support a viable world. In spite of the difficulties one should be optimist for the future, since the resources are there.

The real challenge is to switch from destruction to construction.

Advanced Energy Conversion

Horst Hippler

Universität Karlsruhe (TH), Kaiserstraße 12 - 76131 Karlsruhe. Tel.:
+49(0)721/608-0. Fax: +49(0)721/608-4290.
E-mail: Rektor@verwaltung.uni-karlsruhe.de

Abstract

One of the major challenges for research in the future is to provide energy
on a world wide scale against the background of vanishing resources, climatic changes and economic impact based on novel fuels and new technologies.

1 Introduction

Most of what we have heard in this conference deals with energy conversion for physical appliances. Generally one identifies this problem with the
question of energy resources or energy uses. From a physicochemical point
of view this is not quite a correct approach, since energy is conserved in
any of these processes. We are producing entropy and energy is conserved.

The earth is an open system absorbing and emitting electromagnetic radiation. However, no real stationary state for the energy content of the
earth will be reached since our planet has some complex storage mechanisms, like the formation of crude oil and coal, which humans have decided to burn in large quantities. But this is not the objective of my presentation.

We want to deal with the energy conversion processes which are much
simpler questions to put forward. The form of energy you are going to use
depends on the purpose it is needed for, like transportation, industrial production, chemistry, house heating or cooling, and so forth. This means one
has to look at the whole process of energy conversion in order to optimize

it. Generally, there are two different approaches: (i) one may have an energy source and has to optimize its conversion for a specific application or (ii) one may look back from the end user to that kind of fuel which would be the best for the kind of application. The two answers may be quite different with respect to costs, resources, environment, etc.

Today, fossil resources are the most widely used energy sources, but geographically concentrated at political unstable regions, and in addition it is certain that the total amount of fossil resources is limited. The current prediction is that 50 years from now crude oil and natural gas will be mostly depleted, considering that by the year 2050 the consumption of primary energy is expected to have risen by 50 %. Coal may last somewhat longer. Therefore, we have in particular the demand to increase the efficiency of the actual energy usage. This is also economically very interesting, since it is generally more profitable to improve current processes than inventing and developing new processes. On the other hand, a continuously increased consumption of fossil fuels will accelerate the green house effect (the rise in the concentration of CO_2 in the atmosphere) and because of that, not just nuclear energy, but any other process that reduces CO_2 production will be helpful.

The future energy supply for the next 50 years will continue to rely widely on systems and processes of thermal conversion of chemical fuels to energy for the costumer and we must think about it and look at it in detail. One of the major challenges in future research will deal with providing energy on a world-wide scale against the background of vanishing resources, climatic changes and economic impact, based on novel fuels and new technologies.

2 Activities at KIT

Karlsruhe Institute of Technology (KIT) has been founded by challenging a merger of the University of Karlsruhe, which is a state research institution, with the Forschungszentrum Karlsruhe, which is a federal research institution. Both sites have about 4000 employees, whereof 2000 are researches. KIT is aiming to develop sustainable concepts for energy supply and to generate the scientific basis for innovative engines of the highest efficiency and environmental friendliness, based on novel fuels and new technologies.

As it is presented in Fig. 1, KIT considers three fundamental aspects to develop: (i) Since fuels or engines are controlled by thermo-physicochemical processes mostly, consequently a physicochemical understanding is indispensable for the analysis and validation of these processes. (ii) We also need very developed diagnostics and instrumentations as well as (iii) modeling and computer simulation. These three aspects are mandatory to predict the behavior of the complete physiochemical system in any cyclic process. In this way, the analysis may begin by the consideration of a specific fuel and an engine using this fuel very efficiently. Then analyzing and trying to understand the complete thermo-physicochemical system you may at the end change the fuel or use the same fuel for a better adapted engine.

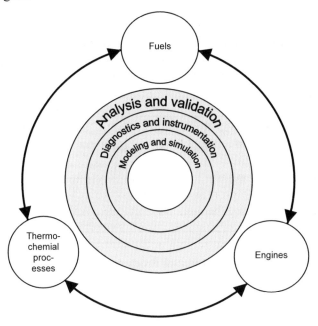

Fig. 1. Methodology of study in the Karlsruhe Institute of Technology

This structure of the energy research as a central research field within the research program of the KIT can be better understood by an energy cascade. It just gives the structure of our energy research which starts on some fuels, for example organic fuels. Then we have some thermo-physicochemical processes of energy conversion linked to some technical energy conversion systems. We obtain the produced energy which then has been to be distributed within some networks. Anyway, you need some kinds of power plants depending on how the energy is produced, maybe decentralized, which then could even be in some of our homes. In the case of operating a decentralized energy conversion there is a strong need for a negotiation system for the prices for selling and buying energy. At the end there is a necessity to look at the final user and what is the user doing with this energy.

For example, for the use of energy in houses like offices or homes we have to ask ourselves what is the future of building houses and how should they be built to efficiently use energy. But this means not only insulation for heating or cooling should be considered but also many other things like the construction process itself, the environment, economic life-time, transportation, how much power, and so on will be necessary at this place. All these questions should be analyzed and validated going back and forth like in figure (Fig. 2). We can follow an example for all these processes from the left to the right in this figure. In going on we shall consider an example of each of the entries in the squares of the figure.

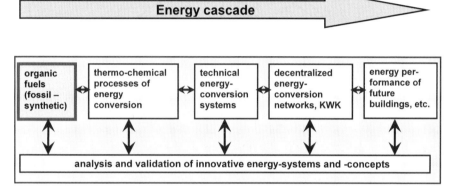

Fig. 2. Structure of the Energy Research Program at KIT

Status of fuel research

At first there is the extraction of raw energetic materials from resources of maybe different nature. It could be crude oil, coal, natural gas, biomass or organic waste. But instead of looking only for combustion processes other kinds of fuels like solar energy, wind, or geothermic energy should be included.

Second, one has to consider the processes for conversion and refinement. For example, for crude oil the following is needed: preparation procedures, distillation, gasification, pyrolysis, syntheses and purification. In any case, it is necessary to investigate the following questions. What are we doing with hot steam coming out of the processes? How can the hot steam be dealt with? What nature are the present impurities? Which machinery is best for dealing with these processes?

Third, the fuel is used, maybe it is incinerated or a propellant is made of for many different things. Then, we have to understand the physical chemistry and how the process can be controlled. It is also very important to know what kind of contaminants have been produced. Or we can also take this basic fuel and try to convert it into other fuels by synthesizing other base chemicals which can be used for different purposes, like the production of food or whatever.

Actual challenges

To begin with, we have scarce and ill pre-conditioned resources, raw materials with high contaminations because we may be forced to use crude oil, for example, with high sulfur contents. However, we also have to comply with high environmental standards. The established methods for resolution of this situation are the synthesis of alternative fuels and propellants for different purposes. One may think of the new airplanes. One synthesizes a special fuel made just for this kind of airplanes in order to have much more efficiency in the turbines, which nevertheless is a great challenge. It is also very important to develop processes which are contamination tolerant, for example to develop fuel cells, a catalytic process to convert chemical energy into electrical energy. However, catalyst may be easily poisoned by impurities. Therefore, most fuel cells are hydrogen based since it is very easy to purify hydrogen. But if instead we use methane or alcohols or hydrocarbons it is almost impossible to purify liquids as well as hydrogen then we have to find impurity insensible catalysts, like high temperature catalysts. The third method of resolution may be the synthesis of clean, efficient, more appropriate and more adapted fuels and propellants.

3 Examples

Production of Synthetic Natural Gas (SNG) from Biomass

In Karlsruhe there is now a pilot plant under construction for the production of synthetic natural gas (Fig.3). The processes within this plant may start from any kind of raw material, like wood, straw, energy crop, organic waste or liquid manure from caws. The process starts with gasification, extraction (as the energy part of crops) or fermentation followed by incineration, esterification or conditioning to produce either a liquid (methanol or ethanol, for example), a gas (synthetic natural gas, for example) or heat which can be used further on as fuel, or even as electricity as the final kind of energy desired. Gasification may be the most promising process for the future which can be applied to almost any kind of carbon containing material.

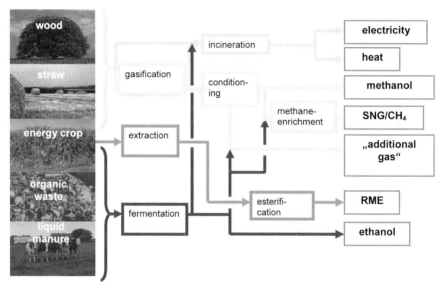

Fig. 3. Example: Production of synthetic natural gas (SNG) from biomass

The Flame: A Thermo-physicochemical Processes of Energy Conversion

The combustion process is an example for a thermo-chemical process of energy conversion and there are several aspects of it which are explored in details in Karlsruhe, as it is shown in Fig. 4. First, we look at the fundamentals of combustion, which means the elementary chemical reactions. Then we go to the technical systems, which means trying to understand what is going on in this flame, why is the heat released, why has it this behavior or what part of the flame is good for conversion and what part of it is bad for that, answers which might depend on the purpose for which you produce the flame. In order to understand these processes we develop some chemical mechanisms, couple them to hydrodynamics and then model the whole process.

It is very important to look into all details of that modeling to figure out what kind of harmful substances have been produced, but it is also important to consider not just the burning process, but also the ignition processes. We aim to understand how to ignite one of these flames and what is going on in the cold process. For example, the important phase in starting a car or engine is in the first two minutes when most of the pollutants are produced. So the starting process until the machine runs steadily is also very important. There are some more technicalities for making the flame to burn in a stable way, which is very important for many new processes. When a flame is unstable however, there is the danger that the flame extinguishes which would be very dangerous if this happens in a flying airplane.

We also consider catalytic combustion systems where the flame burns on a small refined area of a catalyst, so that the energy density is much larger as compared to open flames. We also have to deal with different kinds of flames, like the premixed or the diffusion flames. For all these kinds of processes we have developed competence at KIT.

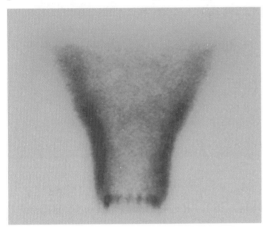

- Fundamentals ⇒ technical systems
- Experiments ⇒ modeling
- Chemical reaction mechanisms ⇒ detailed, reduced
- Formation and reduction of harmful substances
- Ignition processes
- Flame propagation
- Combustion instabilities
- Catalytic systems
- Premixed flames - diffusion flames

Fig. 4. Combustion as the thermo-chemical process of energy conversion for the production of synthetic natural gas

Technical Systems for Energy Conversion investigated at KIT

At KIT there are experts for turbo machines like gas and steam turbines. There is an extensive program for micro-gas-turbines, turbocharger and airplane engines as well as for piston machines. These are essentially car engines or cheap engines, like the Otto (DI), Diesel, and Homogeneous Charge Compression Ignition (HCCI) (Fig.5).

There is also a program on fuel cells. We are studying catalytic processes, avoiding the formation of substances which may poison catalysts. Within this program we are looking for process optimized fuels. Fuels for high temperature catalysts are very promising, and they are very much discussed nowadays. But there are some problems for use in automobiles since these fuel cells are very heavy, and at high temperatures they should be isolated very well. In addition there is a need for pre-heating before running the fuel cell thus starting the car. Maybe high temperature fuel cells are suitable for trucks or for households where even natural gas can be used to produce electric energy.

- Turbo-machines:
    - Gas (and steam) turbines
    - Micro gas turbines, turbocharger, airplane engines
- Piston machines:
    - Otto (DI), diesel, HCCI
- Fuel cells

- Questions:
    - Eficiency/economy
    - Environment friendly (pollutants (harmful)/CO_2)
    - Fuels
    - Operating life
    - Experiment/metrology/modeling
    - Overal system/sub-processes

Fig. 5. Technical systems for energy conversion that should be taken into account

One also has to deal with efficiency and economy as well as whether the process is environmentally friendly, considering the pollutants, in particular the harmful substances. We also do not wish to produce too much carbon dioxide (CO_2). We elaborate fuels, looking in detail on these aspects. Another aspect of importance is the operating life of these whole machines starting from their construction. We have to look at the whole process including fuel, machine, everything together, and to ask ourselves whether this is an efficient process or not. This aspect has been neglected in the past, but in order to produce some of these turbo-machines or some of these fuel cells sometimes you may produce so much waste and other kind of things or you may run into problems because you are producing or using very toxic materials.

We endeavor to develop new experimental technologies. Therefore we need metrology, and so we design measuring systems just for these processes, and we are able to look into running piston machines and turbomachines with optical technologies. We are trying to understand all these processes by modeling with the objective to make them more efficient and better.

Lets discuss decentralized energy-conversion networks, this means that we have to think of the situation that in every household we have some facility to produce electric energy may be from a fuel cell, or natural gas or maybe from the sun. There are, let's say hundred thousand very small electrical energy producers and also electrical energy consumers. Somehow, this fact has to be dealt with, since it may be very non-economical for one household to produce its own energy when it can buy the same amount of energy much cheaper from another household which has an overproduction at the same time. This is not only a free-market economy problem but also a logistic problem (see Fig. 6). We are developing some tools helping to solve this complex problem.

There are many different kinds of household power plants: hydropower on farms, where they have a small river; small wind power plants just for pumping water or similar necessities. These plants may produce energy when wind is present and the resulting energy may be used directly or is produced for use by others. Similarly, photovoltaic energy is not available under cloudy conditions or during the night. So it is indispensable to have a virtual power plant to handle all these kinds of productions and usages which is something very challenging.

As geothermic here we understand the preferentially surface geothermic which can be used by almost every house by heat pumps. We have to link the technologies for regenerative energies with the technical challenges to retain frequency and voltage for the different plants producing electricity because these power plants all have individual processes. Fuel cells produces DC current and geothermic, photovoltaic, micro-gas-turbine and others produces DC or AC, but none of them are at the same voltage and at the same frequency, and we have to develop technology to control all this production of power with different frequency, voltage and phase to link all together.

We have modeled 50 energy producing/consuming units just to show that this can be handled by intelligent software, which has been developed in collaboration between our computer department and our department of economics. This intelligent software negotiates with the individual producer/buyer the price for selling or buying electricity within the system.

Everything is done mostly automatically to optimize the price for each individual energy producer, energy user, and in general to optimize the economy of the whole community in the most efficient way.

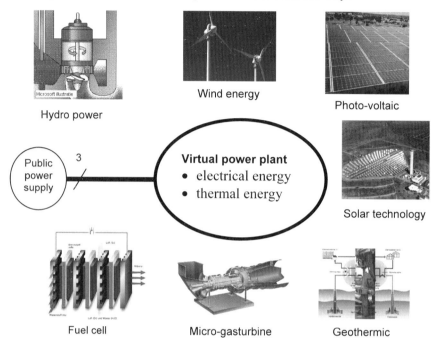

Fig. 6. Energy supply from renewable sources to study

We have also to define a control strategy which is whether deterministic or non deterministic, this means there is somebody deciding about the prices or is an individual negotiating because some small power plant has an specific production of methane or other source of energy and this negotiation is non deterministic. It is then a promising permanent negotiation about price within this virtual power plant. Although it is not a real power plant, it is a power network which may also additionally buy power from a big power production plant.

For a complete system analysis also the machines, maintenance, and other points of view have to be included. The required research for reliable operation of a virtual power plant is summarized in Fig. 7. The big energy producing companies do not really like this vision very much because they may lose their monopolistic situation. However, this may happen anyway within the next 10 years in Germany.

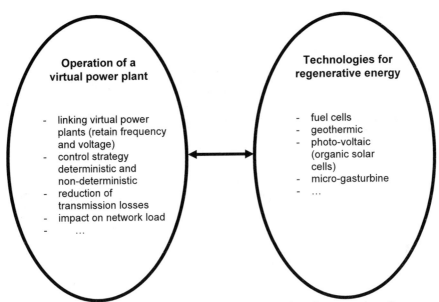

Fig. 7. Required research for reliable operation of a virtual power plant for regenerative energy management

Example of energy performance for future buildings

Finally, as an example of energy performance let us first consider the issue of transportation and traffic. In congested individual transportation, we have engines running all the time, but this is probably not very good. Something should be designed in the car, that when it is in a congested area the engine should stop running, but just running when it is needed. In principle, this is best done by an electric motor and not by a combustion engine relying on hydrocarbons given that speed is not an issue. But hydrocarbons are necessary for power and speed. The complete public transportation system has to be included in this energy research program since you also have to avoid congestion and therefore a traffic guiding system must be developed.

Energy can also be saved by developing construction and modernization of buildings. Sustainability is the superior objective in civil engineering, where the dimensions of sustainability are related to different aspects of the life. In particular, there are social and cultural impacts as well as ecological and economical inputs. These four areas are strongly related to energetic objectives; however, they are difficult to deal with at technical University.

The following example has been developed in the department of Architecture at the University of Karlsruhe. Let us consider the current energy consumption of office buildings within Germany and objectives for the future. This is shown in Fig. 8 for a current building. The energy consumption of a building is given in kWh/m²a, and I have plotted the primary energy needed for operating a building like illumination, air condition, aeration, and heat. However, used is only the end energy as shown in the left column of the figure. There is already a loss of almost 50 % starting from primary energy going to end energy. We have developed in Germany a system, using the heat of the sun to reduce both the primary as well as the end energy. The big difference in numbers is convincing and shows intelligent construction has a high potential for energy efficiency.

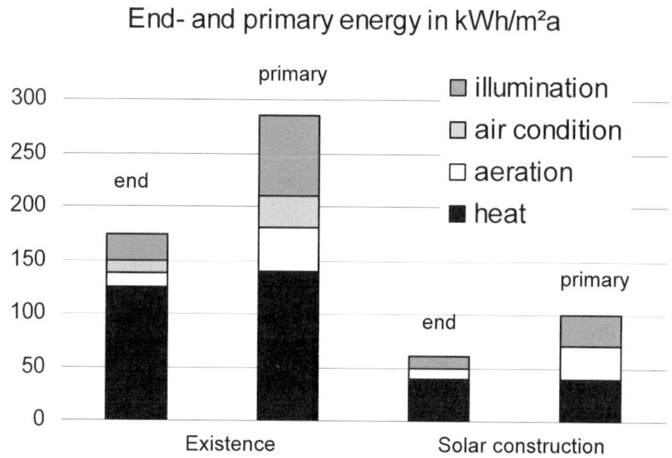

End- and primary energy in kWh/m²a

	Existence		Solar construction	
illumination	25	75	10	30
air condition	11	30	0	0
aeration	13	40	10	30
heat	125	140	40	40

Fig. 8. Current energy consumption of office buildings and objectives for the future

Let me finish by thinking of our children and our future. We have the responsibility to develop for the future for mankind, which means to develop energy concepts of the future.

Energy efficiency in Mexico – a bird's eye view

Mariano Bauer

Instituto de Física, Universidad Nacional Autónoma de México, Ciudad Universitaria, 01000 México, D.F., México. E-mail: bauer@fisica.unam.mx

1 Introduction

> *"Improving energy efficiency is often the cheapest, fastest and most environmentally friendly way to meet the world's energy needs"*

The above sentence can be read immediately below the heading "Energy Efficiency Is Top Priority" in the International Energy Agency report "Energy Technology Perspectives – Scenarios and Strategies to 2050", published in 2006. The report was elaborated at the request of the G8 to provide advice on alternative scenarios and strategies for a clean and secure energy future.

The assertion is not new, but a reiteration of what is been said since the first oil shock of 1973 and its sequel in 1979. The motivation then was without doubt economic. But slowly and fortunately, the consideration and concern for environmental impacts has been added.

Mexico is a country with a growing energy demand, unavoidable certainly if one wants to foster a needed, equitable and on top urgent economic development. Due to its high dependence on hydrocarbons and their derivatives, and the ensuing environmental impacts, energy efficiency can be considered, and in fact is, the main and short term alternative source of energy.

The matter of energy efficiency has many facets. It is not only to achieve the same with less, but in some cases to expend more to obtain much more. An evolution of the economy towards products and services of very high added value justifies greater energy consumption. To improve the cost benefit balance is to make indeed a more efficient energy use.

Today, in the cost benefit balance one cannot leave out the environment. Internalization of environmental costs – local, regional and global – in the energy chains based mainly on hydrocarbons, enhances the value of the alternative energy sources. One, of already significant presence in the world energy supply, nuclear energy, is regaining importance in the development plans of many countries, including Mexico*. This is driven not only by the absence of greenhouse gas emissions, but also by its proven reliability and its technological improvements, and why not, by the volatility in the prices of hydrocarbons and their availability, currently estimated at 60 years at most.

Other alternative sources, marginal today, like solar and wind, but that may provide a substantial amount in the future, are already receiving a larger support in their research, development, demonstration and commercialization phases, although mainly in the developed countries.

2 Some historical developments

At UNAM (the National Autonomous University of Mexico) the subject was broached initially when the Energy Engineering masters program was instituted in 1980 by the Faculty of Engineering.

The University Energy Program of UNAM, created in November1982, devoted two months later it's first Permanent Advisory Forum to the subject of Energy Efficient Use and Conservation. These Advisory Forums, continued periodically until 1997, were directed to review all aspects of the energy scene of Mexico, bringing together specialists and decision makers from academia and the public and private sectors. "Permanent" meant that any subject could and would be taken over repeatedly. The presentations and conclusions of each forum were published. A second round on energy efficiency was indeed carried out in 1988. In the same year the subject of Energy Rationality in the Transport Sector was discussed. In 1986 took place the Symposium on "Energy and Environment", in 1992 the Symposium: "Mexico: the energy-environment relays" and in 1996 the Symposium: "Energy, Environment and Sustainable Development" in collaboration with the Colegio Nacional. The University Energy

---

* Mexico has at present one nuclear power plant at Laguna Verde in the state of Veracruz. It houses two reactors with a total capacity of 1,365 MW. It represents 3 % of the country's installed capacity and supplies 4.75 % of the generation of electricity.

Program also commissioned and supported studies on industrial cogeneration in Mexico and on structural changes for energy saving, which were published. In addition, it integrated a team – engineering faculty professors and students – to carry energy audits in all of UNAM's campuses (over 200 buildings). These resulted in considerable savings and earned twice the National Electric Energy Saving Award – in the category of academic institutions - instituted by the Comisión Federal de Electricidad (CFE – the Federal Electricity Commission) in 1991. The team still operates at the Faculty of Engineering, with continued successes in terms of specialized training of students and of outside projects. It has earned once again the above mentioned award.

Other academic institutions, like the Universidad Autónoma Metropolitana (UAM – Metropolitan Autonomous University) and the Instituto Politécnico Nacional (IPN – National Polytechnical Institute) followed, creating their own programs and integrating the subject in their graduate studies curricula.

With respect to the public sector, a Comisión Nacional de Ahorro de Energía (CONAE – National Commission for Energy Saving) was created in 1989. Its objective was stated as: "… to act as a technical consulting organ for the dependencies and entities of the Federal Public Administration, for the governments of the federal entities, for the municipalities and the private individuals, in matters of saving and efficient use of energy, and of the utilization of renewable energies". Beyond this seemingly advisory only role, it was however assigned many faculties, like preparing national programs, promoting and supporting scientific and technological research, participating in the design of Mexican Official Standards, implementation of labeling of equipments and appliances, among several others (see Appendix).

To date there have been put into effect 18 Mexican Official Standards, that are compulsory for manufacturers and developers. They determine the minimum energy efficiency of household appliances, air conditioners, water heaters, electric motors, compact fluorescent lamps (CFL), and thermal insulations for buildings. CONAE reports that from 1995 to date, the standards related to electricity have resulted in accumulated savings of 16,605 GWh in consumption and 2,296 MW in installed capacity. With respect to household water heaters, the accumulated savings amount to 997,335 m^3 of Liquid Petroleum Gas (LPG). Industrial thermal insulation is credited for an annual saving of 38,600 cubic meters of LPG.

CONAE is also participating in the North America Energy Working Group (NAEWG) together with Canada and the United States, to establish a joint approach to normalization and labeling of energy efficiency in North America.

Also established in August 1990 was the Fideicomiso para el Ahorro de Energía Eléctrica (FIDE – Trust Fund for Electric Energy Saving) whose successful achievements have just received the International Star of Energy Efficiency Award accorded by the NGO "Alliance to Save Energy", being the first time it is given to an entity outside the US. Created at the instance of the CFE, the key of its success has been having from the start an operating fund to finance projects – some by non refundable grants and others with loans at very low or even zero interest rate, to be repaid from the savings obtained. The fund receives yearly allowances from the Mexican Electric Workers Union (SUTERM) and the private sector that are then matched by CFE.

CFE and FIDE launched the first large scale project of energy saving in Mexico in 1993, called ILUMEX. It consisted in procuring and placing 1.7 millions of Compact Fluorescent Lamps (CFL) in households in two major cities, Guadalajara and Monterrey. The cost was estimated at 23 million USD, and was financed by the Mexican Government (ten million), by a grant from Global Environmental Fund (GEF) (ten million) and by the government of Norway (three million).

A similar project was undertaken in 2002 by Luz y Fuerza del Centro (LFC – Light and Power of the Center), the public company that services mainly Mexico City. It was terminated in December 2004, having installed 518,393 CFL. The energy saving was estimated at 3,810 kW in demand and 4,867,083 kWh/year in consumption of electricity.

Through the Program for Financing Electric Energy Saving and other projects, FIDE reports accumulated totals, up to 2004, of thermal insulation of 10,518 households, of replacement of 124,972 air conditioning equipments and 132,243 refrigerators, of substitution of 15.5 million incandescent lamps by CFLs and of carrying out 201,306 energy audits.

FIDE is currently launching an incentives program aimed at substituting conventional electric motors for high efficiency ones, with the purpose of submitting it to the Clean Development Mechanism of the Kyoto Protocol and obtaining funds through the sale of carbon certificates.

The Federal Government accounts for 33 % of the energy consumed in the country, given that it includes the public energy sector, integrated under the Constitution by the State monopolies Petróleos Mexicanos (PEMEX) – oil, gas and basic petrochemicals –, and CFE plus LFC – generation, transmission and distribution of electricity for public service –. Consequently the Energy Secretariat (SENER) instituted a Program for Energy Saving in the Federal Public Administration that contains directives to all sectors. In accordance, they have all initiated internal energy saving programs or reinforced existing ones such as the Program for Energy Saving in the Electric Sector (PAESE - Programa para el Ahorro de Energía en el

Sector Eléctrico) that was also launched by CFE in 1990. To give an example of achievement, Petróleos Mexicanos and its Subsidiaries obtained six 2004 National Awards in Thermo Energy Savings, granted by the Energy Ministry:

- "Petróleos Mexicanos" was granted first place under the category of Integral Programs for the design, development and setting of the Institutional Program of Efficient Use and Energy Savings (PROENER).
- "Pemex Refinación" was awarded first place in the "Best Practices" category, for launching a monitoring and control system in the Refinery "Ing. Héctor Lara Sosa" in Cadereyta, Nuevo Leon, decreasing gas consumption in over 9,000 Million Cubic Feet (MCF) in the last two years.
- The Refinery "Ing. Antonio M. Amor of Pemex Refinación", was awarded the prize in the line of "Facilities Modernization" by increasing efficiency of direct fire heaters in the Combined Plant "AS".
- The PGPB Cactus Gas Processing Facility, was granted two energy national awards:
  - In the "Facilities Modernization" category, for setting two heat recovery units leading to reduce the flaring of gas for over 4 MCF;
  - In the "Best Practices" category, for the solvent change in the gas sweetening process with economic savings of 4.5 million pesos for the use of reagents and feedback steam.
- "The Best Practices Award" was granted to the Ciudad Pemex Gas Processing Center in Tabasco, because of a reduction in gas consumption for over 1,000 MCF per year.

3 The demand management approach

To improve energy efficiency in a society does not consist alone of technological changes in the energy chains. Structural and organizational changes can also have considerable impact on the end use.

One instance of the demand management approach has been the establishment of the Daylight Saving Time or Summer Time since 1996. To note is that most of the country lies in latitudes below the tropic of Cancer, so that daylight hours do not vary very much throughout the year. Nevertheless, it is claimed that up to 2005 it has resulted in accumulated energy and demand savings of 11,133 GWh and 982 MW, respectively, with an accumulated deferred investment of 10,340 million pesos (SENER 2005a). Due to the fact that 75 % of the power generation is carried out with fossil fuels, this results in an avoidance of GHG emissions.

The demand management approach has however not been applied substantially in México. Indeed considerable impacts could be achieved in the transport sector, where an accelerated growth in the vehicle park, in particular private automobiles (Bauer et al. 2003) is resulting in a surge in the demand of gasoline and diesel.

For instance, since 1986, the University Energy Program surveyed the displacements requirements of the population in the Metropolitan Area of Mexico City, the transport modes and the ensuing fuel consumption. It was then pointed out that a modification of the labor laws to allow a Compressed Work Week (CWW) – to comply with the 40 or 48 weekly hours with one day less attendance to the work place (4 or 5 days instead of 5 and 6) and an appropriate extended daily schedule – could result in a reduction in the demand for transport, optimally up to 20 %. This would translate into a reduction of gasoline consumption and of the ensuing environmental pollution that affects Mexico City. It would also result in a benefit for the worker that expends a large amount of time and a significant portion of his income to travel to work and back each day. I estimate that the measure could result in at least a 10 % reduction of the 16 million liters of fuels (gasoline and diesel) consumed daily at the time (Quintanilla et al. 1998). Currently the consumption has risen to about 20 million liters per day, due to the increase in the vehicle park noted above.

Unfortunately, although it would benefit a large amount of the working force that spends over 3 hours daily going to work and back in public and private transport at a significant expense, this has not been implemented in Mexico because it involves changing the labor laws, an issue that would face severe political challenges, not really associated with the issue.

Measures like this have already been implemented in several countries (TDM 2006; CCP 2006) with encouraging results.

4 Conclusions

The energy intensity in Mexico, as well as the emissions intensity, has certainly improved in the last ten years, as shown in the Fig. 1.

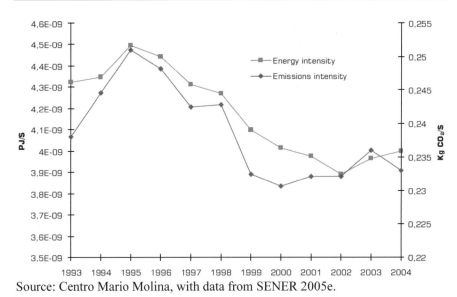

Source: Centro Mario Molina, with data from SENER 2005e.

Fig. 1. Energy intensity (primary energy consumption in PJ/$) and emission intensity (emissions of CO_2 in Kg/$). Pesos ($) in 1993

This is certainly due to the general trend of increasing the participation of the services sector in the economy; the modernization of many industries with up to date equipments; the increased share of the natural gas combined cycle technology in the power sector; the new opportunities for cogeneration; the improvement of the average efficiency in the vehicle park bought by the sales of new cars and in spite of the periodic setbacks created by the regularization of illegal older cars, referred to as "autos chocolate".

But it is also certain that energy saving policies have contributed to the energy and emission intensities.

Appendix

Article 3 of presidential decree on creation of National Commission for Energy Saving (CONAE) as a decentralized member of Energy Ministry (CONAE 1989).

The National Commission for Energy Saving has the following faculties:

1. To issue administrative dispositions in matters of saving and efficient use of energy, in accordance with the applicable legal dispositions;

2. To promote the efficiency in the use of energy by coordinated actions with the diverse dependencies and entities of the Federal Public Administration and with the governments of the federated entities and municipalities and, through concerted actions, with the social and private sectors;

3. To prepare the national programs on matters of saving and efficient use of energy and promotion of the utilization of renewable energies, to submit them to the consideration and, eventually, to the authorization of the Energy Secretariat;

4. To formulate and propose to the Federal Executive, through the Energy Secretariat, the operation, investment and financing programs required in the short, medium and long ranges to comply with the objectives in matters of saving and efficient use of energy and the utilization of renewable energies;

5. To promote and support the scientific and technological research in matters of saving and efficient use of energy, as well as the utilization of renewable energies;

6. To promote, and, eventually, support the activities aimed at obtaining and disbursing the funds from public and private financing sources, for the implementation of actions for the saving and efficient use of energy;

7. To promote the mechanisms that allow the development, manufacturing and utilization of products, gadgets, apparatus, equipments, machinery or systems for the saving and efficient use of energy, as well as the utilization of renewable energies;

8. To facilitate, according to the bases established in this Decree and other applicable judicial ordinances, the participation of specialized consultants and enterprises in the realization of the actions to promote the efficient use of energy;

9. To provide technical, advisory and other services in matters of saving and efficient use of energy;

10. To integrate, analyze and, eventually, diffuse the information on saving and efficient use of energy provided, in accordance with the applicable legal dispositions, by the individuals that carry out these activities within the national territory;

11. To carry out diffusion activities to reach its objectives on energy efficiency, within the time available to the State in radio and television in accordance with the respective dispositions that the law in this matter establishes, and

12. To participate in the elaboration of Mexican Official Standards in coordination with the dependencies that has the faculties to establish

them in the branch of the saving and efficient use of energy, and the utilization of renewable energies.

References

Bauer M, Mar E, Elizal A (2003) Transport and energy in Mexico – the personal income shock. Energy Policy 31:1475–1480

Quintanilla J, Mulás P, Guevara I, Navarro B, Bauer M (1998) Modified urban labor week for energy saving and pollution reduction in the transport sector. In: Proceedings of the 17th World Congress of WEC, Houston, USA, paper 3.3.08, Vol. 5, pp 483–492

CCP (2006) Commuter Challenge Programme "Flexible Work Schedules"

CONAE (1989) Official federal diary, Sept 20, 1989. CONAE webpage

SENER (2005a) Prospectiva del Mercado de Gas Licuado de Petróleo 2005–2014

SENER (2005b) Prospectiva de Petrolíferos 2005–2014

SENER (2005c) Prospectiva del Sector Eléctrico 2005–2014

SENER (2005d) Prospectiva del Mercado de Gas Natural 2005–2014

SENER (2005e) Balance Nacional de Energía 2004

SENER (2006) Energías Renovables para el Desarrollo Sustentable en México

TDM (2006) Transport Demand Management Encyclopedia. "Alternative Work Schedules".Victoria Transport Policy Institute

Information links

Secretaría de Energía (SENER): www.energia.gob.mx

Petróleos Mexicanos (PEMEX): www.pemex.gob.mx

Comisión Federal de Electricidad (CFE): www.cfe.gob.mx

Comisión Nacional para el Ahorro de Energía (CONAE): www.conae.gob.mx

Fideicomiso para el Ahorro de Energía Eléctrica (FIDE): www.fide.cfe.gob.mx

Commuter Challenge Program: www.commuterchallenge.org

Victoria Transport Policy Institute: www.vtpi.org

Energy efficiency and conservation in Mexico

Odón Demofilo de Buen-Rodríguez

Energía, Tecnología y Educación, Calle Puente Xoco 39, Col. Xoco. C.P. 03330 México. D.F. E-mail: demofilo@prodigy.net.mx

Abstract

Based on documental references, this paper enumerates and briefly describes Mexico's main energy conservation programs and institutions, highlighting the main elements and factors for their success. These programs and institutions have been in place since the end of the 1980's, have been generally successful and are certainly the main drivers for a significant reduction of the country's energy intensity in the last twenty five years. The document also describes some of the barriers and challenges for future energy conservation policies.

1 Introduction

Present concerns about the price of conventional energy sources and its environmental impacts have renewed the interest in alternatives to provide energy services at lower costs and reduced impact to the environment. Also, as greenhouse gases in the atmosphere have reached concentrations beyond historical levels and are bringing about Global Climate Change, there is an international drive to promote all possible alternatives.

Among the alternatives to reduce energy costs and environmental impacts is energy efficiency and conservation, as many energy services (such as lighting or comfort) can be obtained with lower levels of conventional energy sources.

After a response to the 1970's oil crisis that focused on increasing its oil production capacity to take advantage of higher prices, Mexican authorities recognized the opportunities for energy conservation in the Mexican economy and began to take action by reducing its energy intensity in the early 1980's. These actions took advantage of the technological developments that followed the oil crisis and of close to ten years of international experiences on energy conservation programs and projects.

More recently, concerns about diminishing oil reserves have renewed interest in energy efficiency and conservation and should be driving authorities to new, improved and larger programs.

2 Mexico's economic and energy context

Mexico is a country with nearly 105 million inhabitants living in, close to 22 million of households in a territory of approximately 2 million square kilometers (INEGI 2006a).

Mexico is one of the most important oil producers in the world, with an average daily production of 3.8 million barrels of oil, of which 47 % are for domestic consumption. Natural gas production averaged 4.8 billion cubic feet per day in 2005, mostly as gas associated with oil production (PEMEX 2006).

Of total domestic energy use in 2004, the energy sector consumed 2,400 Petajoules, while 3,800 petajoules were directed to end-users. This domestic supply was comprised of oil (65 %), natural gas (23.8 %), hydro (3.8 %), coal (3.8 %), nuclear (1.4 %), and others (1.7 %) (SENER 2005).

3 Energy efficiency and conservation in Mexico

The efforts undertaken in Mexico to use energy more efficiently have been driven by a number of factors and are characterized by the continuous institutional development over the past seventeen years. The institutions directly involved in this process have been the National Commission for Energy Conservation (CONAE), the Trust Fund for Electricity Savings (FIDE), the Electricity Sector Energy-Savings Program (PAESE), and the Trust Fund for the Thermal Insulation Program of Households in the Mexicali Valley (FIPATERM).

3.1 The National Commission for Energy Conservation (CONAE)

CONAE is a decentralized administrative agency that is part of Mexico´s Energy Secretariat, with technical and operational autonomy and the objective to serve as a technical advisory body for agencies and entities of Mexico's public and private sectors on issues related to energy savings, energy efficiency and renewable energy use. The most remarkable programs developed by CONAE, in terms of scope and impact, are the mandatory energy-efficiency standards (NOM), the federal buildings program, the technical support to PEMEX internal program, and the programs undertaken in the private industrial and service sectors.

3.2 The Trust Fund for Electricity Savings (FIDE)

The Trust Fund for Electricity Savings (FIDE) is a private, non-profit organization created by the Federal Commission of Electricity (CFE) in 1990 and is supported by a small internal tax on CFE's suppliers and by loans from international development banks. FIDE's objective is to promote actions that encourage and foster electricity conservation and its rational use. FIDE has provided financing for hundreds of energy audits and the purchase of several million units of energy efficient lamps, motors and AC units, thus achieving important energy savings in industrial and commercial installations of the private sector, as well as municipal lighting and pumping systems. As a main result, it has contributed significantly to the development of the energy efficient equipment and systems market, and a solid and highly skilled consultant pool.

3.3 The Electricity Sector Energy-Savings Program (PAESE)

The Electricity Sector Energy-Savings Program (PAESE) was created in 1989 as part of CFE and as an evolution of a previous program operating since 1982 (the National Program for Rational Electricity Use – PRONUREE). PAESE basically operates with a network of specialized professionals which provides support to end-users on issues related to electricity-efficiency improvement. More recently, its work has concentrated in CFE's own installations (mainly buildings).

3.4 Trust Fund for the Thermal Insulation Program of Households in the Mexicali Valley (FIPATERM)

The first systematic effort undertaken to achieve energy savings in Mexico was designed and implemented by PRONUREE to reduce energy consumption from AC use among residential end-users in the city of Mexicali, Baja California[1]. In 1989, based on the PRONUREE's program design, CFE created a trust fund (FIPATERM), which has the objective of providing financing for thermal insulation of high energy-consumption residential users.

4 Main programs and results

Since 1989, a number of programs have been implemented in Mexico by FIDE, CFE and CONAE.

4.1 Energy-Efficiency Standards.

One of the main mandates that CONAE has had from its inception has been the implementation of energy efficiency standards for a number of energy-using equipment and systems. This process started in 2003 and 18 standards for equipment and systems are presently in place and are applied to more than 5 million units which are commercialized in Mexico every year.

The best example of the implementation process of energy-efficiency standards in Mexico is that for refrigerators. The implementation of this standard –that became mandatory in 1996 and had two changes in the following six years–, has meant that refrigerator unit energy consumption has been reduced in more than 60 % in ten years. These efforts crossed an important threshold in 2002, as three of the standards were harmonized with those applied in the United States and Canada.

Energy-efficiency standards have represented significant energy savings. According to a study by Mexico's Electric Research Institute (IIE) done in collaboration with the Lawrence Berkeley National Laboratory (LBNL) more than 52,000 GWh have been saved of electricity in 10 years by CONAE's energy-efficiency standards, which is equivalent to one fourth of total power generation in Mexico in 2005 (Sánchez and Chu 2006).

---

[1] Mexicali is located in the State of Baja California on the Mexico-US border.

4.2 Daylight Savings Time

Daylight Savings Time is a measure by which clocks are moved one hour forward at the beginning of the spring and one hour back in the fall to make better use of natural light. This measure came into effect country-wide in Mexico in 1996. This was possible due to a concerted national consensus process led by FIDE with the support of CFE, CONAE and SENER (Mexican Energy Ministry). Energy conserved by this measure over the past ten years, since it was first applied, amounts to approximately 11,000 GWh, equal to the electricity consumed by all of Mexico's house-holds in a period of more than three months.

4.3 ASI Program

ASI is the program operated by FIPATERM. The program has promoted the widespread installation of thermal insulation on the roofs of the most energy-intensive Mexicali's households. FIPATERM has provided low in-terest loans, allowing the CFE residential customers to pay the cost of the insulation through the electricity bill. By the year 2000, FIPATERM achieved the insulation of more than 60 thousand roof-tops in Mexicali. The program now offers loans for door and window insulation, as well as for the purchase of efficient equipment, such as air conditioners, refrigera-tors, and CFLs. It also includes the performance of energy audits to iden-tify the feasibility of energy-savings measures in the households. By the first semester of 2004 FIPATERM had granted credits for close 60 million US Dollars (USD) for all of the measures.

4.4 ILUMEX

ILUMEX was the first large-scale residential-lighting program designed and implemented in Mexico. The program had the support of a 10 Million USD grant from the World Bank's Global Environmental Facility (GEF). CFE provided an equal amount and the Government of Norway granted 3 Million USD. The program was implemented in 1995 in Guadalajara and Monterrey, the two largest cities in Mexico after Mexico City. More than two and a half million CFLs were installed through ILUMEX up to the year 1999 when the program concluded, resulting in energy savings of more than 300 million kWh and an important reduction of GHG emissions (de Buen 2004).

4.5 FIDE's Incentives Program

This program was operated with resources from a loan granted by the In-teramerican Development Bank (IDB) and additional financial resources from CFE and FIDE. The resources were used for rebates to end-users that purchase high-efficiency technologies. This program included the use of a seal (Sello FIDE) that identified those products that could be subject to re-bates. The program concluded in 2004 and a follow up is being negotiated with the IDB. Through this program, the entire three-phase induction mo-tors markets were transformed, as well as 40 % of the lighting systems market (T-8 fluorescent lamps and low-loses ballasts), and 80 % of the market for compressed-air equipment with capacity higher than 20 HP (FIDE 2006).

4.6 Pemex Energy Conservation Campaign.

CONAE supported energy conservation actions in Pemex for more than ten years, mostly as a process of best-practices implementation. The strat-egy that CONAE followed evolved from the analysis of isolated systems (such as cooling towers and gas furnaces) to the development of an Inter-net-based audit system that led to the implementation of a permanent en-ergy efficiency campaign with significant environmental and economic re-sults (de Buen et al. 2003). According to Pemex, during the 2000–2003 period the oil company reduced its fuel consumption, gas leaks and gas combusted to the atmosphere for an equivalent of 36.6 Million BOE, with an estimated value of 762 Million USD (PEMEX 2004).

4.7 Federal Administration Buildings.

A mandatory energy conservation program for the largest buildings of the Federal Administration was introduced in 1999. By 2002, the program had incorporated 896 buildings, representing 3.8 million square meters of of-fice space. This program, which requires mandatory reports on electricity consumption and the energy-saving measures undertaken, saved 110 GWh from 1999 to 2002 (CONAE 2006).

5 Overall results

Overall, the energy intensity of the Mexican economy has been diminishing at a rate of 1–3 % per year from 1995 to 2004 (SENER 2005), (Fig. 1). Also, per capita CO_2 emissions have practically not changed in the 1990–2003 period, as the economy grew 44 % and population by 17 % (de Buen and Bustillos 2006).

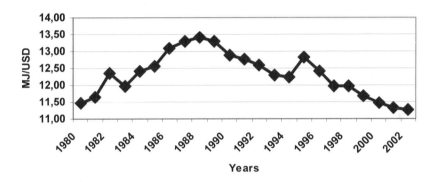

Fig. 1. Evolution of Mexico's energy intensity 1980–2002 (SENER 2005)

6 Barriers and challenges for the future

Mexico's good results in energy conservation are still a fraction of the technical and economic potential for actions of this type. Very general estimates put the potential in 20 % of present consumption.

Existing barriers in Mexico are almost the same to those found in other parts of the world, although these barriers have a different effect on energy-savings project developments.

6.1 The energy sector dominated by a supply side perspective

Even with the clear opportunities and the good results and institutions in Mexico, energy conservation is not seriously considered as an alternative to the expansion of energy supply. This has been reflected by the attempts to close CONAE, which have fortunately failed.

6.2 Energy prices which do not reflect their real cost

The best and foremost policy for energy conservation is real energy prices. In Mexico, the pricing policies which are established by the federal government grant subsidies to certain sectors, particularly the agricultural and the residential. These subsidies represent in some cases more than 50 % of the real cost, which reduces energy-saving measures feasibility for energy users. In 2003, energy subsidies for power customers were greater than 6,000 Billion USD (SENER 2006).

6.3 The growth of the services' sector and its installations

Mexico's service sector has grown faster than the general economy and that of industry (Fig. 2), reaching 70 % of the GDP in 2005. New estimates of energy consumption in buildings (commercial and residential) reveal that it may be as important for electricity consumption as industry (de Buen and Bustillos 2006).

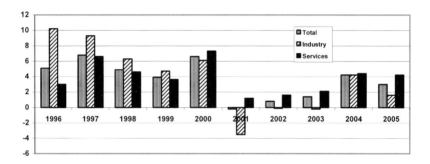

Fig. 2 Annual growth rates of Mexico's GDP, 1996–2005 (INEGI 2006b)

This should drive energy conservation in buildings as a policy priority. Unfortunately, this has not been the case as existing building standards have not been enforced (AEAEE 2005)

7 Conclusions

As primary, fossil-fuel energy prices remain high and environmental concerns may sooner than later drive to stricter measures to the use of fossil fuels, the need for new approaches to fulfill a growing demand for energy services is pushing for alternatives to the dominant energy sources.

Mexico has great challenges in this context as it has a high dependence on oil as reserves have been dwindling in the last decade. Also, as a signatory of the Kyoto Protocol, it is committed to the international efforts to reduce greenhouse gas emissions.

Fortunately, Mexico has a solid institutional structure that has led the design and implementation of successful energy efficiency and conservation programs. However, strengthening of present institutions and programs, and new approaches are needed to respond to the new challenges.

References

AEAEE (2005) Asociación de Empresas para el Ahorro de Energía en la Edificación. Perspectiva y recomendaciones para el confort y el ahorro de energía en la edificación en México. http://www.funtener.org/pdfs/libroblanco.pdf

CONAE (2006) Comisión Nacional para el Ahorro de Energía. Las NOM de Conae para electrodomésticos le ahorraron al país 52, 700 GWh en los últimos diez años. http://www.conae.gob.mx/wb/distribuidor.jsp?seccion=20; Informe de Labores 2005. http://www.conae.gob.mx/work/sites/CONAE/resources/LocalContent/3802/3/informedelabores2005.pdf (Accessed: 2006, Nov 15)

de Buen O, Gutierrez D, Ramos G, Valdivieso E, Gómez S (2003) A Strategy for Energy Efficiency Actions in the Mexican Industrial Sector: The Pemex Experience. In: Proceedings of the ACEEE Summer Study on Energy Efficiency in Industry. July 30 2003, New York

de Buen O (2004) ILUMEX: desarrollo y lecciones del primer proyecto mayor de ahorro de energía en México. http://www.ine.gob.mx/ueajei/publicaciones/libros/437/odon.html

de Buen O, Bustillos I (2006) Energy and Sustainable Development in Mexico. Helio International Sustainable Energy Watch 2005–2006 http://www.helio-international.org/reports/pdfs/Mexico-EN.pdf

FIDE (2006) Fideicomiso para el Ahorro de Energía Eléctrica. Ahorros directos de Energía Eléctrica por Sector y Programa hasta el segundo trimestre del 2005 http://www.fide.org.mx/resultados/index.html (Accessed: 2006, Nov 15)

INEGI (2006a) Instituto Nacional de Estadística, Geografía e Informática Población total annual según sexo. http://www.inegi.gob.mx/est/default.asp?c=119 (Accessed: 2006, Nov 15)

INEGI (2006b) Producto interno bruto anual por actividad económica de origen. Variaciones anuales. http://www.inegi.gob.mx/est/contenidos/espanol/rutinas/ept.asp?t=cuna17&c=1663 (Accessed: 2006, Nov 15)

PEMEX (2004) Pemex a la vanguardia en ahorro de energía http://www.pemex.com/index.cfm?action=content&sectionID=8&catID=40&subcatID=2284

PEMEX (2006) Petróleos Mexicanos. Anuario Estadístico 2006, http://www.pemex.com/files/content/Anuario__2.pdf

Sánchez I, Chu H (2006) Assessment of the Impacts of Standards and Labeling Programs in Mexico (Four Products). Report No. 12933ITF FN LBL 001

SENER (2005) Secretaría de Energía. Balance Nacional de Energía 2004, http://www.sener.gob.mx/wb/distribuidor.jsp?seccion=65

SENER (2006) Secretaría de Energía. Subsidios a Tarifas Eléctricas http://www.sener.gob.mx/wb2/SenerNva/ibEst (Accessed: 2006, Nov 15)

Part II

Traditional Energy Resources

Status of the Mexican Electricity Generation

Gustavo Alonso

Instituto Nacional de Investigaciones Nucleares, Apartado Postal 18-1027, México DF 11801, México. Email: galonso@nuclear.inin.mx

Abstract

Mexico has a major challenge to go from being an emergent economy to an industrialized country. Its electrical installed capacity has been growing during the last 10 years with a 4.5 % annual pace and it is planned to grow for the next 10 years with a 5.2 % annual pace. In 2005 the annual electrical consumption per inhabitant was of 2237 kWh, which is around the world's average. This represents almost one quarter of the average of the industrial countries consumption. The current document shows the prospective for the Mexican electricity sector for the 2005–2014 time frame. It also shows the technologies that will be used to cover the requirements of electricity by region.

1 Introduction

At present Mexico's major challenge is to raise its life standards; part of this problem is to supply electricity to all Mexican regions and for everyone. Mexico is an emergent economy and in 2005 it had an annual average consumption per inhabitant of 2,237 kWh, which is just about the world average. The last year (2006), this average raised to 2,300 kWh (EIA 2004; SENER 2005), in contrast the figure for developed countries is 8,500 kWh.

The installed electricity capacity of the country, from 1995 to 2005, has grown at 3.58 % average annual rate. The plan is to grow at 5.2 % average annual rate for the next 10 years (EIA 2004). Currently the population growth is at 1.44 % annual rate (SENER 2005).

In view of the world's scarcity of primary fuels there is a strong need to consider the diversity of electric generation sources. Mexico, as many other countries has been growing mainly through the use of combined cycle plants that burn natural gas. Recent gas prices volatility makes to wonder about which can be the best option to fulfill energy requirements.

2 Mexican electrical sector

In Mexico, the electric power sector remains largely under state control. However, as result of the 1992 amendment to the Mexican electricity law, some segments of the electric power generation sector are currently open to private participation (WM 2004). Private companies are allowed to generate electricity for areas not considered of "public service." They include generating electricity for export and generating electricity for public service during an emergency. Self- or co-generators and small producers may generate electricity for their own use, and independent power producers are permitted to sell excess power to the Federal Electricity Commission (CFE) under long terms contracts.

From 1995 to 2005 the installed electric capacity has been growing as shown in Table 1, where it can be seen the contribution of the private companies (PIE's). It is also shown in that table, that the generating capacity has been growing at a 4.37 % average annual rate due to better generating practices and higher availability power plants factors.

Table 1. Evolution of the Mexican Electrical Sector

Year	Installed capacity [MW]			Generation [GWh]		
	CFE	PIE's	Total	CFE	PIE's	Total
1995	32,166		32,166	140.82		140.82
1996	33,920		33,920	149.97		149.97
1997	33,944		33,944	159.83		159.83
1998	34,384		34,384	168.98		168.98
1999	34,839		34,839	179.07		179.07
2000	35,385	484	35,869	190	1.21	191.21
2001	36,236	1,455	37,691	190.88	4.04	194.92
2002	36,855	3,495	40,350	177.05	21.83	198.88
2003	36,971	6,756	43,727	169.32	31.62	200.94
2004	38,422	7,265	45,687	159.53	45.86	205.39
2005	37,325	8,251	45,576	170.07	45.56	215.63

Figure 1 shows the contribution by source type to the total installed capacity at December 2004 and Fig. 2 illustrates how power plants are distributed through the whole country.

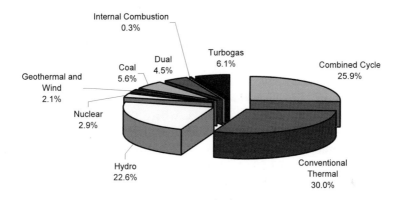

Fig. 1. Installed Capacity Contribution by source type

Fig. 2. Electrical Power Plant Type Distribution

3 Forecasting scenarios

To estimate the consumption and demand of electricity for the next 10 years several macroeconomics assumptions must been taken into account along with the recent evolution of the Mexican electricity sector and regional studies to determine electricity requirements.

The prospective of the Mexican Electricity sector is an annual planning exercise. The Mexican Ministry of Energy defines three economic scenarios –planning, high and low– to be used in the electricity consumption estimations. The planning scenario considers the most likely economic forecast for the next 10 years (2005–2014), where the starting point for the annual planning cycle is to make a new electricity consumption estimation.

For the planning scenario the gross domestic product (GDP) annual growth rate scenario is 4.3 % (4.7 % in 2004). In the high and low scenarios the projected rates are 5.2 % and 2.8 % respectively (5.6 % and 3.2 % in 2004, respectively).

On the other hand, Mexican GDP has not growth as it was expected. This is shown in Table 2. Thereby, some programmed projects have been delayed. Since in the short term it is not possible to delay the projected new power plants, just 26 generating plants have been delayed in the medium term (more than 5 years) because of the mentioned slower GDP pace.

Table 2. Real Gross Domestic Product annual growth rate evolution

Year	Gross Domestic Product annual growth rate [%]
2000	6.60
2001	-0.16
2002	0.83
2003	1.41
2004	4.36

The ways to expand the installed electricity and generating capacity is done trough optimization models. Some of the main parameters involved in these models are: the fuel price forecast along with the investment cost for the different technologies. The values for these variables must be estimated.

In the latter procedure, in particular, the recent high volatility of natural gas price and the gas supply uncertainty, create a risk if taking the decision for future expansion through this source. As a measure to guarantee the gas supply, the Mexican government has decided to build a LG plant in Altamira (Gulf of Mexico) and a second project in the Pacific area. The forecasted gas price for the next 30 years has been estimated as 4.87 USD/mmBTU.

Also other types of power plant are under study to diversify the sources; among them is the nuclear option. The optimal power plant mix is the one that can satisfy electricity demand under a minimal global cost (investment plus generating cost).

4 Electricity gross consumption

The forecast for the gross consumption for 2005–2014 can be found out using the macroeconomic assumptions and regional studies. This is shown in Table 3.

Table 3. Gross Consumption in the Planning Scenario

Year	Gross Consumption [GWh]	Growth rate [%]
2000	184,194	6.38
2001	187,661	1.88
2002	192,307	2.48
2003	197,242	2.57
2004	203,398	3.12
2005	214,160	5.29
2006	224,158	4.67
2007	235,357	5.00
2008	246,809	4.87
2009	260,452	5.53
2010	275,929	5.94
2011	291,105	5.50
2012	306,749	5.37
2013	322,263	5.06
2014	338,890	5.16
Average		5.2

In addition, the gross demand by region for 1995–2004, without considering cogeneration and exporting, is shown in Table 4 and Table 5 shows the demand forecast by region for 2005–2014 period. Also shown the average growth rate (AGR) in percents.

As it was mentioned the Mexican electricity prospective is an annual feedback exercise that is corrected through the time. As an example, in 2002 the electricity demand forecast for 2009 was of 41,440 MW, while in 2005 the current forecast was of 37,922 MW.

Table 4. Gross Electricity Demand 1995–2004 [MW]

Region	1995	1996	1997	1998	1999	2000	2001	2002	2003	2004	AGR [%]
Central	5,819	6,347	6,447	6,884	7,181	7,439	7,700	7,737	7,874	8,047	3.2
East	4,352	4,463	4,528	4,797	4,954	5,058	5,291	5,373	5,434	5,425	3.6
West	4,688	4,837	5,209	5,472	5,702	6,062	6,157	6,345	6,632	6,523	3.7
North-West	1,911	2,041	2,182	2,195	2,217	2,365	2,496	2,457	2,491	2,606	3.7
North	1,790	1,887	1,937	2,163	2,231	2,421	2,516	2,660	2,720	2,853	5.2
North-East	3,693	4,005	4,307	4,662	4,759	5,245	5,558	5,676	5,688	6,148	5.7
Baja California	1,388	1,458	1,329	1,393	1,491	1,695	1,698	1,699	1,823	1,856	3.5
Baja California South	153	164	170	181	186	204	224	215	214	234	4.8
Peninsular	671	702	737	805	839	908	971	985	1,043	1,087	5.0
Small Systems	16	17	19	19	20	21	22	22	22	24	4.8

Table 5. Gross Electricity Demand Planning Scenario 2005–2014 [MW]

Region	2005	2006	2007	2008	2009	2010	2011	2012	2013	2014	AGR [%]
Central	8,347	8,762	9,120	9,453	9,852	10,298	10,744	11,196	11,615	12,069	4.1
East	5,788	6,187	6,533	6,890	7,305	7,743	8,171	8,617	9,057	9,558	5.8
West	6,928	7,412	7,882	8,302	8,801	9,344	9,871	10,423	10,961	11,493	5.8
North-West	2,747	2,873	3,003	3,140	3,295	3,473	3,650	3,821	3,969	4,135	4.7
North	3,042	3,216	3,365	3,542	3,726	3,941	4,163	4,382	4,607	4,832	5.4
North-East	6,396	6,673	7,000	7,363	7,776	8,293	8,814	9,352	9,904	10,511	5.5
Baja California	1,969	2,060	2,164	2,294	2,435	2,599	2,774	2,965	3,153	3,338	6.0
Baja California South	248	263	283	304	327	352	379	405	431	461	7.0
Peninsular	1,160	1,229	1,308	1,391	1,492	1,604	1,709	1,816	1,938	2,072	6.7
Small Systems	25	26	29	30	32	33	34	36	38	39	5.0

5 Electricity prospective

As result of the prospective for the period 2005–2014, the amount of additional electricity that is required by the grid system is 22,126 MW.

That capacity is divided in two types, the first one is call committed capacity and it means that the project is committed or it is in the construction process. The second one is non-committed projects, in this category there are two classes, the first class is the one where the technology is already decided and for the second class the technology needs to be selected (free).

According to the prospective 2005–2014 there are 6,184 MW as already committed capacity and 15,942 under the non-commited capacity. From these there are 6,178 as free technology.

Table 6 shows how the additional capacity will be covered by the different technologies. As it can been seen in that table, the main contribution is trough combined cycle power plants. This technology has a high efficiency in the energy conversion process and low emissions.

However, because the high volatility gas prices or limited gas supply, other scenarios need to be considered. Less number of natural gas combined cycle facilities will be participate in the scenario while those based on coal gasification, nuclear, hydro and wind will be considered as viable options.

Currently, Mexico can supply only 70 % of its natural gas demand; the other 30 % comes from abroad. As already mentioned, to guarantee the national gas supply, one Liquid Gas plant has been built in the Gulf of Mexico. This plant will start commercial operation in October 2006, it has a daily capacity of 3000 millions of cubic feet (MCF) and it will increase its capacity to 500 MCF by January 2007. In addition, there is another LG plant in the pacific coast with a daily capacity of 500 MCF and one more in Baja California with a daily capacity of 230 MCF.

Table 6. Power Plants Additional Capacity by Technology [MW]

Technology	Committed	Non-Committed	Total
Combined Cycle	4,555	7,128	11,683
Hydro	754	1500	2,254
Coal	700	0	700
Geothermal	0	125	125
Gas Turbine	0	479	479
Internal Combustion	90	25	115
Wind	85	506	592
Free	0	6,178	6,178
Total	6,184	15,942	22,126

Figure 3 shows the projects that will deal with the committed capacity according the 2005–2014 Mexican Electricity prospective. These projects are plants already built, in construction process or already given.

Figure 4 shows the electricity requirements by region that are under the non-committed capacity category. This figure shows where the electricity requirements are. Also shown where the technology has not been chosen. Table 7 shows the share percent changes of different technologies as participants in the electricity generation, as forecasted by the Mexican electricity prospective for 2005–2014. It can be seen that the combined cycle plants based on natural gas are the main sources to be used.

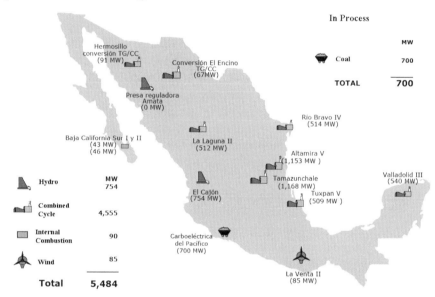

Fig. 3. Electricity committed capacity 2005–2014

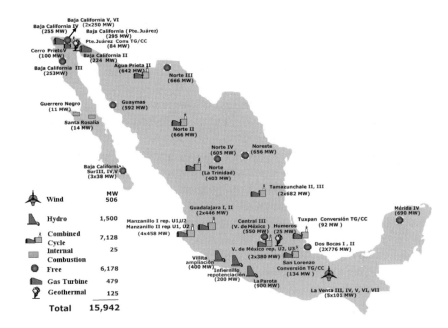

Fig. 4. Electricity non-committed capacity 2005–2014

Table 7. Technology participation in the Mexican Electricity Prospective 2005–2014

Fuel	Technology	2004 (46,552 MW) [%]	2014 (64,210 MW) [%]
Fossil	Combined cycle	25.9	39.7
	Gas turbine	6.1	5.1
	Conventional thermal	30	12.3
	Internal combustion	0.3	0.4
	Total	62.3	57.5
Coal or petroleum coke	Dual	4.5	3.3
Coal	Coal	5.6	5.1
Renewable	Hydro	22.6	19.9
	Geothermal and wind	2.1	2.5
	Total	24.7	22.4
Fissile	Nuclear	2.9	2.1
	Free		9.6

6 Discussion

If the gas relative price increases over 25 % with respect of today's cost, then other technology options become competitive and a new planning needs to be done.

Coal technology is restrictive because Mexico does not have large coal resources thereby this resource needs to be imported and facilities need to be build to import it and manage it.

Hydro has high investment cost along with social and environmental issues to be solved, but it is a renewable and clean technology.

Use of combined cycles based on synthesis gas coming from fossil fuel is another option to be considered.

Recent advances in nuclear technology and construction improvements have made this technology an option economically sound and very competitive.

Thus, a diversification strategy can give greater protection against the volatility prices of primary fuels. It also eliminates the international dependence to only one natural gas provider among other benefits.

Renewable Energy will have in 2014, 22.4 % participation according to the current prospective. On the other hand, free technology has a participation of 9.6 % and it gives the opportunity to use another technologies. It also can create a better diversification.

References

EIA (2004) Energy Information Administration. International Energy Outlook. US Department of Energy
SENER (2005) Secretaria de Energía. Prospectiva del Sector Eléctrico 2005–2014, México
WM (2004) World Market Research Centre. Country Report - Mexico (Energy) Electricity (January 21, 2004). http://www.worldmarketsanalysis.com

Thermoeconomic Study of CCGT Plants

Dolores Duran[1], Salvador Galindo[2]

[1]Departamento de Ingeniería Mecánica, Facultad de Ingeniería, UAEM, Cerro de Coatepec s/n, Toluca, Estado de México, México. [2]Instituto Nacional de Investigaciones Nucleares, Apartado Postal 18-1027, México DF 11801, México. E-mail: mddg_2210@hotmail.com

Abstract

This work presents a thermo-economic study of combined–cycle gas turbine (CCGT) facilities based on a flexible genetic algorithm search technique. The results here presented will maximize the cash flow by pointing out the correct parameter values for plant design when the turbines have already been chosen among the existing commercial options.

1 Introduction

Today's economic rhythm of development has considerably increased the world's growing demand for energy, this fact along with the growing costs of fuels and the problem of greenhouse emissions require the best possible design for ecologically and cost efficient power plants, to be built in short time.

In this perspective, in order to avoid the risk of environmental havoc in the future, we should concentrate on using and producing energy more efficiently and at the same time substituting, where possible, fossil fuel (coal, oil, and natural gas) for non-carbon renewable or nuclear energy sources.

Meanwhile, in the move toward more efficient energy production by combustion of fossil fuels, combined–cycle facilities show great advantages over others because they are capable of surpassing a 60 % efficiency, have short construction periods and are cost-effective, besides yielding low emission levels of CO_2 and NO_x, primarily if they burn natural gas.

Not surprisingly, since 1990 the number of combined–cycle facilities has significantly increased. In the past 5 years 19 combined-cycle facilities have been built in Mexico. This growth is also owed to the changeover or modifications of coal-fired and gas power plants to combined–cycle technology. It is expected that in the course of 1997 to 2007, world power production by the latter plants will rise to about 150 GW (Boyce 1997).

In response to the demand for a methodology that allows us to obtain the best possible design for ecologically and cost efficient combined–cycle power plants, there are several approaches nowadays. Some are centered on the system simulation itself, and others on the search for the most favorable thermodynamic settings or thermo-economic conditions of the system. However, the majority of these methods are either loose or apply to very particular circumstances and as consequence they put on view forecasts in a difficult scheme to be valued for power company planners, investors and decision makers. For these reasons it is the purpose of this work to present comprehensible thermo-economic information based on a flexible genetic algorithm search technique. These results will maximize the cash flow by choosing the correct parameters for plant design when the turbines have already been chosen among the existing commercial options.

2 Combined-Cycle Gas Turbine Plant (CCGT)

Simple heat power plants are only able to use a portion of the energy their fuel generates (usually less than 30 %). The remaining heat from combustion is generally wasted. This is not the case of a CCGT, because it produces energy employing more than one thermodynamic cycle. A CCGT power plant is a facility where a gas turbine generator produces electricity and the waste heat from the gas turbine is used in turn to make steam, via a boiler and a heat exchanger, known in the field as a heat recovery steam generator (HRSG). This steam is then used to breed a steam turbine to produce additional electricity; this last thermodynamic cycle enhances the efficiency of electricity generation.

The thermal efficiency of a combined cycle power plant is normally defined in terms of the net power output of the plant as a percentage of the lower heating value or net calorific value of the fuel. In the case of generating only electricity, power plant efficiencies of up to 60 % can be achieved. (Horlock 1991; Rapún 1999).

3 The CCGT systems

The power generated by CCGT plants is primarily established by the gas turbine. This component is normally purchased in the market. For this reason, one of the most important elements of a CCGT plant design is the HRSG because it links the gas cycle with the steam cycle in such way that the extra power enhanced by the latter will depend on the heat exchange and its exergetic efficiency (see Fig.1). Therefore the thermodynamic HRSG design parameters must be carefully selected in order to maximize the heat exchanged in the HRSG to reach the best performance in the combined cycle (Duran 2006). There are different types of HRSGs configurations depending on the number of sections they have. They may have normally one or more sections (low pressure, intermediate pressure, high pressure section). Each section in turn may encompass an economizer, in which water raises to the saturation temperature; a steam drum where gas and liquid phases are separated; an evaporator that transforms water into steam; super-heaters, to raise the temperature up the saturation point. Some operate under a single pressure cycle others have multi-pressure cycles. Moreover there are also once-trough HRSGs that are capable of working at supercritical pressures. More information about them and about HRSGs in general may be found in Horlock (1991).

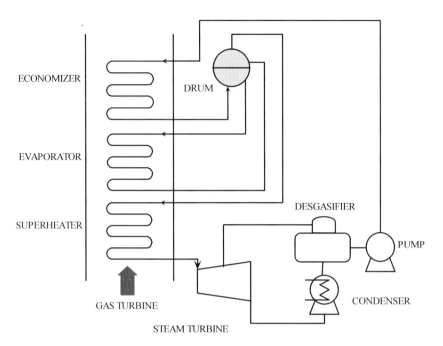

Fig. 1. Schematics of a CCGT system

Therefore there is a wide variety of selections of CCGT systems, due to the ample number of components involved in their design, such as the number of pressure levels, the distribution of the economizers, vaporizers, and over-heaters inside the HRSG and the use of pre-heaters and super-heaters.

To facilitate the analysis of this ample number of CCGT systems we will apply our thermo-economic study to the following types of HRSG:

- Single pressure steam cycle type (1P)
- Double pressure steam cycle type (2P)
- Double pressure and reheat steam cycle (2PR)
- Triple pressure levels and reheat steam cycle. (3P)

4 Gas turbine trends

Gas turbines are standardized elements whose design is determined by the makers. However, analyzing different commercial turbines, it was observed that in spite of the diverse companies, the design parameters (the compression ratio in the compressor, the entrance temperature to the turbine and the air mass flow) as well as the turbine price, follow certain trends depending on the turbine power just as it is shown in Fig. 2. In this figure, the value of the compression ratio and the inlet gas temperature in the turbine are plotted as function of its power. The points shown in this graph correspond to 165 existing turbines available in the world market. It is observed that these points follow clear tendencies. Two regression lines were fitted to these data, specifically one line for those in the low power range and another for those in the high power range. The thicker dots in the plot show those turbines that don't adjust to none of the regression lines. They correspond to sequential combustion turbines and with cooled blades, such as the popular models: ABB GT 24 and ABB GT 26.

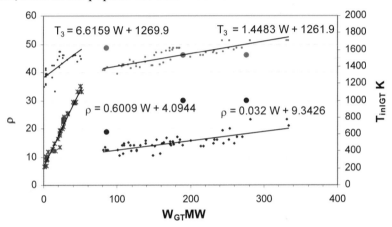

Fig. 2. Gas turbine design parameters tendencies. Left hand axis depicts the compression ratio (ρ), right axis the turbine entrance temperature (T_{inlGT})

From the mentioned regression lines it is possible to find, for a given power, the compression ratio in the compressor, and the inlet turbine temperature. In the same way, Fig. 3 shows that the mass flow also follows a clear trend for the commercial gas turbines and its amount can be obtained from the compression ratio and turbine inlet temperature.

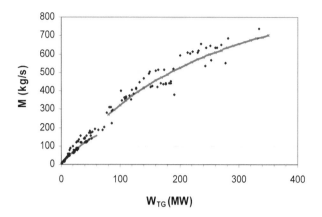

Fig. 3. Gas mass flow value trend

In addition, Fig. 4 shows the costs of gas turbines as a function of their power output. One must notice that sequential combustion turbines are still fitted to a regression line. The relevance of these previous results for the present work is that it is possible to select a commercial gas turbine for each specific power, whereupon the optimum parameters for coupling a HRSG design can be found.

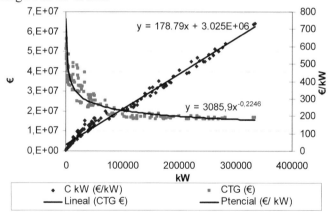

Fig. 4. Commercial gas turbine cost. Left hand axis shows the turbine cost, right axis shows the production cost per kW. Both scales are in Euros

To establish the guide lines obtained in this work, four hypothetical power values for the gas turbine were chosen, and starting from the adjustment functions shown in Figs. 2 and 3 the design parameters corresponding to each one of them were calculated. These data are shown in Table 1. Also shown are those corresponding to the ABB GT 24 and ABB

GT 26 sequential combustion turbines in view of the fact that they are of a different type. After that, a CCGT design was proposed for each of the six gas turbines, to continue with the optimization of each one of them.

Table 1. Turbine design parameters

GT	W_{GT} [MW]	T_{inTG} [K]	ρ	m_a [kg/s]
1	80	1377	11	271
2	150	1479	14	435
3	250	1624	17	595
4	350	1768	20	701
GT24	191	1533	30	581
GT26	277	1533	30	549

5 Thermo-economic balance search

In the design of power plants and thermal systems in general, both thermo-dynamic and thermo-economic studies are important. Through a thermo-dynamic study one can find the dominant factors for cycle performance and the main inefficiency sources. However, an efficiency rise in a plant usually increases the cost. For that reason, thermo-economic models were developed to find a compromise between efficiency and cost. Several works exist about this topic. There are some which are based on a break-down of the system elements into smaller subsystems to subsequently apply a cost balance to each exergy flow. (Valero et al.1994; Tsatsaronis 1993 and Agazzani 1997). Other authors limit their studies only to the HRSG elements (Franco and Ruso 2002) while Dechamps develops a method to find the middle ground between performance and cost genera-tion. (Dechamps 1995). In a similar vein, a model based on the HRSG de-sign was developed by one of us (Duran et al 2005).

The method used in the present work is the one proposed by (Rovira 2004), this maximizes the cash flow or minimizes the generation cost by choosing the correct parameters for the plant design subject to the restric-tion that hypothetical, but realistic turbines have already been chosen. The method used is that of genetic algorithms (GA) proposed by Duran (Duran 2004). GA methods are well established computer search techniques (Goldberg 1989). In particular, the method used here is described in extent by Duran, where specific details can be found (Duran 2004; Valdés et al. 2003). However it is important to mention that the thermodynamic pa-rameters being considered in the GA -taken as independent variables - were, for each considered pressure level: the drum pressures, pinch points, approach points and temperature differences at the super-heaters outlets. In addition, and in order to obtain a realistic results, some of the mentioned design parameters were constrained to manufacture and operational range values, this in order to avoid, for example, having high humidity rates at the turbine outlet and low temperature values at the HRSG exit.

6 Results

6.1 Efficiency

The analysis was done for each one of the turbine power values given in Table 1 combined with each of the four HRSG configurations mentioned in the section describing the HRSG. By this means it can be established which of all possible arrays provides larger dividends and smaller costs.

Figure 5 shows the variation of the efficiency as function of the power for the four configurations under study. One can observe that for the triple pressure steam cycle type with reheat (3PR), efficiency is always above the others, for approximately power rates higher than 250 MW. This means that if power installations above 250 MW are considered, the best installation is the mentioned one. However if lower power rates are considered, the best selection should be a double pressure steam cycle type (2P). One must notice that single cycle type has always the worst performance.

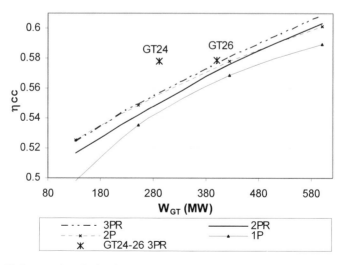

Fig. 5. Efficiency of optimized CCGTs

6.2 Cash flow

Figure 6 compares the yearly cash flow as a function of the CCGT power. It is seen that there is not perceptible difference between the cash flow of a triple pressure type (3PR) from a double one type with reheat (2PR). This is probably due to the high cost of a triple pressure level steam cycle type. It is also observed that, for powers lower than 340 MW, the double cycle type without reheat (2P) is the one that presents higher cash flow; however, for higher powers it is surmounted by the double pressure cycle with reheat and also that of three levels of pressure with reheat (3PT).

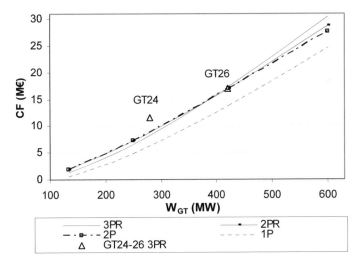

Fig. 6. Yearly cash flow of optimised CCGT

6.3 Fuel Cost Impact

The above results allow decisions on which configurations will provide the best economic benefits. However it is necessary to take into account other aspects like the increment in the price of the fuel.

Therefore, this work also shows what is the effect that fuel price has on the generation costs for the configurations 2P and 3PR. This is shown in Fig. 7, where it can be seen that after a fuel price rise, the 3PR optimum generation costs increase less than those corresponding to the 2P system. In the same figure it may be seen that the optimization model selects a configuration with poorer efficiency in order to compensate the fuel cost increment. It can be seen that a 3PR system operates well again after eventual fuel price rises.

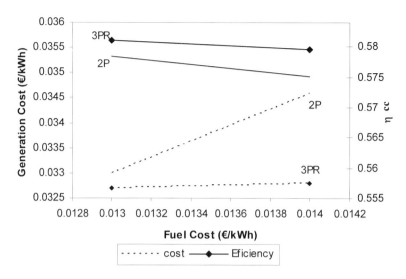

Fig. 7. Optimal generation costs as function of fuel price rise. Also shown is the obtained efficiency for each optimized system

7 Conclusions

As mentioned in our introduction, with the aim of avoiding an ecological disaster in the future, we must substitute fossil fuels for non-carbon renewable, or nuclear energy sources. In the meantime, while hopefully this conversion takes place, we must focus on producing conventional energy more efficiently. This is the case for combined cycle power plants (CCGT). The present study is a comprehensive analysis of the thermo-economic parameters of CCGT facilities.

Our study was derived from an analysis of 165 existing turbines available in the world market and so we would like to stress its importance. The analysis here presented was vast enough to be able to specify the behavior of the turbine compression ratio and the inlet gas temperature as function of a selected power.

In the same way we were able to show that the mass flow also follows a clear trend for the commercial gas turbines and its amount can be obtained from the compression ratio and turbine inlet temperature. In addition we found the trend cost of gas turbines as a function of their power output. The relevance of these findings is that it is possible to select a commercial gas turbine for each desired specific power and, as a result of this, the optimum parameters for coupling a HRSG design can be found.

Finally a genetic algorithm method was applied to maximize the cash flow by minimizing the power generation cost as a result of choosing the correct parameters for the plant design subject to the restriction that hypothetical, but realistic turbines have already been chosen. The scenario of a fuel cost rise in power producing profits was also pondered. Attention to this point in the future will be unavoidable.

References

Agazzani A, Massardo A (1997). A Tool for Thermoeconomic Analysis and Optimization of Gas, Steam and Combined Plants, Trans ASME, J of Eng for GT and P 119(4):885–892

Boyce M (1997) Handbook for cogeneration and combines cycle power plants

Dechamps P (1995) Incremental cost optimization of heat recovery steam generators. ASME COGEN-TURBO, Viena, Austria

Duran M (2004) Estudio de Calderas de Recuperación de Calor de Ciclos Combinados de Turbinas de Gas y Vapor Empleando la Técnica de Algoritmos Genéticos. Ph.D Thesis (In Spanish). Universidad Politécnica de Madrid. Madrid

Durán M, Valdés M and Rovira A (2004). Aplication of the Genetic Algorithms on the thermoeconomic optimization of CCGT Power Plants, ECOS 2004, 3:1335–1350

Franco A, Russo A (2002) Combined Cycle Plant Efficiency Increase Based on the Optimisation of the Heat Recovery Steam Generator Operating Parameters. Int J of Thermal Sciences 41:843–859

Goldberg D (1989) Genetic Algorithms in Search, Optimization, and Machine Learning. First Ed, Addison-Wesley, Michigan

Horlock J (1992). Combined power plants, First ed., Oxford, Pergamon Press

Rapún J (1999).) Modelo matemático del comportamiento de ciclos combinados de turbinas de gas y vapor. PhD (In Spanish). ETSII, UPM, Spain.

Rovira A (2004) Desarrollo de un Modelo para la Caracterización Termoeconómica de Ciclos Combinados de Turbinas de Gas y de Vapor en Condiciones de Carga Variable. Ph.D. (In Spanish). ETSII, UPM Spain

Tsatsaronis G (1993) Thermoeconomic analysis and optimization of energy systems. Prog Energy Combust Sci 19:227–257

Valdés M, Durán M and Rovira A (2003) Thermoeconomic optimization of combined cycle gas turbine using genetic algorithms, Applied Thermal Engineering 23:2169–2182

Valero et. al. (1994) CGAM Problem: Definition and Conventional Solution. Energy 19(13):279–286

CO$_2$ Capture for Atmosphere Pollution Reduction

Rosa-Hilda Chávez[1], Javier de J. Guadarrama[2], Jaime Klapp[1]

[1] Instituto Nacional de Investigaciones Nucleares, Apartado Postal 18-1027, México DF 11801, México. [2] Instituto Tecnológico de Toluca, Av. Tecnológico S/N, Metepec, 52140 México. E-mail: rhch@nuclear.inin.mx

Abstract

Carbon dioxide is considered to be the major source of greenhouse gases responsible for global warming; man-made CO$_2$ contributes approximately 63.5 % to all greenhouse gases. Efforts towards reducing greenhouse gas emissions have increased in the past few years, offering promising alternatives in power generation and better fuel efficiency. However, the incorporation of these new technologies to our daily lives represents a big challenge to be solved in the mid- to long-term, leaving separation and CO$_2$ sequestration to be an immediate priority for researchers. CO$_2$ capture and storage can support the transition of our fossil fuel based energy supply towards a sustainable energy system, based upon nuclear and renewable sources. Our present energy infrastructure will largely remain the same during this transition period. For example, electric power plants will be equipped with CO$_2$ capture units, but will produce the same electricity, transported and distributed over the same grid. The first step in the CO$_2$ capture and storage chain is to capture carbon in a high concentration. This can be done before or after combustion of the fuel. Capture is best carried out at large sources of emissions, such as power stations, refineries and other industrial complexes. There are several ways to capture CO$_2$, some of the main methods are absorbents using solvents or solid sorbents, which have been used in industry for several years and seems to be the most feasible solution at this time; membranes have also become an interesting alternative and although extensive research is in progress, new materials for membranes have yet to be discovered. Other methods like pressure- and temperature-swing adsorption using various solid sorbents, cryogenic distillation and new emerging technologies show promising results in the bench testing scale. This report presents a comparison between these different methods.

1 Introduction

Climate change is one of the greatest and probably most challenging of environmental, social and economical threats the world is facing. The issue of climate change is not about air quality or smog but it is about global warming. Human activities have altered the chemical composition of the atmosphere through the build up of significant quantities of greenhouse gases, which remain in the atmosphere for long periods of time and intensify the natural greenhouse effect. Increasing concentrations of greenhouse gases are likely to accelerate the climate change rate. Concerns are growing about how increases in atmospheric greenhouse gases caused by human activities are contributing to the natural greenhouse effect and raising the earth's average temperature (UNEP 2001; IPCC 2001).

Several CO_2 separation methods have been developed with relative success, due to the fact that they all have advantages and disadvantages. Some methods offer CO_2 high-purity streams but their components are prone to degradation and damage. Others prove very durable but cannot be used with high flow volumes. Others methods seem to be very promising, but their economical cost or energy penalties requires further investigation. This report presents a summary of the present status of the methods as well as their advantages and disadvantages. Furthermore, it offers a comparison regarding the feasibility of retrofitting them to actual applications.

Prior to CO_2 absorption from flue gas, it is recommended to remove other compounds such as SO_x for which several methods are used. One recent method that has been studied uses the separated SO_x compounds to form a solvent that absorbs CO_2. Laboratory experiments show that up to 0.125 m^3 CO_2/kg of flue gas desulphurization product can be absorbed. Limestone based compounds are being used to separate SO_x from flue gas streams prior to absorption. These compounds are then slurried, becoming the solvent for absorption. An attractive feature to this design is that many power plants already have flue gas desulphurization units. Some small amount of additional transport equipment for the slurry to be moved to the absorber would be necessary, minimizing capital cost for retrofitting an existing plant (Taulbee et al. 1997).

2 Absorption by Monoethanolamine (MEA)

A typical system is illustrated in Fig. 1, which consists of two main elements: an absorber where CO_2 is removed, and a stripper, where CO_2 is released and the original solvent recovered. The performance of a MEA based CO_2 capture, a key feature of an amine system, is the large amount of heat required to regenerate the solvent. This heat is typically drawn from the steam cycle and significantly reduces the net efficiency of the power plant. Substantial electrical energy is also needed to compress the captured CO_2 for pipeline transport to a storage site. The overall energy penalty of this process has a major impact on system performance and cost. From a multi-pollutant perspective, there are also important interactions between the CO_2 capture system and the control of other air pollutants, especially SO_2 and NO_x emissions. Acid gases like SO_2 and NO_2 react with MEA to form heat-stable salts that reduce the CO_2 absorption capacity of the solvent. Thus, very low concentrations of these gases (on the order of 10 parts per million) are desirable to avoid excessive loss of (costly) solvent. Additionally, the flue gas temperature needs to be reduced to approximately 45°C (MEA ideal temperature) since high temperatures of at least 100°C related with flue gases can degrade solvents and lower the solubility of CO_2 (Chapel et al. 1999; Yeh and Pennline 2001).

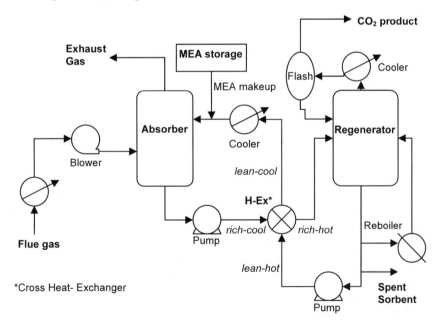

Fig. 1. Flowchart for CO_2 capture from flue gases using amine-based system

MEA is a CO_2 well proven method that has been used for several decades, thus offering a good level of knowledge in its mechanisms and involved thermodynamics. Another advantage is the capability of being regenerated by raising its temperature in order to release the absorbed CO_2. Some MEA costs are the addition of new solvent and other operating and maintenance at about \$40–\$70/ton CO_2 separated and the energy penalty of the entire absorption process of approximately 0.341 kWh/kg CO_2 (\$13.95/ton CO_2), (Gottlicher and Pruschek 1997). Neither estimate includes capital nor installation costs.

The elementary steps for the reaction can be represented by the following equilibrium reactions:

$$2H_2O \Leftrightarrow H_3O^+ + OH^-$$

$$2H_2O + CO_2 \Leftrightarrow H_3O^+ + HCO^{-3}$$

$$H_2O + HCO^{-3} \Leftrightarrow H_3O^+ + CO_3^{-2}$$

$$C_2H_8NO + H_2O \Leftrightarrow H_3O^+ + C_2H_7NO$$

$$C_3H_6NO^{-3} + H_2O \Leftrightarrow HCO^{-3} + C_2H_7NO$$

Possibly the most promising absorption process is based on the KS-1, KS-2 and KS-3 solvents being developed by Mitsubishi Heavy Industries (Iijima 2005). This family of solvents shows higher CO_2 loading per unit solvent lower regeneration conditions, and almost no corrosion, degradation, or amine loss. A novel packing material, KP-1, has also been developed that will further improve this process. Development has reached the pilot plant stage at Mitsubishi Heavy Industries.

We have constructed structured package columns in the Mexican National Institute of Nuclear Research (ININ) with the aim of offering low cost systems capable of capturing different types of contaminants from liquid and gas mixtures and, at the same time, to retrofit existing industrial plants (Fig. 2).

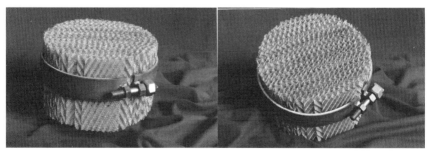

Fig. 2. Structured packing "ININ 18" developed at ININ

The geometric characteristics of our structured packing design are presented in Fig. 3 and Table 1, and Fig. 4 shows our Pilot test separation column research facility. A mathematical model of the CO_2 capture system has been developed (Chávez et al. 2004, 2005; Chávez and Guadarrama 2006; Chávez 2006).

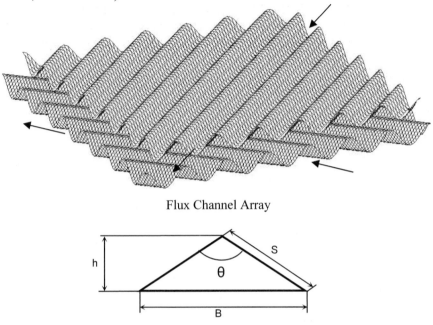

Flux Channel Array

Triangular Transverse Section

Fig. 3. Flux channel geometry of a structured packing

Table 1. Geometric parameters of "ININ 18" versus commercial structured packings

	Structured packing			
Variable	ININ 18	Sulzer BX	Mellapak 250Y	Units
Construction material	Stainless steel mesh	Stainless steel mesh	Polypropylene	
S	0.012	0.00889	0.0225	m
B	0.0165	0.0127	0.035	m
h	0.009	0.006	0.012	m
θ	35	30	45	°

Fig. 4. Pilot test separation column at the ININ facilities

3 Adsorption

Adsorption is a CO_2 separation method that uses the interactivity between sorbent and guest molecules. In this way, CO_2 molecules are attracted into small cracks, pores or external surfaces of the sorbent under specific temperature and pressure. However, the sorbent must be selective with the component tried to be separated since flue gases normally contain other components such as N_2, CO_2, H_2O, NO_x, SO_x, CO, O_2 and particulate matter. The two main adsorption methods are pressure and temperature swing. For both methods, adsorption rates depends upon the temperature, partial pressures of CO_2, surface forces (interaction energy between sorbent and CO_2), and pore size of available surface area of the sorbent. Results show that pressure swing adsorption has lower energy demand and higher regeneration rate than temperature swing adsorption (Li et al. 2003; Meisen and Shuai 1997).

Three main disadvantages of adsorption methods are the difficulty to deal with large CO_2 concentrations, that sorbents are not selective enough for CO_2 separation from flue gases and the low velocity of the process. Sorbents usually can deal with CO_2 concentrations between 0.04 % and 1.5 % while a value usually found in power plants is approximately 15 %. The issue with selectivity is that smaller particles such as N_2 fill pore space easier in sorbents, making the process less efficient. Finally, adsorption velocity is low, allowing maximum adsorption approximately in 20 min, making it only suited for small-scale application (Li et al. 2003; Meisen and Shuai 1997). However, adsorption can be considered as a good alternative in hybrid systems. It can be placed after another separation process, due to its low CO_2 concentration requirements.

4 Cryogenic distillation

For cryogenic separation all components of the flue gas must be removed except for the N_2 and CO_2 prior to be sent into a cryogenic chamber. The temperature and pressure are manipulated in order to liquefy CO_2. To achieve this, the triple point of CO_2 (56.6°C and 7.4 atm) is achieved where CO_2 will condense while N_2 remains a gas. This distillation allows N_2 to escape through an outlet at the top of the chamber while the highly concentrated liquid CO_2 can be collected at the bottom of the chamber (Schussler and Kummel 1989).

The CO_2 separation produces CO_2 liquid. Besides, the resulting liquid CO_2 has high levels of purity (up to 99.95 %) making it ready for transport. Nevertheless, the process requires the use of liquid nitrogen which implies high energy costs. The fact that H_2O, NO_x, SO_x and O_2 must be removed from the flue gas before the cryogenic process takes place represents another important disadvantage (Meratla 1997).

5 Membrane diffusion

The use of membranes is not new, has been used to separate CO_2 from light hydrocarbons. They are also used to separate hydrogen gas from various other gases. Several types of membranes are commonly used: inorganic, metallic, polymeric, and solid-liquid (Baltus and DePaoli 2002). The selectivity process of membranes is based on the ability to interact with the target and to diffuse it across, by solution-diffusion or absorption-diffusion mechanisms.

Ceramic and metallic membranes have certain porosity allowing only the pass of gases of a certain size through its pores. They are usually located into a chamber in which flue gases pass; CO_2 passes through the membrane into a low pressure chamber, where it is collected.

Other types of membranes under study are gas absorption membranes, which can be considered a hybrid system between solid membranes and liquid absorption. The mechanism is the selective diffusion of CO_2 gas in the membrane, to be later captured and removed by the liquid absorbent.

With a two-stage separator, the same process applies, except the gases that permeated the membrane in the first chamber are again separated in a second separation tank. This approach provides greater separation and a gas stream that is more suitable for carbon sequestration. The unpermeated gases from both tanks are sent to the same receiver for further treatment. While double separation yields a much higher purity stream of CO_2 (89.1 % instead of a 46.4 % at 25°C with a single stage) (Tokuda et al. 1997), it generally cost as twice as much as a tradition amine separation processes (Tam et al. 2001).

The requirements necessary to implement membranes are simple: the only necessary equipment is the membrane and fans, thus keeping the installation and maintenance costs low. Their main drawback is that they are either very selective or very permeable to CO_2, but cannot have both properties at the same time. High permeability allows other gases besides CO_2 to permeate, requiring extra separation steps. Other drawback is that some membranes are not able to work under the high temperatures usually found in flue gases. They are also sensitive to chemical attack from some components in the gas (Park and Lee 1995; Kovvali and Sikar 2002).

New research is being done towards finding membranes that: withstand the high temperature of the entering gases (350°C ideally, see Mowbay 1997), resist the pressure changes associated with the gas flow through the chamber, and to increase chemical resistance to other components in flue gas.

6 Other separation methods

The hydrate formation separation takes advantage of its ice-like structures in which water forms a cage with cavities where small gases such as CO_2 can be trapped. Due to the higher affinity of CO_2 to form a hydrate, it is more likely to be trapped in these cavities. A maximum of eight CO_2 molecules can be trapped in a cage of 46 water molecules, giving a mole fraction of 0.148 but a weight fraction of 0.31 g CO_2/g H_2O (Pruschek et al. 1997). The high pressures (~88.8 atm) and low temperatures (0°C) required for hydrate formation makes it to date a cost-ineffective method (Brewer et al. 1999).

The electrical desorption method uses an electric current to liberate the CO_2 from a physical sorbent (quinine carrier). By means of electricity the CO_2 trapped in the sorbent is liberated. A low energy-penalty is involved in the process, since the sorbent used is electrically conductive (Judkings and Burchell 1997).

Another method is to use a redox active carrier to bind the CO_2 to itself at high pressure and then release it at low pressure. The carrier ability to bind with the CO_2 is determined by whether the carrier is reduced or oxidized; reduction allows the carrier to pick up CO_2 while oxidation causes it to release the CO_2.

CO_2 can also be separated by reacting the CO_2 in flue gas with ammonia gas and water vapor in a gas-phase reaction. The reaction proceeds according to the following equation:

$$CO_2 + NH_3 + H_2O \Leftrightarrow NH_4HCO_2$$

The product of this reaction is solid CO_2; N_2 and other gases encountered in the flue gas stream continue through for release or treatment. Some experiments show that the reaction can take place at room temperatures and ambient pressures, suggesting that the energy penalty for this method would be lower than most current methods. Other advantage is that the solid product can be used as fertilizer (Li et al. 2003; Meisen and Shuai 1997).

7 Conclusions

Due to its simplicity and low energy requirements, membrane diffusion looks like a very promising method. In addition, retrofitting membrane separation units to existing power plants seems to be simple. However, under some operation conditions like high temperatures and chemically aggressive flue gases, actual membrane materials show poor performance. The biggest challenge facing membranes is the development of new polymeric materials able to reach up to 350°C.

Absorption is already a well-established process and through research and development could still improve its results, becoming the most feasible separation technology. For example, easier regeneration, faster loading and solvents resistant to degradation and less corrosive to equipment are areas to be improved in the near future. Retrofitting existing power plants with chemical absorbers are predicted to be relatively easy, based on the fact that some power plants already have chemical absorbers incorporated in their designs. Worth mentioning are the results that Mitsubishi and Econamine FG (Chapel et al. 1999), that could make absorption the preferable separation method. Apparently, they are able to reduce the energy requirements, making this process as efficient as membrane separation for a traditional power plant system.

Another way to make the efficiency increase of absorption methods is to maximize interaction between the solvent and the flue gases by using structured packed material in the exit column of flue gases. Several types of packing material have been used, both random and structured. With a solvent utilization of 43.9 +/– 0.5 % (fraction of solvent that absorbed CO_2) versus random packing with 28.6 % utilization, the structured packing shows better results. The use of both types of packages relies on the principle of increasing the interaction of solvent with entering gases by increasing the surface area available of the solvent.

Our experience in the capture of liquid and gas contaminants by using separation columns with structured packing has shown positive results. As our target was to develop a solution specifically directed to Mexican industrial needs, commercial costs have been kept low. Other aspects like reliability and maintenance meet the required standards.

Diluting the MEA solution with organic solvents such as methanol is also a possibility under study. The concept is that if the organic solutions have lower heating capacities, the heat of regeneration should be decreased, thus reducing the energy penalty for regeneration. However, the use of methanol shows that the solution evaporated while heated under atmospheric pressure. This type of solutions may be more adequate in high pressure systems such as the Integrated Gasification Combined Cycle.

For the pressure swing adsorption process, conditions necessary for its operation are not as difficult to manipulate or achieve as for cryogenic distillation, though they are more difficult than absorption or membrane diffusion. Among the main drawbacks are that materials do not show a high enough CO_2 selectivity to make adsorption cost effective, and the retrofitting of existing plant would be more expensive than with absorbers and membranes. Thus, the development of new sorbents with better selectivity to handle large-scale separation will tell if this technology is going to be effective.

Due to its high energy penalty and high equipment cost, cryogenic distillation is the process with least promising future. Although the final product is liquid CO$_2$, ready for transport and sequestration, this process is more complex than the other ones, and retrofitting an existing plant would not be a simple upgrade.

Hydrate formation, electrocatalysis, and ammonium bicarbonate formation are all very new concepts for CO$_2$ that are still in lab testing. These methods, upon further investigation, may prove efficient enough to progress to bench scale testing in the near future.

A final option may be hybrid systems. These systems offer high degrees of CO$_2$ separation through multiple methods combined (some systems use chemical absorption with hydrate formation to have liquid CO$_2$ as the final product). However, the most promising hybrid system seems to be Integrated Gasification Combined Cycle. This system offers high levels of efficiency at a lower energy penalty, although its degree of CO$_2$ separation is not as high as in other hybrid systems.

An important consideration to take into account is the capacity of flue mixture to be treated as seen on Table 2. If the separation results of a certain method prove positive but it cannot cope with a proper separation capacity, an increase in costs for its implementation must be considered.

Table 2. Approximate Single-Line Maximum Capacities for Selected Separation Processes

Alternative Processes	Upper commercial scale for single-line operation
Distillation	No limit
Extraction	No limit (certain types of column have been used only up to 1.83 m) diameter
Crystallization	10–70×10^6 kg/yr
Absorption	No limit
Reverse osmosis	0.45×10^6 kg/yr water flow per module, with many modules per unit
Membrane gas separation	0.9×10^9 kg/yr gas permeated per module, with many modules per unit
Ultrafiltration	0.45×10^6 kg/yr water flow per module, with many modules per unit
Ion exchange	450×10^6 kg/yr water flow
Electrodialysis	0.45×10^6 kg/yr water flow
Electrophoresis	1000 kg/yr carrier free basis
Adsorption	0.0025×10^6 kg/yr gas
Chromatographic separation	0.24×10^6 kg/yr gas, 0.09×10^6 kg/yr liquid

With further research, hybrid systems utilizing hydrate formation, ammonium bicarbonate formation and electrocatalysis may be the choice for power plant design due to their streamlined integration of processes, high CO_2 separation and capability to utilize the other components of the separated gas streams.

Acknowledgement

We thank Ing. José Guillermo Cedeño Laurent for preparing the figures and evaluations. This work has been partially supported by the Mexican Consejo Nacional de Ciencia y Tecnología (CONACyT) under contract U43534-R.

References

Baltus R, De Paoli D (2002) Separation of CO_2 using room temperature ionic liquids. Oak Ridge National Laboratory. Environmental Sciences Division Carbon Management Seminar Series: Oak Ridge, TN, June 20

Brewer PG, Friederich G, Peltzer E, Orr FM Jr (1999) Direct experiments on the ocean disposal of fossil fuel CO_2. Science 284:943–945

Chapel D, Ernest J, Mariz C (1999) CO_2 Recovery from flue gases: commercial trends, Canadian Society of Chemical Engineers Annual Meeting. http://www.netl.doe.gov/publications/proceedings/01/carbon_seq/2b3.pdf

Chávez RH (2006) Study of CO_2 Capture Process by Reactive Absorption, Technical Report 66/2006 at Institute Technische Thermodynamic and Thermische Verfahrenstechnik (ITT), University Stuttgart, Germany, 30 April 2006 and IT.G.C.AMB/DFR/09/2006 at Instituto Nacional de Investigaciones Nucleares, July 2006

Chávez RH, Guadarrama JJ (2006) Natural gas sweetening using structured packing, Information Technology Journal 5(2):285–289

Chávez RH, Guadarrama JJ, Hernández A (2004), Effect of the structured packing on column diameter, pressure drop and height in a mass transfer unit, International Journal of Thermodynamics 7(3):141–148

Chávez RH, Guadarrama JJ, Segovia N (2005) Removal of sulfur dioxide from exhaust gases using hazardous and structured packing, International Journal of Environment and Pollution 23(1):81–91

Gottlicher G, Pruschek R (1997) Comparison of CO_2 removal system for fossil-fuel power plant processes. Energy Convers Manage (Suppl) 38:173–178

Iijima M (2005) CO_2 Recovery from Flue Gas Using Hindered Amines CO_2 Amines, Mitsubishi Heavy Industries, Ltd. 3-3-1, Minatomirai, Nishi-Ku Yokohama 220–8401, Japan http://www.co2management.org/proceedings/Masaki_Iijima.pdf

IPCC (2001) Intergovernmental Panel on Climate Change. Climate Change 2001: The Scientific Basis, Contribution of Working Group 1 to the Third Assessment Report of the Intergovernmental Panel on Climate Change, Cambridge University Press

Judkings RR, Burchell TD (1997) A novel carbon fiber based material and separation technology. Energy Convers Manage (Suppl) 38:99–104

Kovvali AS, Sikar KK (2002) Carbon dioxide separation with novel solvents as liquid membranes. Ind Eng Chem Res 41(9):2287–2295

Li X, Hagaman E, Tsouris C, Lee JW (2003) Removal of carbon dioxide from flue gas by ammonia carbonation in the gas phase. Energy Fuels 17(1):69–74

Meisen A., Shuai X (1997) Research and development issues in CO$_2$ capture. Energy Convers Manage (Suppl) 38:37–42

Meratla Z (1997) Combining cryogenic flue gas emission remediation with a CO$_2$/O$_2$ combustion cycle. Energy Convers Manage (Suppl) 38:147–152

Mowbay J (1997) Ceramic membrane to combat global warming. Membr. Technol. 92:11–12

Park YI, Lee KH (1995) The permeation of CO$_2$ and N$_2$ through asymmetric polyetherimide membrane. Energy Convers Manage, 36(6–9):423–426

Pruschek R, Oeljeklaus G, Haupt G, Zimmerman G, Jansen D, Ribberink JS (1997) The role of IGCC in CO$_2$ abatement. Energy Convers Manage (Suppl) 38:153–158

Schussler U, Kummel R (1989) Carbon dioxide removal form fossil fuel power plants by refrigeration under pressure. In: Proc of the 24th Intersociety 6–11 Aug 1989. Energy Conversion Engineering Conference, vol 4 pp 1789–1794

Tam SS, Stanton ME, Ghose S, Deppe G, Spencer DF, Currier RP, Young JS, Anderson GK, Le LA, Devlin DJ (2001) A High pressure carbon dioxide separation process for IGCC plants. http://www.netl.doe.gov/publications/proceedings/01/carbon_seq/2b3.pdf

Taulbee DN, Graham U, Rathbone RF, Robi TL (1997) Removal of CO$_2$ from Multi-component Gas Streams using Dry-FGD Wastes. Fuel 76(8):781–786

Tokuda Y, Fujitsawa E, Okabayashi N, Matsumiya M, Takagi K, Mano H, Haraya K, Sato M (1997) Development of hollow fiber membranes for CO$_2$ separation Energy Convers Manage (Suppl) 38:111–116

UNEP (2001) United Nations Environmental Program and UN framework convention on climate change: Climate change information kit

Yeh JT, Pennline HW (2001) Study of absorption and desorption in a packed column. Energy Fuels, 15(2):274–278

Fossil Fuels Pollution and Air Quality Modeling

Darío Rojas-Avellaneda

Centro de Investigacion en Geografía y Geomatica "Ing. Jorge L. Tamayo". Contoy No. 137, Lomas de Padierna, Tlalpan, C.P. 14740. México
D.F., México. E-mail: dariorojas@centrogeo.org.mx

1 Introduction

World energy demand has been increasing continuously with human development and the increase of world population. An increase of energy
consumption is forecasted to rise by 50 % in the next three decades. Fossil
resources – natural gas, coal and oil – are more widely used to supply consumers with this energy. At present time, fossil fuels continue to be the
dominant energy source. Fossil resources supply almost 88 % of the total
energy consumed in the world, followed by hydrodynamic (6.3 %) and nuclear (6 %). In spite of the development and use of other sources of energy
such as nuclear, hydrodynamic and renewable sources, future energy supply will continue to rely on fossil resources, although with a lower relative
utilization.

On the other hand, the burning of fossil resources, coal, oil and natural
gas are the principal sources of several major air pollution problems. These
problems include local and regional pollution and also global effects. Local and regional effects, like the generation of locally unhealthy air, indoor
air pollution and urban smog are some of the causes of damage to the environment, human health and quality of life.

Air pollution can expand beyond a localized region to cause global effects. Two global issues have recently become major concerns. The greenhouse effect, which could cause an increase of the earth's average temperature as a consequence of increasing concentrations of carbon dioxide,
and the depletion of the stratosphere's ozone layer, a natural protective
shield from harmful solar radiation.

How can humanity satisfy its energy requirements in spite of the increasing population and the need of more energy resources, without irreparable damage to the environment? The only known mechanism of limiting air pollution, aside from willingness, is government intervention. Intervention can take the form of setting up economic markets for the rights to emit pollution, limiting emissions from specific sources, requiring emission control technologies, or setting limits on pollutant concentrations and allowing the use of emission reduction methods to achieve those goals.

In order to understand air pollution problems and to establish an air quality legislation to implement air quality and emission standards, modeling as well as monitoring is required. It will be necessary to develop air pollution modeling techniques based on methods that rely upon available data and limited computational resources.

Monitoring and modeling studies constitute a relatively inexpensive activity whose results, in the best case, provide useful information for possible future implementations of much more expensive emission reductions and control strategies.

We consider here the main efforts to measure and to estimate the concentration levels of the principal pollutant in the atmosphere of the Mexico City region (ZMVM) and the actions performed to identify the types of emitting sources and estimate their emission rates. The data obtained as a result of these actions and measurements are used as input information and validation data to develop mathematical models.

Air quality modeling is an indispensable tool for most air pollution studies. Air quality modeling is required for establishing emission control legislation in the determination of maximum allowable emission rates that will achieve fixed air quality standards. It is also required for the evaluation of emissions control techniques and strategies, and applied to select locations of future sources of pollutants in order to minimize their environmental impacts. To avoid severe episodes in a given region, it is very useful to plan the control of air pollution episodes.

2 Major pollutant sources

Most air pollution comes from burning fossil fuels to power industrial processes and motor vehicles. Among the harmful chemical compounds that this burning emits to the atmosphere are carbon dioxide, carbon monoxide, nitrogen oxides, sulfur dioxide and particles. When the burning of fuels is incomplete, volatile organic chemicals (VOCs) also enter the air.

Pollutants also come from other sources. For instance, decomposing garbage and solid waste emit methane gas.

Once in the atmosphere, pollutants often undergo chemical reactions that produce additional harmful compounds. Ozone, for example, formed via photochemical reactions. Pollutants in the atmosphere are the cause of the major air pollution problems on different spatial scales. These problems include photochemical urban smog, indoor air pollution, acid deposition, ozone depletion in the Antarctic and globally, and global warming. Many cities, including Mexico City, exhibit photochemical smog. Photochemical smog is the air pollution that is a result of intense sunlight and high levels of emissions from fossil-fuel combustion.

3 Air pollution measurements

There are two kinds of air pollution measurements: ambient measurements and source measurements.

3.1 Ambient measurements

The measurement of pollutant concentrations is important in the study of urban and regional smog problems to determine zones where the concentration is high enough to affect the health of humans, animals and vegetation.

Air pollution regulations in Mexico City were first enacted in 1971, aimed to prevent and control environmental pollution. A full set of monitors was implemented in 1986 to measure outdoor pollutant mixing ratios. Today the Atmospheric Monitoring System for Mexico City, SIMAT, consists of 32 ambient monitors to measure the concentrations of pollutants in the air that everyone breathes in real time; this network of monitors is named the "Red Automática de Monitoreo Atmosférico de la Ciudad de Mexico", RAMA (http://www.sma.df.gob.mx/simat/homecontam.php). The SIMAT also measures meteorological parameters, radiation, temperature, humidity and wind velocity.

The RAMA net provides hourly values for the main air pollutants: carbon monoxide CO, sulfur dioxide SO_2, nitrogen oxides NO_2 and NO_x, ozone O_3 and suspended particulate matter PM10 and PM2.5. Ozone concentrations are strongly linked to meteorological conditions and are also influenced by winds due to transportation of both ozone and its precursors. Ozone, after suspended particulate matter, is the second most important pollutant found in Mexico City's atmosphere. In fact, the maximum allowed ozone concentration level (0.11 ppm, one hour per year) was exceeded in year 2000 in around 84 % of the days. This percentage of days has been continually reduced thanks to enacted control norms and to improvement in the fuel quality, gasoline particularly. In year 2005 the maximum allowed ozone concentration level was exceed in 60.5 % of days (SMA 2006). Although these pollution levels are far from ideal, these regulations have improved notably the air quality in the Mexico City region.

3.2 Source measurements

In order to control pollutant concentrations, we must regulate the time, place and amount of their emissions. Thus concentrations and emission rates from all sources of air pollutants, that is, factories, power plants and automobiles, among others, must be measured.

The desire of all, concerning air pollution, is to have a completely unpolluted environment at no cost to anyone. That appears to be impossible, so our logical goal is to have an appropriately clean environment, obtained at an appropriate cost, and with this cost appropriately distributed amongst industry, car owners, homeowners and others sources of pollutants.

Emission testing is expensive. For single, well-defined source, such as a power plant stack, testing can be tedious but is not difficult. For a poorly defined source, as for example, road dust from an unpaved road or CO from forest fire, reliable test results are difficult to obtain. Furthermore, such testing is only possible after the source is in place; it is often necessary to know in advance what will be the emissions from a new source before it is built.

To achieve these goals, the Environmental Protection Agency of the United States, EPA, has produced a very useful set of emissions factor documents. These are summaries of the results of past emission tests, organized to make them easy to apply. (US Office of Air Quality Planning and Standards 1991).

4 Emissions inventories in the ZMVM

In order to know the quantity and content of emissions from different industrial processes and human activities in the ZMVM, the federal and local authorities have realized several emission inventories since 1988.

The emissions inventory 2004 of the ZMVM contains not only criteria pollutant emissions (PM10, CO, NO_x and SO_2) but also pollutant emissions of PM2.5, total organic compounds, TOCs, volatile organic compounds, COVs, ammonia NH_3 and methane CH_4. Furthermore the emissions data are spatial and temporally discriminated (SMA 2004a; SMA 2004b).

The estimation of emissions data also considers the national and international recommendations for reducing uncertainties, which have been made by international research organizations (Molina et al. 2000; ERG 2003)

The first Toxic Pollutant Inventory in the Air of the Mexico City Region was undertaken in year 2004 in order to identify the more contaminating industrial activities, the toxic pollutant types, their emission sources and emission factors. This inventory estimates that each year more than a hundred thousand tons of toxic pollutants are emitted to the atmosphere as a result of anthropogenic activities. The main sources of pollution are the area sources that emit 53.9 % followed by mobile sources with 35.2 % of total emissions (SMA 2006).

5 Developed Research Programs

The emission inventories have been in part the support for developing programs to improve the air quality in the ZMVM.

Among these programs it is worthwhile to mention the Integral Program for Controlling the Atmospheric Pollution (PICCA), and the programs to improve the quality of the air in the Valley of Mexico (PROAIRE 1995–2000), and in the ZMVM (PROAIRE 2002–2010) (CAM 2002).

In the year 2006, with the goal to measure local and global impacts of atmospheric pollution with the support of several national and international organizations the program MILAGRO was initiated, "A Mega city Initiative: Local and Global Research Observations". Some of the International organizations, which supported and participated in the initial part of this program, are: NASA; National Science Foundation, NSF; National Center for Atmospheric Research, NCAR; and the Department of Energy (DOE) among others. As initial part of this program, between 1 and 29 March were realized four campaigns: MIRAGE-MEX – Mega city Impacts on Regional And Global Environments; MCMA – Mexico City Metropolitan Area-2006; MAX-Mex – Mega city Aerosol eXperiment; and INTEX-B part 1 – the Intercontinental Transport Experiment (Madronic 2006).

During these campaigns were measured gases, particulates and radiations in the atmosphere and meteorological parameters such as wind velocity, temperatures and humidity.

The results of these measurements will be released publicly in March 2008. These can be used for the photochemical characterization of particles and gases, for emission inventory validation, a major knowledge of contamination processes and pollutant dispersion, climate impacts of air pollution and improvement in climate, meteorological and air quality models.

6 Mathematical models

A mathematical model is a set of numerical or analytical algorithms that describe physical or chemical aspects of the pollutants in the atmosphere.

Two types of mathematical models are mainly used in urban, regional and global air quality studies: deterministic and statistical. Deterministic models are based on the fundamental description of atmospheric chemical and physical processes (Seinfeld and Pandis 1998, pp 1193–1240), while statistical models are characterized by their direct use of air quality measurements to infer semi-empirical relationships (Seinfeld and Pandis 1998, pp 1245–1283).

6.1 Deterministic models

The concentration of a pollutant satisfies the advection-diffusion equation (Jacobson 2000). Because the flows of interest, in air pollution studies, are turbulent, actual numerical solutions of this equation require that variables in the equation be averaged in space, over grid volumes and in time. Solutions of the equation for the average term are found only through several semi empirical assumptions (Seinfeld and Pandis 1998).

The application of mathematical models of this kind is often limited due to their complexity, the requirement of precise knowledge of some values (such as initial value and boundary conditions of concentrations, and inventories and emission factors of sources), and the need for detailed meteorological information.

6.2 Statistical models

On the other hand, statistical models are based on the fact that air pollutant concentrations are inherently random variables, because of their dependence on the fluctuations of meteorological variables (such as wind direction and speed) and emission variables (such as emission rates and types of

sources). Several types of statistical methods are frequently used in air pollution studies. Frequency distributions are mostly used to assess the probability density function of the air pollutant concentration. Time series
analysis methods are used in the analysis of ordered data in a time sequence. Spectral analysis techniques allow the identification of cycles in
meteorological and air quality time-series measurements and regression
analysis techniques are used to study the concentration of a pollutant as a
function of meteorological conditions and of others pollutants concentration.

7 Methods of spatial interpolation

Many air pollution studies have employed spatial interpolation methods for
estimating levels of pollutant concentrations in regions that contain a number of monitoring stations, and to produce maps of air pollution concentrations. These studies are based primarily on distance-weighting methods
(Phillips et al 1997) and kriging (Liu and Rossini 1996; Yi and Prybutok
1996; Phillips et al 1997; Mulholland et al 1998). The Inverse Distance-
Weighting method (IDW) typically assigns more weight to nearby points
than to distant points (Phillips et al 1997). Kriging is a regression-based
technique that estimates values at unsampled locations using weights re-
flecting the correlation between data at two sampled locations or between a
sample location and the location to be estimated.

Both the IDW method and kriging directly use coordinate information
of sample points to perform interpolation and kriging's performance specially is dependent on the presence of spatial autocorrelation (values at
nearby points are more similar than are values at distant points).

We have recently considered these two techniques to predict ozone pollution levels across the Mexico City region (Rojas-Avellaneda and Silvan-
Cardenas 2006). Although linear interpolation processes such as IDW and
kriging normally require relatively high sampling densities and uniformly-
spaced sample locations, the relative accuracy obtained with the application of these methods for estimating ambient ozone concentrations in the
Mexico City region was stimulating.

8 Conclusions

Air has historically been polluted without limit. The burning of fossils fuels is the major cause of atmospheric pollution. These problems include

local and regional pollution and also global effects. Local and regional effects, as the generation of locally unhealthy air quality, indoor air pollution and urban smog, for example are the cause of damage to the environment, human health and quality of life. Two global issues have recently become a major concern: the greenhouse effect (global warming) and the depletion of the stratosphere ozone layer.

The only effective mechanism for limiting air pollution is government intervention. National governments did not act aggressively to control global air pollutions problems until the 1970's and 1980's. Actions were not taken earlier because lawmakers were not always convinced of the severity of air pollutions problems or because industries opposed government intervention.

Although work still needs to be done, government intervention has proved to be effective in mitigating some of the major air pollution problems facing the humanity, despite the opposition to government intervention.

In order to understand air pollution problems and to establish legislation to implement air quality and emission standards, modeling as well as monitoring is required. Several air pollution modeling techniques based on methods that rely upon available data and limited computational resources have been developed.

In spite of all these efforts some of the air pollution problems have been mitigated but not eliminated. The mitigated problems in the air of Mexico City include urban air pollution and acid deposition. Other major global air pollution problems, such as global climate change, have not yet been controlled.

References

CAM (2002) Comisión Ambiental Metropolitana. Programa para mejorar la calidad del aire de la Zona Metropolitana del Valle de México 2002–2010, México

ERG (2003) Eastern Research Group Inc. Evaluation of the 1998 Emissions Inventory for the Metropolitan Zone of the Valley of Mexico. Prepared for Western Governors' Association. Denver, Colorado

Jacobson MZ (2000) Fundamentals of Atmospheric Modeling. Cambridge University Press, New York

Liu LJS, Rossini AJ (1996) Use of kriging models to predict 12-hours means ozone concentrations in metropolitan Toronto- a pilot study. Environment International 22:677–692

Madronic S (2006) Conference given in "Foro de Monitoreo Atmosférico y Taller de Gestion Ambiental del Aire". México City

Molina MJ, Molina LT, Sosa G, Gasca J, West J (2000) Análisis y Diagnostico del Inventario de Emisiones de la Zona Metropolitana del Valle de México. Technological Institute of Massachusetts

Mulholland JA, Butler AJ, Wilkinson JG, Rusell AG (1998) Temporal and Spatial Distributions of Ozone in Atlanta: Regulatory and Epidemiologic Implications. Air and Waste Manage Assoc. 48:418–426

Phillips DL, Tingey DT, Lee EH, Herstrom AA, Hogsett WE (1997) Use of auxiliary data for spatial interpolation of ozone exposure in southeastern forest. Environmetrics 8:43–61

Rojas-Avellaneda D, Silvan-Cardenas JL (2006) Perfomance of geostatical interpolation methods for modeling sampled data with non-stationary mean. Stoch Environ Res Risk Assess 455–467

SMA (2004a) Secretaria del Medio Ambiente, Gobierno del Distrito Federal Inventario de Emisiones ZMVM

SMA (2004b) Secretaria del Medio Ambiente, Gobierno del Distrito Federal Inventario de Contaminantes Tóxicos del Aire en la ZMVM

SMA (2006) Secretaria del Medio Ambiente, Gobierno del Distrito Federal. Informe Ejecutivo de la Calidad del Aire ZMVM, march 2006

Seinfeld JH, Pandis SN (1998) Atmospheric chemistry and physics. Wiley-Interscience, New York

US Office of Air Quality Planning and Standards (1991) Compilation of Air Pollutant Emission Factors. Volume I: Stationary Point and Area Source. Volume II: Mobile Sources for Autos and other Vehicles. AP-42

Yi J, Prybutok R (1996) A neural network model forecasting for prediction of daily maximum ozone concentration in an industrialized urban area. Environmental Pollution 92: 349–357

Fundamentals of Boiling Water Reactor Safety Design and Operation

Javier Ortiz-Villafuerte, Rogelio Castillo-Durán, Héctor Hernández-López, Enrique Araiza-Martínez

Instituto Nacional de Investigaciones Nucleares, Apartado Postal 18-1027, México DF 11801, México. E-mails: jov@nuclear.inin.mx, rcd@nuclear.inin.mx, hhl@nuclear.inin.mx, earaiza@nuclear.inin.mx.

Abstract

Nuclear power is a viable, economically competitive and safe option to contribute to the high electricity demand expected for the next decades. Nuclear energy is currently the largest source of electricity without emission of greenhouse gases. In several countries, for baseload electricity generation, the capacity factors of NPPs are the highest for any type of fuel and, at same time, except for hydroelectric power, production costs from nuclear power are the lowest. These two factors, among others, have led the nuclear power industry to become competitive today in the electricity generation market. However, public acceptance is still a major issue to overcome before nuclear power can be exploited to its fullest. Issues as nuclear plant security, waste management and reactor safety are constantly being debated by society, even when such issues have been shown to have technically sound solutions. In particular, NPP safety will be discussed in this work.

Although safe operation of current nuclear plants is at its highest level, misinformation and lack of understanding of the physical fundamentals on nuclear reactor design and operation have led general public opinion to still show concerns regarding nuclear reactor safety. In this work, it is presented a short description of the fundamental physical principles behind the safety core and plant design and operation of BWRs, with the intention of helping to better understand and clarify concepts of frequent use in the nuclear engineering and safety areas.

1 Introduction

It is commonly accepted that there exists a direct relationship between energy consumption and wellbeing of the population in a country. Secure and continuous access to energy sources is therefore a fundamental aspect to consider for the future progress of a country. Near and mid-term projections show a clear increment in energy consumption, worldwide. An average of about 2.0 % per year growth is expected to 2030 (DOE 2006). In particular, energy requirements for electricity generation follow a similar growing tendency. Worldwide, electricity generation has shown an about 3.0 % annual growth during the last decade, and an expected demand increase of about 2.9 % per year to 2015 (SENER 2006). Currently, coal is the main fuel used for generating electric power, followed by natural gas. This tendency is still expected to be the same to 2030. Consequently, carbon dioxide emissions are also expected to increase (DOE 2006).

Under the scenario described above, electric power generation by nuclear energy has regained worldwide attention in the energy industry. Nuclear energy is the largest energy source without emission of greenhouse gases for electricity generation. The perspectives for nuclear power have also improved because of the high performance operation of current nuclear power plants (NPP), and the current high and volatile price of natural gas and oil. Excluding hydro, production costs (fuel plus operation and maintenance) of nuclear power are the lowest for any baseload generation technology in several countries (WNA 2005). Further, capacity factors of NPPs are also the highest for any type of fuel in several countries.

During 2005, there were 443 nuclear reactors in operation all over the world. These produced 2,625.9 TWh of electricity, representing 15.5 % of the world total electricity generation (IAEA 2006a). At the end of 2005, there were still 27 new nuclear reactors under construction. It is expected that the nuclear share in power generation will continue being about 16 % until 2020. In particular, in México there is one NPP with two nuclear reactors, generating about 5 % of the total electricity.

Although the importance of commercial nuclear power is clearly supported by the NPPs either planned or already under construction (mainly in the Pacific Rim), and safe operation of current nuclear plants is at its highest level, general public opinion still shows clear concerns regarding nuclear reactor safety. To the end of helping in understanding the safety issues during design and operation of commercial nuclear reactors, it is presented here a short description of the main design and operation parameters related to nuclear reactor safety.

2 Nuclear Reactor Designs

Grossly, a nuclear rector is a complex system in which the energy produced by fission nuclear reactions and the chain reaction are kept under control. Nuclear reactors can be divided is several different categories, as for example the purpose or function of the reactor; type of moderator (material used to slow down or decrease the neutron energy to the range suitable for causing fissions); fuel type; energy of the neutrons causing the chain reaction; etc. For example, there are nuclear reactors for research purposes, production of radioisotopes, or power generation. Further, according to the type of coolant, reactors can use light (ordinary) water, heavy water (D_2O), other liquids (Na), or gas (CO_2 or He).

In a general view, the main difference between a typical large NPP and those using fossil fuels is the energy source: nuclear fission versus chemical combustion. The rest of major components of the power plants are basically the same, as a steam supply system, turbine and condenser, and the electrical generator, as shown in Fig. 1.

In the case of a NPP, the steam supply system is referred to as nuclear steam supply system (NSSS). Other major difference between fossil-fueled and nuclear reactor plants is that the later are required by regulation to have redundant safety systems.

For electric power generation, nuclear reactors have as main objective to find an optimal balance between safety and economic issues. The main parts of a nuclear power reactor are the core, the reactor pressure vessel (RPV), internals, structures and instrumentation, and a control rod drive system. All these components are sometimes called the reactor assembly. The reactor core is composed of the fuel (the fuel itself that is uranium dioxide, UO_2, and structural materials for the fuel rods and assemblies), the moderator and/or coolant. The reactor pressure vessel contains all the main components of the reactor assembly. Technically, reactor design comprises neutronic (nuclear) and thermalhydraulic aspects and the feedback issues between them.

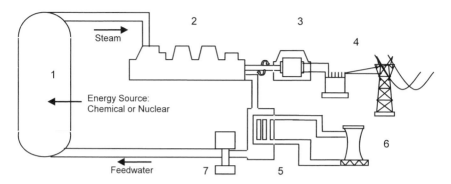

1) Steam Supply System, 2) Multi-stage Turbine, 3) Generator, 4) Transformer,
5) Condenser, 6) Cooling System, 7) Feed-water Pump

Fig. 1. Typical main components of power plants

The vast majority of the current operating reactors are cooled either by gas or water. If heavy water is used as coolant, this type of nuclear reactor is referred as to Heavy Water Reactor (HWR), whereas the term Light Water Reactor (LWR) is applied to the nuclear reactor cooled by ordinary water. Light water reactors are thermal reactors – those in which most neutrons causing fissions have energies below about 0.5 eV (although 0.625 eV is also taken as the boundary between thermal and epithermal neutrons). Two types of LWR exist: Pressurized Water Reactor (PWR) and Boiling Water Reactor (BWR). In a PWR water is prevented from boiling by keeping a very high pressure (14 MPa) inside the RPV. Currently, about 81.5 % of all nuclear reactors in the world are of LWR type and they produce about 88 % of the total nuclear power. Thus, in here, the safety aspects discussed will be focused on LWR NPPs, and more specifically in BWR plants.

3 Nuclear Safety

Before mentioning technical aspects of nuclear reactor safety, it is important to define *nuclear safety*, and how it is achieved. Nuclear safety is a set of actions taken to protect individuals, society (in general), and environment against radiological risks. These actions can be divided into three general groups: a) safe normal operation of nuclear facilities; b) prevention of transient events and accidents; and c) mitigation of the consequences of the transients and accident events should they occur.

For a NPP, the set of actions related to safe operation imply that normal operation must be performed within specific limits and conditions. Besides normal operation, it also includes maneuvering during reactor startup, power increase and decrease, shutdown, maintenance, test, and refuel (reload of nuclear fuel). This group also includes the anticipated operational transients, which are events expected to occur during the life of a nuclear facility. These events must be considered in order to take actions to prevent significant damage to reactor components or to avoid reaching accident conditions.

The prevention of transient events and accident conditions in a nuclear power reactor is accomplished by the use of components, systems and procedures, related all to safety. Accident prevention is the top priority for reactor designers and operators. Operation personnel are required to have strong commitment with the culture of safety. Means of accident prevention include: 1) technical aspects, as emergency systems used to control conditions that could lead to an accident scenario, 2) the *in-depth defense* strategy, which prevents the release of radioactive material by a series of physical barriers, 3) inspections and tests, which are regularly performed to systems and components to reveal any possible malfunction or degradation, and 4) operator training, which is mainly focused on recognizing conditions leading to accident scenarios, so the response is fast and appropriate.

4 Engineering Safety Barriers

The nuclear aspects of the design of a nuclear reactor core are highly dependent on other areas of design of the power plant, as thermalhydraulics, structural analysis, economic performance, etc. Thus, the throughout design of a large commercial NPP is an enormously complex task that involves coordination among several and quite diverse disciplines. The design is not at all a one-time, static process, but an iterative one, since the design is refined through several steps to identify and satisfy constraints, safety issues, and economical performance.

The major safety concern for a commercial NPP is to avoid the release to the environment of the large inventory of radioactive fission products accumulated in the nuclear fuel, for any foreseeable accident. *Fission products* are the nuclei (fission fragments) formed by the fission of heavy elements, plus the nuclide formed by the fission fragments radioactive decay. To avoid such fission product escape –and the much less likely escape of highly- radioactive solid material in the fuel-, several safety engineering barriers exist: first, the fuel pellet itself keeps the solid and some of the gaseous fission products in the fuel matrix. Long-lived actinide and tran-suranic isotopes produced by neutron capture or radioactive decay also stay into the fuel matrix. Then, the next barrier is the fuel rod cladding, which keeps those fission products accumulated in the fuel rod gap from reaching the core coolant. Figure 2 shows a sketch of a typical BWR fuel element, and a fuel assembly and its corresponding part of a cruciform control rod.

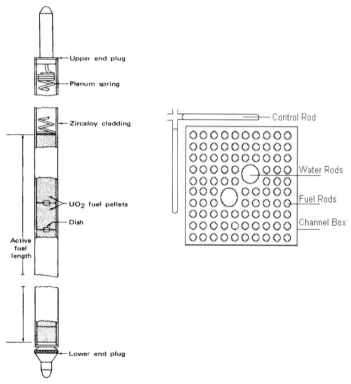

Fig. 2. Typical BWR fuel element (first safety barrier) and assembly

If some fission products could leak out through the fuel rod clad (*fuel failure*) to the coolant system, the next barrier is the RPV. This is made of forged steel, with dimensions of about 21 m height, 5.3 m diameter and thickness between 15 and 20 cm. Some radioactive material could also be transported with the coolant in the coolant piping loops, but it is not prob-able it can leave the coolant system. If a catastrophic situation is considered, the pressure vessel could fail, however the radioactive material can still be contained in the next safety barrier, referred to as the *primary containment*. This is a hermetic building, surrounding the RPV. This containment structure has 1.5 m thick reinforced concrete walls, which in addition have a steel liner of about 1 cm thick. Although the probability of leak out from the primary containment for radioactive material is quite low, there exists an additional barrier: the secondary containment, also known as *reactor building*. This is the last barrier that the radioactive inventory in the core needs to pass through to, finally, reach out to the power plant surrounding environment. However the reactor building structure has walls 0.6 to 1.2 m thick and the interior atmosphere is kept to a negative pressure, thus avoiding gaseous radioactive material to escape. Figure 3 shows all the typical safety barriers in a Mark II containment design, as the one employed in the Laguna Verde NPP for the two BWR Units. Several types of containment designs have been developed over time as consequence of an evolutionary process to better protect the environment (Gavrilas 2000; Lahey and Moody 1993).

The design of the above mentioned safety engineering barriers involves choosing the correct construction materials for each of the barriers, since the environment inside a nuclear reactor is characterized by very high pressures, large thermal gradients, and an intense nuclear radiation field. Therefore, those materials employed for the safety barriers are required to have *nuclear quality*, since nuclear radiations alter the properties of such materials, besides the demanding thermomechanical stresses.

Although the engineered safety barriers are in charge of physically contain the fission products and other radioactive material, there are additional operational measures and systems designed to take preventive action, in case of abnormal behavior of the nuclear reactor. Separate safety systems have as primary function to keep the reactor core cooled, and fully covered at all times, in case of accident. Even when the reactor is shut down, the remaining decay heat needs to be removed from the core to avoid core meltdown. Safety systems include therefore a reactivity control system, as the control rods, an Emergency Core Cooling System (ECCS), and a Residual Heat Removal (RHR) system to accomplish the safety features required. The ECCS comprises reactor system components (pumps, valves, heat exchangers, tanks, and piping) that are specifically designed to remove residual heat from the reactor fuel rods should the normal core cooling system (reactor coolant system) fail. The ECCS mainly includes high and low pressure coolant injection systems to keep the core fully covered. An additional Automatic Depressurization System (ADS) is included to decrease the vessel pressure, so the low-pressure cooling systems can provide more coolant, if required.

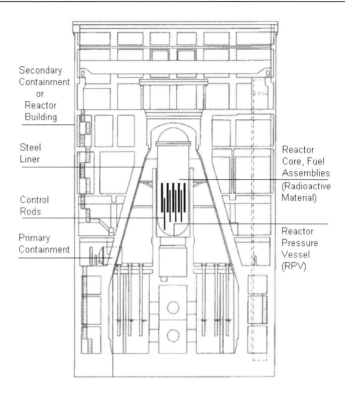

Fig. 3. Multiple layers of Safety in the Mark II containment design. Taken and adapted from www.cfe.gob.mx/en/LaEmpresa/generacionelectricidad/ nucleoelectlagverde/nucleoelectricidadenelmundo/

5 Reactivity and Feedback

The previous section showed measures considered in the reactor design to avoid or at least minimize radiological risks to the environment and population, in case of accident. In following sections, design features for reactor control and safe operation are presented. All such control and safety features are based in basic physical principles. In order to safely operate a nuclear power reactor, the chain reaction must be stable. A measure of such stability is the multiplication factor k, which is defined as:

$$k \equiv \frac{\text{neutron population in one generation}}{\text{neutron population in the preceeding generation}}. \tag{1}$$

A more practical definition of k is

$$k \equiv \frac{\text{rate of neutron production}}{\text{rate of neutron removal}}. \tag{2}$$

Neutron production is achieved by nuclear fission, where for each neutron captured and causing fission, two or three new neutrons are generated. Neutron removal is due to absorption (it can or cannot cause fissions) or leakage from reactor core. It is clear that k is a time dependent function.

A nuclear power reactor is designed to operate when there is a balance between neutron production and removal, that is, $k = 1$. When reaching such condition, the reactor is said to be *critical*. *Criticality* thus means that the chain reaction is steady; it does not grow up, nor it dies out. The fractional change of k, or fractional deviation of core multiplication from criticality, is defined as the *reactivity* ρ, that is,

$$\rho = \frac{k-1}{k}. \tag{3}$$

As k, ρ is consequently time dependent too. To measure ρ, there exist three main units of reactivity: (a) *inhour*, related to the inhour equation. One inhour of reactivity is the magnitude that results in a reactor stable period of one hour; (b) percent (%) reactivity or percent excess k; and (c) dollars and cents, where one dollar of reactivity corresponds to the value of the delayed fraction of fission neutrons (β), which is $\beta \approx 0.0064$ for a U235 fueled reactor.

Changes in power in a reactor lead to changes in the core reactivity, which in turn lead to new changes in reactor power. This feedback effect then needs to be considered in reactor design. The reactivity management thus has two objectives: control and safety. For the control function, control rods, burnable poison or flow control can be used. Control rods are not only designed for abnormal or emergency situations. Control rods are an important aspect of fuel management -economics of the NPP- because they are normally employed to keep the reactor core critical. Control rods, therefore, are needed to compensate the excess reactivity of fresh and low-burnup fuels, so they are used during normal operation to achieve the desired power output and profile. Reactor power control using the recirculation flow is a unique feature of the BWRs.

Regarding the safety function of reactivity management, during an emergency situation, the rapid insertion of all control rods (*scram*) stops the chain reaction. For abnormal situations, control rods are designed to diminish or kill the neutron chain reaction in the core, by absorbing more neutrons than those produced during the fission reactions. In BWRs, the control device is a cruciform blade containing stainless steel tubes with the reactivity control material, as shown in Fig. 2. This is compacted boron carbide powder. There is a cruciform control rod for each four fuel assemblies. In a BWR, the control rods are inserted from the bottom of the reactor vessel. For modern BWRs, the scram time is from 1.5 to 2.5 seconds (Gavrilas et al. 2000).

In a nuclear reactor, changes in power involve changes in temperature throughout the core. Such temperature change affects the neutron flux distribution, since the macroscopic cross sections (probability of different kinds of interactions between neutrons and nuclei) are temperature dependent, in general. Consequently, if there is a change in power, k changes, and then ρ changes too, leading to a new power change, and so on. In order to accurately account for the feedback effects from power to reactivity and vice versa, it would be necessary to solve the fundamental transport equations of mass, momentum and energy for the coolant, coupled to the neutron transport equation (or to use the diffusion approximation where permitted), which describes the neutron distribution in the core, and consequently the heat generated (power) profile. Since such strategy leads to a formidable computational effort even today, state-of-the-art computational tools are currently being designed and coded to solve the problem, but there still exists fundamental issues, as physical models for certain phenomena (turbulence and inter-phase exchange in multiphase flows, etc.), before the problem can be attacked to its fullest.

5.1 Reactivity coefficients

An alternative way to account for the feedback effects on reactivity is through the *reactivity coefficients*. These coefficients are then inserted in simple models of the transport equations mentioned, which can be solved easily and fast. In a BWR, both the effects of temperature and void fraction are important for total change of core reactivity during normal operation, and especially during transient conditions. Reactivity coefficients simply represent the rate of change of the reactivity respect to some variable, for example temperature, that is:

$$\alpha_T \equiv \frac{\partial \rho}{\partial T} \, . \tag{4}$$

This is called the temperature coefficient of reactivity. Note that if α_T is positive, it implies that an increase in temperature leads to an increase in ρ, and thus the power level also increases, causing a new increase in temperature, and so on. This situation is clearly unacceptable, because the reactor would be unstable in such conditions. α_T depends on several different physical processes –each within its own temperature range- in the reactor core, which must be considered to accurately calculate the total change in ρ. One way to do this is to introduce different temperature coefficients of reactivity. Each coefficient represents the contribution of a major component of the core, as fuel, moderator/coolant, and structures, that is:

$$\alpha_T = \sum_j \alpha_{T_j} \equiv \sum_j \frac{\partial \rho}{\partial T_j} \tag{5}$$

One important issue regarding temperature reactivity coefficients is that the time scale for each different physical process involved in reactivity feedback can be quite different one to the other. Thus, temperature reactivity coefficients can be divided into prompt and delayed. Effects related to instantaneous phenomena occurring in the fuel, as the Doppler effect or physical changes in the fuel matrix, are considered as prompt; while effects due to changes in the moderator or coolant, as density or neutron energy spectrum, are taken as delayed (Duderstadt and Hamilton 1976). It should be clear that if $\alpha_T < 0$, the reactor is *inherently stable* to changes in temperature; otherwise the reactor will be *inherently unstable* to the same phenomenon, as explained above.

It is customary then to refer to the *fuel temperature coefficient* as the *prompt temperature coefficient*, α_{prompt}. Technically, this coefficient is the derivative of reactivity with respect to the fuel temperature, while keeping the coolant temperature and void fraction constant. For this coefficient, the *nuclear Doppler effect* is the governing process. Shortly, in the Doppler effect more neutrons are absorbed in some nuclei's resonances –no fission occurs- as the temperature increases, thus negatively affecting the neutron economy for the chain reaction. In LWRs, the term *Doppler reactivity coefficient*, α_D is, for the reasons mentioned, also used when referring to the fuel temperature reactivity coefficient. Although the magnitude of this coefficient depends on several parameters, such as coolant density, presence of control rods, etc., for modern BWRs, actual values are in the range from -1×10^{-5} to -3×10^{-5} $(\Delta k / k) / °C$ (Gavrilas et al. 2000). Since $\alpha_{prompt} (\approx \alpha_D)$ determines the first response to a change in fuel temperature and/or power, no NPP will be licensed, in practically any country in the world, if $\alpha_{prompt} \geq 0$.

The (delayed) moderator coefficient, α_{mod}, governs the final state of the reactor for a temperature change. As with α_{prompt}, $\alpha_{mod} < 0$ assures reactor stability, for normal operation and during transient events. For BWRs, α_{mod} has two important components: one due temperature and the other due to void fraction, ε, content. Both components are related to coolant density, and thus α_{mod} is also known as the *moderator density reactivity coefficient*. Actual values of the steam void reactivity coefficient, α_V, fall in the range from -80×10^{-5} to -140×10^{-5} $(\Delta k / k) / \varepsilon$, for beginning and end of cycle, respectively, and taking ε=40 %. These quite large values are of primary importance in the BWR operation, and allow for automatic load-following capability.

Once the reactivity coefficients are determined, they can be used in models to study feedback effects during transients or to evaluate consequences during accident scenarios (Ash 1965; Lewis 1977; Lewins 1978).

6 Power Peaking Factors

The principal goal of the thermalhydraulics safety design of a reactor core is to ensure that set-by-design limiting operating, transient and expected accident conditions are not exceeded. The design analysis is commonly performed for a single coolant channel of the reactor core, where limiting conditions are assumed. This channel is referred as to the *hot channel*. By ensuring that this hot channel does not surpass the thermal safety design limits, thus, the rest of the channels in the core will most certainly fall within such safety constraint values.

The *hot spot* is the point of maximum heat flux or lineal power density in the core, and the hot channel is generally defined as that coolant channel in which the heat flux and/or enthalpy rise reach their maximum value in the core. Is it important to note that under certain conditions, control rod movement can cause a hot spot to show up, in one of the assemblies around the control device.

Nuclear hot channel factors or *power peaking factors* are parameters related to the power distribution generated in the reactor core. Peaking factors are no longer directly used as *thermal limits*, which are the actual limit values set to maintain fuel cladding integrity during normal and transient events, and the capability of cooling the core for postulated accidents. Peaking factors are however used to calculate the thermal limits by the process computer in the control room of a NPP. These parameters are defined in terms of the generated power, or the heat flux on the fuel elements and assemblies.

Although several different peak factors exist, according to each different reactor design, the most common ones are the radial peaking factor, F_R, also called relative assembly power, the axial peaking factor, F_A, and the total peaking factor, F_q. These are defined, respectively, as:

$$F_R = \frac{\text{average heat flux of the hot channel}}{\text{average heat flux of the channels in the core}}, \qquad (6)$$

$$F_A = \frac{\text{maximum heat flux of the hot channel}}{\text{average heat flux of the hot channel}},$$

$$F_q = F_R \times F_A = \frac{\text{maximum heat flux in the core}}{\text{average heat flux in the core}}.$$

Clearly, any other of the channels in the core can be used in the definitions given above. To determine the point of maximum heat flux in a fuel rod (hot spot), it is introduced the local peaking factor, F_L, which is defined as:

$$F_L = \frac{\begin{array}{c}\text{maximum fuel rod average heat flux}\\ \text{in an assembly at a particular axial position}\end{array}}{\begin{array}{c}\text{assembly average fuel rod heat flux}\\ \text{at the same axial position}\end{array}}. \qquad (7)$$

The total peak factor is therefore now constructed introducing the local peak factor, that is,

$$F_q = F_R \times F_A \times F_L. \qquad (8)$$

In both cases, F_q can be considered as a three-dimensional parameter of the core, relating the highest power produced by a fuel rod, in a specific axial position, to the average power generated in all the fuel rods in the core, that is,

$$F_q = \frac{\text{highest fuel rod power at a particular axial position}}{\text{core average fuel rod power}} . \qquad (9)$$

For a BWR/6 reactor, for example, typical values for the peak factors are: $F_R = 1.40$, $F_A = 1.40$, $F_L = 1.13$, and $F_q = 2.22$ (Lahey and Moody 1993). All these peaking factors change all along an operation cycle, even at steady-sate nominal power, since they are affected by the control rod movements made to compensate the fuel burnup.

7 BWR Core Thermal Limits

Normal NPP operation implies that certain physical parameters need to be kept under control. The core thermal limits are limit values set to maintain fuel cladding integrity during normal and transient events. For postulated accidents, the thermal limits are set to ensure the capability of cooling the core, by avoiding gross cladding failure and thus maintaining the core ge-ometry. The *Design Basis Accident (DBA)* is the postulated accident that a nuclear facility must be designed and built to withstand without loss of systems, structures, and components necessary to assure public health and safety. For a modern BWR, the *Loss Of Coolant Accident (LOCA)* is the postulated accident that results in a loss of reactor coolant at a rate in ex-cess of the capability of the reactor makeup system from breaks in the re-actor coolant pressure boundary, up to and including a break equivalent in size to the double-ended rupture of the largest pipe of the reactor coolant system. During normal operation and transient events, the parameters to monitor in a BWR are the *Minimal Critical Power Ratio (MCPR)* and the *Linear Heat Generation Rate (LHGR)*. For the LOCA DBA, it has been established the *ECCS/LOCA limit*, because during this unlikely event the ECCS core re-flooding action could lead to gross fragmentation failure, due to the quenching effect on the highly heated up fuel rod cladding.

7.1 The linear heat generation rate thermal limit

In order to satisfy the thermal and mechanical limits of basis design, there exist limits on the power level that can be generated. Specially, limits on the power generated by unit length in the axial direction, the LHGR, are set to prevent cladding cracking that eventually could lead to cladding fail-ure. Cladding cracking could occur if a certain power level (in kW/m) is exceeded. In this case, the fuel pellet and cladding material could come in contact, because of the difference on the thermal expansion rate of the materials. If the stress placed on the cladding by the fuel pellet touch would exceed the yield stress of the cladding, then the cladding cracks. In this case, the power level, that is the LHGR limit, is limited such that will not induce a 1 % plastic strain on the cladding. This is a quite conservative value, since there is evidence that 1 % does not imply cladding cracking. This 1 % plastic strain value is known as the *mechanical limit*.

Another limit on the peak fuel pin power is that preventing the melting temperature of the fuel pellet centerline (2800°C at zero exposure for UO2). The limit value for the power level in this case is known as the *fuel pellet thermal limit*. These two peak power level limits are grouped in one single value, referred as to the *thermal-mechanical limit*. The same LHGR value serves for satisfying both the thermal (peak fuel pellet centerline temperature) and mechanical (cracking) design limits for the cladding.

During a LOCA DBA event, the power level (in kW/m) is limited such that will not induce a cladding temperature of 1200°C (2200 F) (Lahey and Moody 1993), when Zircalloy is the material used for cladding. This is the *DBA LOCA thermal limit* or the *ECCS/LOCA limit* mentioned above. If the cladding temperature stays above 1200°C for an extended time, the increasing rate of Zircalloy oxidation could lead to cladding brittleness, and thus increasing the probability of cladding fragmentation, due to the much lower temperature of the sub-cooled liquid injected by the ECCS.

The LHGR depends both on the type of fuel and exposure. Generally, the LHGR thermal limit refers to the LHGR value at zero exposure. A typical value for the LHGR thermal limit of the fuel in a BWR is 14.4 kW/ft, although in new fuel rod and assembly designs this value is lower to provide a greater safety margin.

7.2 Boiling heat transfer in a nuclear reactor core

The MCPR limit is directly related to the thermalhydraulics of a BWR. Since boiling naturally occurs in the core of a BWR, it is given next a short introduction to the heat transfer and flow regime characteristics of a boiling system. *Flow regime* is the topology or geometry of the flow. When a liquid is vaporized in a heat channel the liquid and vapor generated take up a variety of configurations known as flow regime. The particular regime depends on the conditions of pressure, flow, heat flux, and channel geometry. Similarly a *heat transfer regime* is the different flow and thermal patterns that occur in a heat transfer system.

The heat amount that can be transferred through the cladding towards the coolant depends on the thermalhydraulics conditions of the flowing coolant itself. Several heat transfer regimes can occur in a boiling system (Collier and Thome 1996). Starting with *single phase subcooled liquid*, the heat flux corresponds to the natural or forced convection heat transfer regime. In here, the heat flux increases about the same as the temperature difference between the cladding and liquid saturation temperature increases too.

Then, the *subcooled boiling* regime occurs. This type of boiling occurs on the cladding surface, which has a temperature equal or slightly higher that the coolant saturation temperature, while the bulk flow is still below saturation temperature. The small bubbles formed on the clad surface collapse quickly when going into the still subcooled bulk flow. However, the agitation and mixing generated by this bubbles nicely enhance the heat transfer from the cladding to the coolant. The heat transfer in this regime is difficult to compute, and corresponds to a mixture of single and dilute bubbly flows. The heat flux increase in this regime is still somehow proportional to the temperature difference increase.

Next, *fully developed nucleate boiling* occurs. This a very efficient heat transfer mode, since the turbulence is enhanced by the movement of the bubbles formed, inducing a better mixing of the whole fluid. The heat transfer in this regime is even more difficult to compute than in subcooled boiling, because it corresponds to a bubbly, slug, and annular flows. In all these three flow regimes, the cladding surface is practically always wet and well cooled. Bigger and bigger bubbles form in the middle of the coolant flow, until the annular flow occurs. At this point, small liquid droplets travel in high velocity steam, while a liquid film covers the cladding surface. The highly turbulent flow nicely enhances the heat transfer. In this heat transfer regime, heat flux greatly increases with little temperature difference occurring between the cladding and the coolant saturation temperatures.

Transition boiling follows the nucleate boiling. In this mode, cladding temperature is unstable since patches of steam and liquid alternatively occur on the cladding surface. Large differences between cladding and coolant saturation temperatures can occur here, with little or none heat flux increase. These fluctuations occur because the dry regions are generally unstable, existing momentarily at a given location before collapsing and allowing the surface to be rewetted. The *onset of transition boiling (OTB)* marks the point where the cladding temperature becomes unstable.

Finally, the *film boiling* regime can occur. In this regime, cladding surface is covered by a steam film, and thus clad temperature is high. Film boiling is a flow regime similar to an inverted annular flow, where liquid flows in the centre and a thin vapour film flows adjacent to the heating surface. It occurs at low qualities and mass flow-rates.

LWRs, both PWR and BWR, are designed to operate below the starting of the transition boiling regimen, the OTB point also known as the *Departure from Nucleate Boiling (DNB)*. This means that all the heat generated in the fuel can be transferred to the coolant flow, and the temperature of the cladding will be only a little higher that the coolant saturation temperature. Thus, in normal operation conditions and during transient events caused by a single operator error or single equipment malfunction, a BWR will present only regions of single phase subcooled liquid at the bottom of the core, and then a region of subcooled boiling, nucleate boiling, and finally annular flow at the core top.

Because of its importance to reactor safety, several terms have been employed to refer to the physical phenomena occurring at DNB, although they do not describe the exact same conditions. Such terms are: (a) *Boiling crisis*: The point at which the heat transfer coefficient deteriorates, resulting in a temperature excursion of the heated surface. It characterizes the rapid temperature excursions typical of low-quality DNB. (b) *Burnout*: Condition in which a small increase in heat flux on a particular surface leads to an excessive increase in wall temperature. (c) *Critical heat flux (CHF)*: Another term to describe the boiling crisis, although they are not necessarily equal. It is the local heat flux that determines the OTB. CFH is not generally descriptive of the phenomena involved in BWR conditions. (d) *DNB*, as already explained, it refers to the point at which the heat transfer from a fuel rod rapidly decreases due to the insulating effect of a steam blanket formed on the rod surface when the temperature continues to increase. It is more frequently used to describe the high-pressure, high-flow phenomena that can occur in pressurized water reactor bundles. (e) *Dryout*: The point where the processes of evaporation, deposition, and entrainment lead to a condition in which the film flow rate becomes zero. See also boiling crisis. It is considered the term that better describes the physical mechanism that occurs at conditions of interest to BWR technology.

7.3 The minimum critical power ratio thermal limit

The other thermal limit used in normal and transient operation is the minimum critical power ratio (MCPR). In this case, the goal is to avoid transition boiling in the reactor core, in order to prevent cladding damage due to *overheating*, which is defined, conservatively, as the onset of transition from nucleate boiling. Furthermore, significant weakening of the fuel cladding by overheating is not expected until the flow enters into the film boiling regime zone. As mentioned above, overheating is avoided by staying below the thermal limits associated to the critical power ratio. *Critical power* is the power of an assembly that produces the *critical quality*, which is the flow quality corresponding to the DNB. That is, the critical bundle power is the minimum fuel bundle (assembly) power that, under the conditions of interest, will produce transition boiling at some point in the bundle. Thus, the *critical power ratio* is defined as the ratio of the critical power to the actual operating power of the assembly considered. The minimum CPR (MCPR) is the minimum value of the CPR among all fuel assemblies in the reactor core. Thus, the reactor operational limits are based on terms of the MCPR.

Based on statistical analysis, there exists the possibility, however small, that a combination of a transient event and the uncertainties and tolerance associated to different parameters could cause transition boiling somewhere in the fuel bundle for some time period. Thus the MCPR limit has a design basis requirement that transients caused by a single operator error or single equipment malfunction, must be limited, considering uncertainties in the monitoring of the state of operation of the core, such that more than 99.9 % of the fuel rods are expected to avoid transition boiling.

7.4 Operational limit and margin

By definition, MCPR = 1.00 at OTB. Because of the uncertainties associated to monitoring the actual state of operation and boiling regime at certain axial position of a fuel assembly, a statistical margin is applied to the MCPR. That is, instead of considering 1.00 as the transition boiling onset value, the MCPR limit value used is set, typically, at 1.06, and naturally, it is not allowed to operate below such limit value.

When considering the worst transient expected event, there is another increase to the MCPR limit value used for operation. This limit value is referred as to the *operation limit*. Furthermore, at full power, the MCPR operation limit value used is set even higher that the MCPR operation limit. The difference between the actual operation MCPR limit value used and the MCPR operation limit is called the *operation margin*.

8 Nuclear Industry Accidents

Two are the major accidents in the whole history of the nuclear power industry: Three Mile Island (1979) and Chernobyl (1986). The latter is worst catastrophe in the industry; while the former showed that the engineering barriers properly implemented could perform their containment duties and thus avoid radiological risk to population and environment. None of the nuclear reactors involved in these accidents were BWRs, although at the Three Mile Island (TMI) site the accident occurred to a PWR, which is a LWR as BWRs are too. These two accidents are briefly described next, with the intention of showing the lessons learned from them, and measures taken to avoid such kind of problems to happen again. The material presented in previous sections can help understanding the causes leading to the accident scenario.

8.1 TMI

The major accident in nuclear industry for a LWR, occurred in Unit 2 (PWR) at the TMI NPP, in Pennsylvania. Half of the core has been melt down (USNRC 2004). The accident occurred on March 28, 1979. In this case misleading signals from instrumentation led operators to perform erroneous actions that actually aggravated the original transient. After the core meltdown, actions were taken to ensure the core was properly cooled, but a large hydrogen bubble formed in RPV dome. It was feared that a hydrogen explosion could eventually lead to a breach of the containment. This did not happen because of the absence of air in the RPV and preventive actions.

Although small amounts of radioactivity were released, partly because of a pressure relief action, no harm to human health or environment has been corroborated. In TMI, the four major causes identified leading to the accident were: (a) inadequate operator training; (b) mechanical problems and faulty instrumentation; (c) poor control room design; and (d) communication failures at the facility and in information exchange across the industry.

8.2 Chernobyl

The Chernobyl NPP, located in Ukraine, had at the time of the accident four units of the RBMK type in operation. The accident occurred on April 26, 1986, in Unit 4. The reactor was completely destroyed and large amounts of radioactive material were thrown out to the surrounding areas. *RBMK* is an acronym for *Reaktor Bolshoy Moshchnosti Kanalnly* (high-power channel reactor), a type of light water (as coolant) reactor using graphite as moderator. This type of reactor was never built outside the, at the time, Soviet Union (NucNet 2006), and certainly it would have not possibly be licensed in western countries, because, for example, this reactor design has a positive void coefficient and it did not have a full containment.

Nowadays it is considered that both the reactor design and the operational errors were decisive for the accident to occur. For example, the control rod design in RBMKs could actually increase reactivity when fully inserted, instead of killing the chain reaction. Many of the mistakes performed by the operators could be due to lack of safety culture and poor training. It is important to mention that for the intended tests to be performed the ECCS was deliberately disabled and the protection system that would have tripped the reactor was also turned off. Details of the accident evolution and aftermath consequences can be consulted elsewhere (NucNet 2006; IAEA 2006b).

There still exist 15 RBMKs in operation, 14 of them in Russia. The one in Lithuania is scheduled to shut down by 2009. All these units have incorporated safety changes to eliminate deficiencies as those leading to the Chernobyl accident.

8.3 Lessons and actions

The two accidents mentioned led the industry to look deeper into the safety issues of power plants and to develop and improve the safety culture and in-depth defense strategy. To eradicate or minimize the problems that could lead to severe accident situations nowadays exists extensive training to operators, and key craftsman and technician positions, are required. The training is provided in classrooms, laboratories and on the job itself, by accredited institutions. Human performance thus has been recognized as critical in reactor safety.

Inspections are performed by regulators and plant personnel closely collaborating with the IAEA. Such inspections have helped in the upgrading and strengthening of plant design, components and instrumentation. Also, NPPs have grouped together in organizations to address common issues. This allows sharing information among different NPPs and countries.

The arrival of digital instrumentation has greatly helped on the way information is presented to the reactor operators. Further, detection of faulty behavior of equipment or components is an important step in prevention of accident occurrence. *Online monitoring* systems are currently being used and implemented in NPPs to monitor and provide a diagnosis of the state of the equipment or component under surveillance (Hines and Uhrig 2005; Ortiz-Villafuerte 2006). One major issue in online monitoring is *signal validation*, which is defined as detection, isolation and characterization of faulty signals. Monitoring thus must include signal validation for use in safety related components.

9 Fuel Failure

Recent advances in competitiveness in the nuclear industry come, among other reasons, from power uprates. This is an economical and effective way to generate more power without the need of constructing more power plants. New fuel rod and assembly designs along with core management strategies are an important part that makes possible the power uprates in NPPs. However, *fuel failure* is still an issue in reactor safety. Thus, the recent advantages in competitiveness of nuclear energy can be challenged by public opinion, thus forcing the regulatory entities to restrict the use of the new core management strategies, particularly on power peaking and linear heat generation rate (LHGR) operation limits.

In general, BWR fuel has shown a quite good performance. For example, the number of failed boiling water reactor (BWR) fuel rods has decreased considerably since the 1970's, from about 1,000 to less than 5 failed rods per million fuel elements in operation (Edsinger 2000). However, after more than a decade of a continuous decreasing fuel failure rate, in the last few years BWR fuel presented a noticeable increase failure rate (NEA 2003). The cause is considered to be a combination of very diverse areas, such as water chemistry, new cladding materials and manufacturing procedures, and higher fuel duty (Yang 2004).

During the 1990's, the major cause of BWR fuel rod failure in the United States of America (USA) was crud-induced localized corrosion (CILC) (Yang 2004). In comparison, during the 1970's the principal cause of fuel rod failure in BWRs was the pellet-clad interaction (PCI) mechanism (Billaux 2004). Three main reasons for reducing the BWR fuel rod failure rate because of PCI were: more restrictive fuel preconditioning operational procedures; the introduction of the Zr-liner, or barrier, during the 1980's; and better fuel assembly designs (9×9 and 10×10), which reduced the average fuel rod LHGR. Although the Zr-liner has shown great success resisting failure by PCI, it does not eliminate the problem completely. Moreover, the Zr-liner has been associated with fuel failure as the secondary degradation that causes the axial splits, leading to high off-gas and coolant activity. New enhanced Zr-liners have been introduced to prevent and reduce this new problem. Currently, both fuel fabricants and utilities work together towards a *zero failure goal*, taking into account materials, production and operational strategies.

10 Other Issues Related to Nuclear Safety

The safety of a NPP requires today considering into account the plant security. Another major issue is waste management and transportation. These issues are beyond the scope of this work, but it can be said that after September 11, 2001, NPP security has been strengthened by regulation. Regarding waste management, low level repositories exist in several countries with no major safety issues. No final repository exists for high level waste and spent fuel, but temporary sites are already receiving waste in several countries. The final repository issue is more a political and public opinion problem that a technical one.

11 Next Generation Reactor Designs

So far, the safety features described are related to the operating nuclear power reactors. The vast majority of these working reactors belong to the Generation II (GII). Few of the nuclear power reactors of the next generation, III, are already in operation, but several are under construction. Generation III (GIII) refers to an evolutionary design, which is based on proven technologies, but with important economical benefits to operation and improved safety systems. In Japan, for example, there are two Advanced BWRs (ABWRs) in operation for almost a decade now. Regarding safety, some GIII reactors include passive safety systems, and not only those active systems in GII reactors. Passive safety systems rely on the forces of nature, as gravity, and not on human operation. The current use of probabilistic safety analysis shows that GIII designs present important advances respect GII reactors. Some GIII designs have included features in their containment design that avoid the radioactive material in the core to reach the NPP ground, in case of RPV failure, and thus no underground water sources could be affected.

12 Conclusions

A brief description of the safety engineered systems and operational procedures of typical commercial NPPs were presented here. Concepts such as thermal and mechanical limits, reactivity, power peaking factors, critical power, linear heat generation rate, and their relation to safety of nuclear reactor were introduced to end of helping understanding the fundamental physical principles behind the safety core and plant design and operations of BWRs.

References

Ash M (1965) Nuclear Reactor Kinetics. McGraw-Hill, New York

Billaux M (2004) Modeling pellet-cladding mechanical interaction and application to BWR maneuvering. International Meeting on LWR Fuel Performance; Orlando, September 19–22

Collier JG, Thome JR (1996) Convective Boiling and Condensation. Clarendon Press, Oxford

DOE (2006) Department of Energy. Energy Information Administration, International Energy Outlook. DOE/EIA-0484(2006)

Duderstadt JJ, Hamilton LJ (1976) Nuclear Reactor Analysis. John Wiley & Sons, New York

Edsinger K (2000) A review of fuel degradation in BWRs. 2000 International Topical Meeting on Light Water Reactor Fuel Performance; Park City, Utah; April 10–13

Gavrilas M, Hejzlar P, Todreas NE, Shatilla Y (2000) Safety Features of Operating Light Water Reactors of Western Design. CANES MIT, Cambridge MA

IAEA (2006a) Energy, Electricity and Nuclear Power Estimates for the Period up to 2030. IAEA-RDS-1/26

IAEA (2006b) The Chernobyl Forum: 2003–2005, Chernobyl's Legacy: Health, Environmental and Socio-Economic Impacts and Recommendations to the Governments of Belarus, the Russian Federation and Ukraine. IAEA/PI/A.87 Rev.2/06-09181

Lahey RT Jr, Moody FJ (1993) The Thermal-Hydraulics of a Boiling Water Reactor. American Nuclear Society, La Grange Park IL.

Lewis EE (1977) Nuclear Power Reactor Safety. John Wiley & Sons, New York.

Lewins J (1978) Nuclear Reactor Kinetics and Control. Pergamon Press, New York

NucNet (2006) Chernobyl Fact File

Hines JW, Uhrig RE (2005) Trends in computational intelligence in nuclear engineering. Prog. Nucl. Energy 46(3/4):167–175

Ortiz-Villafuerte J, Castillo-Durán R, Alonso G, Calleros-Micheland G (2006) BWR online monitoring system based on noise analysis. Nucl Eng Design 236:2394–2404

NEA (2003) NEA/Committee on the Safety of Nuclear Installations. Fuel Safety Criteria in NEA Countries. NEA/CSNI/R10

SENER (2006) Secretaria de Energía: Prospectiva del Sector Eléctrico 2005–2014

USNRC (2004) United States Nuclear Regulatory Commission. The Accident at Three Mile Island. Fact Sheet

WNA (2005) World Nuclear Association. The Economics of Nuclear Power. WNA Report

Yang R, Ozer O, Edsinger K, Cheng B, Deshon J (2004) An integrated approach to maximizing fuel reliability. 2004 International Meeting on LWR Fuel Performance; Orlando, FL; September 19–22

General Overview of the Current Situation of Nuclear Energy

Raúl Ortiz Magaña, Enrique García Ramírez

Instituto Nacional de Investigaciones Nucleares, Apartado Postal 18-1027, México D.F. 11801, México. E-mail: rortizm@nuclear.inin.mx, jenrique@nuclear.inin.mx

Abstract

Nuclear energy has a half of a century contributing to the generation of electricity in the world. The way has not been easy, since risk perception on part of the public opinion has not remained totally favorable to this technology, as some issues like accidents and waste management have been misused by detractors. However, as security of energy supply, climate change, and profitability have raised as concerning issues in energy planning, the generation of electricity by nuclear means has started to be on the table again, and a renaissance seems to be close. This article intends to provide with a brief description on the current situation of Nuclear Energy, and their perspectives, pointing out its main characteristics that can be advantageous in order to be included in the fuel mixings of the future.

1 The first years

The production of electricity using nuclear fuel began during the decade of 1950's, in the post war era, when few nations had available this technology and strong efforts were made to dedicate the nuclear power to peaceful uses. Historically, the first nuclear reactor that was connected to the grid, in order to supply with electricity one town, was sited in the former Soviet Union village of Obninsk in 1954. This reactor was a RBMK (acronym in Russian language of *Channelized Large Power Reactor*), which used graphite as moderator and water as coolant, with thermal capacity of 30 MW, delivering to the grid only 5 MWe. Ever considered as an experimental issue, it was operational until 1959.

In the UK Queen Elizabeth II inaugurated officially one nuclear reactor in the location of Calder Hall in October 1956. This reactor was one Magnox type (called so due to the content of non oxidable Magnesium in the alloy used in fuel rods cladding), also graphite moderated but cooled by CO_2 gas. This reactor, owned by British Nuclear Fuels PLC, had a capacity of 50 MWe and was maintained in commercial operation until 2003.

Since December 1953, US president D. Eisenhower in his famous *Atoms for Peace* speech, in the United Nations General Assembly, had put on the table the necessity to found one international atomic energy agency devoted to watch and monitor that this arising technology were used in the world as much as possible with peaceful purposes (Eisenhower 1953).

One conspicuous paragraph of President Eisenhower's speech stated that: "The more important responsibility of this atomic energy agency would be to devise methods whereby this fissionable material would be allocated to serve the peaceful pursuits of mankind. Experts would be mobilized to apply atomic energy to the needs of agriculture, medicine and other peaceful activities. A special purpose would be to provide abundant electrical energy in the power-starved areas of the world".

The Soviet Union and UK developments achieved in nuclear electricity production were in complete alignment with this US initiative, fact which in terms of propaganda urged the government of this country to commission its Navy force to work jointly with electrical industry to adapt one reactor, originally designed for ships propulsion, to be used for electricity production. In 1957 it was connected to the grid in Shipping Port, Pennsylvania, one Westinghouse PWR with 60 MWe, which was operated until 1982 under the ownership of Duqesne Light Co.

The significance of these milestones is based on the fact that the military purposes of nuclear energy were for the first time conspicuously overcome, provided that the main blocks in the world, i.e. the Soviet Union, Europe and US, showed the will to use this powerful technology for the benefit of civilian society.

2 Nuclear industry: a growing experience

Since the late 1950's, electricity production by nuclear means was consistently growing, with more and more nuclear power plants (NPP) being constructed in the world in the first stages of this industry. According to the records (BP 2006), nuclear electricity consumption grew from 25.7 TWh in 1965 to 2771 TWh in 2005.

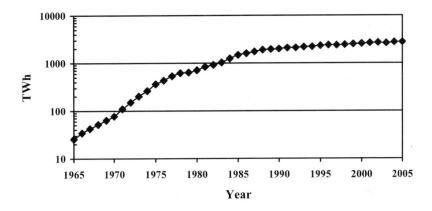

Fig. 1. The growing of nuclear electricity consumption in the world [TWh]

A careful analysis of Fig. 1 reveals how the slope of this tendency has varied in time. For instance, between 1970 and 1985 the growing rate was about 85 TWh per year, whereas between 1987 and 2005 such a rate decreased to 52 TWh per year, mainly due to the fact that after the Chernobyl accident in 1986 the construction of new plants suffered a severe diminishing. Then, the worldwide nuclear share of electricity generation has been kept in around 16 % in the last 20 years (IAEA 2006a).

Since its beginning, 50 years ago, nuclear industry has gained almost 12,000 reactor-years of cumulative operational experience. Nowadays 442 reactors are being operated in 31 countries in the world (IAEA 2006b), totalizing 369.6 GWe of installed capacity. Table 1 shows the countries with operating reactors ranked by nuclear share, i.e. the percentage of relative contribution to the total electricity generation recorded during 2005. It is also shown the nuclear installed capacity by country.

France is the country where nuclear energy plays the most important role in the national electricity, since 78.5 % of its generation is nuclear. By the way, France is the first exporter of electricity in the world. On the other hand, USA is the country that has the greatest value of nuclear installed capacity (over 98 GWe), with 103 reactors that delivered 19.9 % of total electricity generation during 2005.

After the Chernobyl accident in 1986 which caused many concerns in the public opinion respecting to the safety of the nuclear industry, many countries mainly in western Europe, established the so called phase-out policies on their nuclear programs, which meant to stop the construction of new NPP, and to close the running units at the end of their life times. However, security of energy supply and climate change which are current issues in the international agenda, have been moderating extremist positions respecting to the refusal to use the nuclear technology for the electricity generation.

Table 1. Countries with operating reactors and their nuclear shares

Country	Nr. Of Units	Installed Capacity [MWe]	Nuclear Share [%]
FRANCE	59	63363	78.5
LITHUANIA	1	1185	69.6
SLOVAK REPUBLIC	6	2442	56.1
BELGIUM	7	5801	55.6
UKRAINE	15	13107	48.5
SWEDEN	10	8916	44.9
REP. OF KOREA	20	16810	44.7
BULGARIA	4	2722	44.1
ARMENIA	1	376	42.7
SLOVENIA	1	656	42.4
HUNGARY	4	1755	37.2
FINLAND	4	2676	32.9
SWITZERLAND	5	3220	32.1
GERMANY	17	20339	31.1
CZECH REPUBLIC	6	3373	30.5
JAPAN	55	47593	29.3
UNITED KINGDOM	23	11852	19.9
SPAIN	8	7450	19.6
USA	103	98145	19.3
TAIWAN	6	4884	17.6
RUSSIAN FED.	31	21743	15.8
CANADA	18	12584	14.6
ROMANIA	1	655	8.6
ARGENTINA	2	935	6.9
REP. OF SOUTH AFRICA	2	1800	5.5
MEXICO	2	1360	5.0
NETHERLANDS	1	450	3.9
INDIA	16	3483	2.8
PAKISTAN	2	425	2.8
BRAZIL	2	1901	2.5
CHINA	10	7587	2.0

As a matter of fact, the same table shows that, with the exception of China and India, every country with 10 or more reactors has a nuclear share over 14 % of generation, which makes difficult to some countries to continue with their phased out policies.

3 Nuclear electricity: greenhouse emissions free

One analysis made by the International Atomic Energy Agency (IAEA) showed that a comprehensive comparison regarding emission factors for greenhouse gases (GHG) from whole electricity production chains, resulted for the nuclear option, in values of carbon equivalent emissions per kWh in the order or below of those of the renewable options (i.e. solar, hydro, biomass and wind) (Spadaro et al. 2000). It is well known, that emissions of GHG during the generation phase is cero for nuclear generation (same case as renewable energies), which already represents an enormous advantage with respect to fossil fuel options. Table 2 shows typical emission factors in terms of CO_2 equivalent values for fossil fueled, nuclear and renewable options, during the generation phase, and also the share of each fuel in the worldwide electricity mixing.

Table 2. Emission factors (generation phase) and share by fuel

Type of fuel	t CO_2/GWh[a]	World Share (2004)[b] [%]
Coal	1019	41
Oil	788	6
Gas	575	19
Nuclear	0	16
Hydro	0	16
Other (Ren)	0	2

[a] Calculated from Spadaro et al. (2000) considering conservative values for fossil fuel.
[b] Taken as rounded off numbers from IEA (2006).

Considering that the total electricity generation in 2004 was 17,450 TWh (IEA 2006), using the values of table 2 can be easily demonstrated that nuclear energy avoided the emission of 2,300 millions of metric tons of CO_2 equivalent, in that year. In 2005 the CO_2 avoided raises to 2,466 millions of metric tons. The significance of such figures can be better understood if it is perceived that in 2004 the total emissions from fossil fuels were around 27,000 millions of metrics tons of CO_2 (EIA 2006a), see figure 2. This 9 % of emissions avoided each year can give a good idea on why nuclear energy can not be abandoned nowadays as a sustainable option.

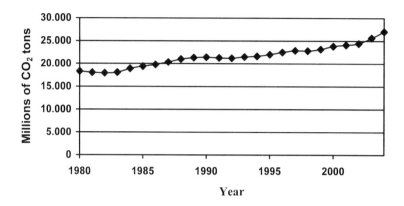

Fig. 2. CO_2 emissions in the world, 1980-2004 (EIA 2006a)

4 Chernobyl and TMI: the toughness of western safety design

The two more important accidents in the history of nuclear industry have been indeed the Three Mile Island (TMI) accident in US in 1979 and the Chernobyl accident in the former Soviet Union in 1986. Comparing them in terms of root causes and consequences, big differences are revealed in terms of design safety features, and operational procedures. The Chernobyl's RBMK reactor was designed with positive void coefficient which essentially means that the power will be increased instead of diminishing in case of an increasing of reactor temperature, exactly at the reverse of western design light water reactors (LWR). The root cause event of Chernobyl's accident was initiated when the operator started an experiment which included the blocking of certain automatic-shut-down safety signals, whereas in western LWR, experiments are rigorously forbidden and the operator has not access to block safety signals. In addition to the above causes, a containment structure was not a part of the design in the Chernobyl reactor, which permitted that the fission products released during the heat explosion were widely dispersed into the environment.

The TMI accident, whose exact initiating event never was known but whose sequence was aggravated by certain errors in instrumentation design that prevented the operator to perceive that the capacity of heat removal was compromised, was however well mitigated by the engineered safety features such as the automatic reactor shut-down, and the conservation of the containment structure integrity up to the end. Even though the core reactor suffered an important melting and fission products were released from fuel cladding rods to the coolant, most of dangerous radioactivity (i.e. iodine and particles) was contained inside the plant, and only small doses were delivered to the public. Although there were no consequences in workers' health nor the public's, the accident of TMI meant a sort of inflexion point in the safety regulations, as well as in the design and the procedures of operation not only in the USA, but in all the western nuclear power stations, where important programs of safety improvement were implemented.

Summarizing, a comparison of the factors behind each of these accidents shows how the resistance of western design to severe events could contain and mitigate serious consequences, or to say in other words: safety engineered features worked as expected according to the design.

5 Nuclear Performance: the longest periods in the grid

There are several concepts created to measure the performance of nuclear power reactors. It has been recognized that regarding the operation on nuclear power plants, performance and safety are almost synonyms. Good maintenance programs in NPP mean more guaranteed safety and more hours delivering maximum power to the grid. The capacity factor is the ratio of the energy actually generated to the maximum energy that could have been generated in a given period.

Table 3 shows the countries that averaged over 80 % in their capacity factors during 2005. Excellent performances of countries such as USA, France and Korea operating 20 or more reactors are of particular worth, since the more the number of operating reactors the more difficult to maintain an even high performance in all of them.

The 21 countries listed in Table 3 represent 68 % of the countries generating nuclear electricity, and the number of implicated NPP is 305 (i.e. around 70 % of the total). The world average capacity factor in the same year was 84 %. This figure is significantly high if compared with capacity factors for other sources, for instance in the US, during 2005 (Table 4).

Table 3. Capacity factors over 80 % during 2005 (IAEA 2006b)

Country	Units	Capacity Factor [%]
SLOVENIA	1	98.5
NETHERLANDS	1	95.9
FINLAND	4	95.6
MEXICO	2	92.9
KOREA, REP. OF	20	91.6
USA	103	90.9
BELGIUM	7	90.3
ROMANIA	1	89.6
LITHUANIA	1	89.3
SWEDEN	10	88.6
GERMANY	17	88.0
CHINA	9	86.9
HUNGARY	4	84.7
SPAIN	8	84.1
CANADA	18	83.5
FRANCE	59	83.4
SLOVAK REP.	6	82.2
UKRAINE	13	81.9
BULGARIA	4	80.6
SOUTH AFRICA	2	80.5
INDIA	15	80.3

Table 4. Capacity factors during 2005 for non-nuclear electricity options in US (EIA 2006b)

Fuel	Capacity Factor [%]
Coal (Steam Turbine)	72.6
Gas (Combined Cycle)	37.7
Oil (Steam Turbine)	29.8
Hydro	29.3
Wind	26.8
Solar	18.8
Gas (Steam Turbine)	15.6

6 Nuclear electricity in Mexico

The Laguna Verde NPP is located 70 km to the north of the port Veracruz, 290 km to the east of Mexico City, over the southern coast of the Gulf of Mexico. It has two BWR twin units, totalizing 1365 MWe of installed capacity. Nuclear Steam Supply System is General Electrics's and the turbogenerator is Mitsubishi Heavy Industries's. Unit-1 started its commercial operation in 1990 and Unit-2 in 1995. In terms of installed capacity, nuclear represents about 3 % of the total in the country.

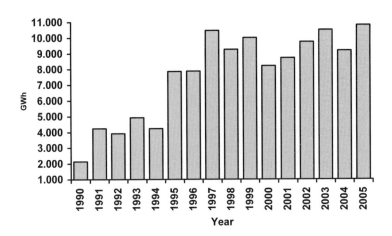

Fig. 3. Nuclear generation produced in Mexico

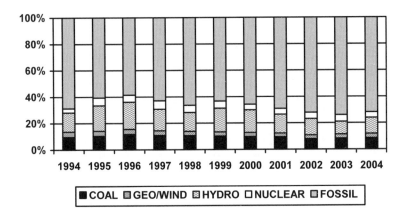

Fig. 4. Nuclear share in Mexico

The history of nuclear power generated in Mexico by nuclear means is represented in graph Fig. 3 in terms of GWh per year for the two units, where about 9,300 GWh have been averaged per year in the last 10 years. As of September 2006, U-1 had generated 75.1 millions of MWh, and U-2 55.6 millions of MWh in their lifetimes, totaling over 130 millions of MWh. Based on a high standard performance, whose lifetime average reaches 82 % in terms of capacity factor, nuclear generation has been maintained with a share of around 5 % in the last 10 years (Fig. 4). Due to this successful records, government officers of energy sector made a proposal at the end of 2006 to reopen the nuclear option for future plans of electricity generation.

7 Towards the future: now and then

As of September 2006, 28 new NPP are being constructed in the world which will be connected to the grid in the next few years: Japan, Argentina, Iran, Finland, South Korea, Romania and Pakistan are constructing one reactor each; Taiwan, Ukraine and Bulgaria, are constructing two; Russian Federation and China are constructing four; and India is constructing seven reactors. As a matter of fact, construction of new units in the world never stopped, neither in the hard days following the Chernobyl accident.

In USA, where no new NPP have been connected to the grid since the one of Watts Bar-1 in 1996, there has been however a tendency to increase the use of nuclear power in two different ways: to extend the original lifetime, and to rise the original power of operating NPP. Regarding lifetime, as September of 2006, 44 renewal licensee applications have been approved by the National Regulatory Commission (NRC), which means an advanced permission to operate for 20 years longer than the original lifetime (NRC 2006). At the same date, NRC was evaluating 8 more applications and was expecting 20 more to be received in the next 9 years. Regarding increasing of power, NRC has approved 17 *extended* power uprates (i.e. 7–20 %), and is evaluating 4 more. In addition, this agency is expecting to receive 7 more applications in the next 5 years. An installed capacity of 1.7 GWe has been increased in this way. As a matter of fact, since 1998, polls indicate a higher bias in "favor" position of public opinion, revealing that as September 2006, around 70 % of US people supports Nuclear Energy (Bisconti 2006). In the middle of this favorable atmosphere, NRC has received applications for four early site permits in: North

Anna, Clinton, Grand Gulf and Vogtle, where new NPP will be constructed.

In Europe, in spite of the prevalence of the phased out policies that have been expressed recently with the final shutdowns of Obringheim in Germany, Barseback-2 in Sweden (both during 2005), and Jose Cabrera-1 in Spain (2006), some openness has been noted in recent years. For instance, in Sweden there have been approved two extended power uprates and four more are being evaluated; in Netherlands it has been approved one license renewal for 20 additional years for the only NPP in the country (in Borssele). In UK there were two license renewals granting 10 additional years for two of the four units in Dungeness.

In Asia, Nuclear Energy has kept a great driven force. Twenty four out of the last thirty four units connected to the grid, have been connected there, and sixteen out of twenty eight units are being constructed in Asian countries. Japan is going to increase its nuclear share up to 40 %, through the commissioning of 10 new units in 2014. South Korea, where an interesting self reliance program permitted them to turn from importers to exporters of nuclear technology, is planning to add another additional 9,600 MWe through the construction and connection of 8 new reactors in 2016. Japan and Korea are probably the most efficient constructors of NPP, since their total constructions times (since first concrete to connection) are in the range of 46–54 months. This mention is interesting because the impact of construction time in overnight costs. China and India in their own, have announced huge expansion programs for the next 15 years: China will augment its installed capacity from 6.7 GWe to 40 GWe in 2020, reaching then 4 % in nuclear share; India will increase from 3.5 GWe to 29 in 2022.

Nuclear Industry has been taking advantage of its great experience, and some of the reactors currently planned, in construction or even already operating, have actually evolutionary design concepts: for instance, recent connected units of Hamaoka-5 and Shika-2 in Japan, are ABWR type reactors (from Advance Boiling Water Reactor), which are the successor of the former General Electric's Boiling Water Reactors. In fact, Japan had put in operation ABWR units in Kariwa-5 and Kariwa-6 previously, in 1996–1997; the Korea's shin-Kori units 3 and 4, and shin-Ulchin units 1 and 2, planned to be connected to the grid between 2013 and 2015, are APR-1400 type reactors (from Advanced Power Reactor), which are the successor of the previous ABB Atom's PWR design; and France's planned a Flammaville-3 EPR reactor, which is the successor of the previous Frammatome's PWR reactor. This EPR is the same type than the Finland's one currently under construction. All of the above mentioned models are one generation more advanced than the reactors currently in operation, so they are called the Generation III (or III *plus* in some cases), which have a standardized design for each type, reducing construction time and hence capital cost. They have also a more robust design that makes them easier to operate and less vulnerable to operational events; higher availability and longer lifetime by design (60 years); less probability of core melt accidents; minimal effects into the environment; higher burn-up rates which means better advantage taking of the fuel, etc.

Besides of the above described efforts, an international task force called Generation IV International Forum (GIF) has been formed to work in the design of the next generation of reactors (i.e. beyond Generation III plus), expected to be deployed in 2030. At the time, six different technologies have been selected as conceptual models to perform research and development on them (GIF 2006). Generation IV reactors are planned to be highly economical, with an enhanced safety design, very resistant to the proliferation, and with a minimum production of wastes.

8 Radioactive waste: Achilles heel or political issue?

Many words have been said about the production of radioactive waste in the nuclear industry. Detractors claim that radioactive waste is a problem with no solution that makes unfeasible the generation of electricity using nuclear fuel. The true is that nuclear industry has been very responsible in the management of its produced wastes and none has ever suffered any harmful from radioactive waste in produced in NPP which is not the case of some other industries. For management purposes, radioactive wastes are divided in two groups: low and intermediate level waste (LILW), which can be, according to their half life, short or long lived, and high level waste (HLW), that for practical purposes is the spent fuel unloaded during refueling activities in NPP. LILW are managed according to a general process which includes conditioning, immobilization, volume reduction, and preparation for definitive disposal. Facilities for disposal of LILW can be in surface, shallow sub-surface or even geological repositories, and shall be designed with enough integrity to remain under institutional control for hundreds of years, when they enter into closure, time enough to reach a complete radiological stability. Currently, more than one hundred of LILW disposal facilities are operating in the world and some other forty are under some stage of development (Han et al. 1997).

Regarding HLW, there is an international consensus in the sense that the deep geological repositories are the most feasible option to dispose this kind of wastes. The concept itself, which can be used also for long lived LILW, consists of a facility constructed hundreds of meters underground in stable geological formation whose design is based in a multibarrier system considering natural and engineered barriers. The host rock and surrounding geological formations with their structural, hydrogeological and geochemical characteristics are the main components of the natural barriers whereas the engineered barrier system (EBS) includes among others, the conditioned waste form, the waste package and its potential overpack, the backfill, buffer and seal materials. This system of barriers will provide over a very long period of time, adequate containment and isolation of waste, long term retardation of any released radionuclides and ultimately, dilution of any remaining radionuclide concentrations.

Since the sixties and seventies last century, underground research laboratories (URL), were founded in several countries to do research addressed to characterize and assess the potential of geological media to contain radioactive waste. The results produced have proved to be valuable both in generic terms, i.e. developing and assessing the disposal concept and building confidence in it, and in specific terms as an essential means for detailed characterization, design and assessment of potential repository systems (IAEA 2001). Belgium, Canada, Switzerland, Sweden, UK and USA have formed, under the IAEA auspice, a network of URL, devoted not only to the exchange and sharing of the extended information collected among IAEA country members, but also to provide with training and space for researching to countries with less experience and development in the field.

At present, most of the spent fuel is being stored temporarily either in spent fuel pools of reactors or in some other temporary fashions while final decisions are made regarding the sites to be selected and the correspondent studies are finished. This is in part because, although the above described efforts have demonstrated that technological solutions exist to the safe disposal of radioactive wastes, public acceptance and its political use remain as the main obstacles to the actual implementation of HLW disposal. That is why in many countries, the agencies devoted to radioactive waste management are enhancing transparency in information and promoting public participation in the making decision processes regarding radioactive waste disposal, factors that are being fruitful towards a change in the traditional trends of public acceptance. Some of these are join initiatives at a regional-international level, as the project COWAM with the participation of 14 European countries (COWAM 2003).

9 Conclusions

Nuclear Energy is a technology with more than 50 years of experience contributing to the worldwide generation of electricity. In spite of its detractions, nuclear generation of electricity is a competitive option in terms of sustainability and financial feasibility, since it has evolved in terms of safety and operational efficiency, achieving high standards of performance. Technological solutions there exist for the issue of radioactive waste, and a change in public perception is being achieved based on transparency and participation in decision making. Due to its many achievements nuclear energy shall be considered as a valuable option, at the same level of renewable energies, in the future electricity generation plans.

References

Bisconti Annual (2006) U.S. Public Opinion About Nuclear Energy; available in: http://www.nei.org/documents/PublicOpinion_0906.pdf

BP (2006) British Petroleum. BP Statistical Review of World Energy (Workbook), June 2006; available in http://www.bp.com/liveassets/bp_internet/globalbp/globalbp_uk_english/publications/energy_reviews_2006/STAGING/local_assets/downloads/spreadsheets/statistical_review_full_report_workbook_2006.xls

COWAM (2003) Community Waste Management 2 – Improving the Governance of Nuclear Waste Management and Disposal in Europe, Annex 1, "Description of Work". http://www.cowam.org/dav/web/Workdescription.doc

Eisenhower D (1953) Atoms for Peace Speech; available in: http://www.eisenhower.archives.gov/atoms.htm

GIF (2006) Generation IV International Forum. http://gif.inel.gov/

Han KW, Heinonen J, Bonne A (1997) Radioactive waste disposal: Global experience and challenges. IAEA Bulletin 39-1, March 1997. http://www.iaea.org/Publications/Magazines/Bulletin/Bull391/bonne.html

IAEA (2001) The use of scientific and technical results from underground research laboratory investigations for the geological disposal of radioactive waste. IAEA Tecdoc-1243; available in: http://www-pub.iaea.org/MTCD/publications/PDF/te_1243_prn.pdf

IAEA (2006a) Nuclear Technology Review 2006. Report by the General Director. http://www.iaea.org/About/Policy/GC/GC50/GC50InfDocuments/English/gc50inf-3_en.pdf

IAEA (2006b) Power Reactor Information System Data base, September 2006; available in http://www.iaea.org/programmes/a2/index.html

IEA (2006) International Energy Agency. Key World Energy Statistics; available in http://www.iea.org/Textbase/nppdf/free/2006/Key2006.pdf

EIA (2006a) Energy Information Administration. International Energy Annual 2004 http://www.eia.doe.gov/pub/international/iealf/tableh1co2.xls

EIA (2006b) Global Energy Decisions/Energy Information Administration Quoted as preliminary; available in http://www.nei.org/documents/ U.S._Capacity_Factors_by_Fuel_Type.pdf

NRC (2006) Nuclear Regulatory Commission. NRC Web Page http://www.nrc.gov/reactors/operating/licensing/renewal.html

Spadaro JV, Langlois L, Hamilton B (2000) Assessing the Difference: Greenhouse Gases Emissions of electricity generation chains http://www.iaea.org/Publications/Magazines/Bulletin/Bull422/article4.pdf

The Clean and Safe Nuclear Reactors of the Future

Karl Verfondern

Research Center Juelich. Institute for Safety Research and Reactor Technology (ISR), D-52425, Juelich, Germany.
E-mail: k.verfondern@fz-juelich.de

1 Energy as global challenge

Despite energy saving efforts and improved efficiency of energy production, projections of the World Energy Council indicate a significant increase in global energy consumption in the medium and long term due to a further growing world population and rising prosperity. On the other hand, the energy supply is connected with a threatening of the earth's environmental and climate system. The present share of fossil fuels of 80 % in the world's energy consumption is hardly expected to change in the short and medium term. Continuation of the extended use of natural resources of coal, oil, and natural gas, however, carries the danger of over-stressing the environment. In addition, strongly fluctuating energy prices, the dependence of many countries on energy imports from politically unstable regions, and the uncertainty about how long oil and gas reserves will last, are raising fears of supply security. This all urgently calls for political and technological counteractions.

World summits in Toronto (1988), Rio de Janeiro (1992), and Kyoto (1999) have shown the awareness of most countries of a need for change, but little was successful to curb the trend of present energy consumption (see Fig. 1). In contrast, the impact of the mechanisms of the Kyoto protocol and the policy beyond the year 2012 remain highly uncertain. The worldwide demand for more energy, particularly in emerging economies like China or India, appears to be without limit, predicting a further increase by 50 % in the next two decades.

The situation in the European Union as predicted for the next 30 years is characterized by a growing demand for energy and, at the same time, a decreasing domestic energy production (Martin Bermejo 2004). This development will push CO_2 emissions in the EU countries, if no additional measures are taken, to a plus of 14 % by 2030 compared to the 1990 level, far off the Kyoto commitment of an 8 % reduction.

Satisfying the need of future generations for a more "sustainable development", a balanced mix of diversified energy sources must be applied to meet ecological, economic, and social requirements. It is the objective of climate change policies to enhance the trend towards renewable energies and other non-fossil alternatives. Nuclear energy as one of the significant carbon-free sources may contribute a robust element to a more secure energy world.

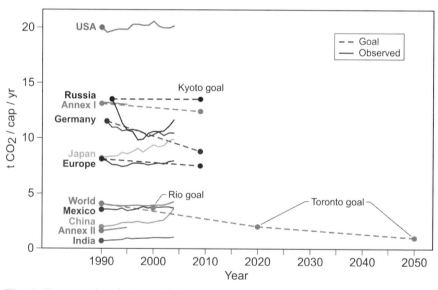

Fig. 1. Energy-related CO_2 emission per capita for selected countries and groups of countries (EIA 2006)

A long-term strategic response to the global energy challenge is the intensified research of energy technologies. It may lead to innovative concepts with improved safety characteristics, and services of energy supply, and whose success may be determined by market processes. It will, however, be dependent on a continuation of the long-term intensive cooperation among the nuclear vendors, utilities and research organizations.

With the recent worldwide increased interest in hydrogen as a clean fuel of the future, Europe has also embarked on comprehensive research, development, and demonstration activities with the main objective of the transition from a fossil towards a CO_2 emission free energy structure as the ultimate goal. The near and medium term, however, due to the growing demand for hydrogen in the petrochemical and refining industries, will be characterized by coexistence between the energy carriers hydrogen and hydrocarbons. Due to the increasing share of "dirty fuels" such as heavy oils, oil shale, tar sands entering the market, the need for both process heat and hydrogen will also increase significantly.

2 Three generations of nuclear reactors

2.1 Nuclear electricity production in the world

It is more than 50 years ago that in 1954 the first commercial nuclear power station, the Obninsk reactor near Moscow, started its operation producing 5 MW(e). Five decades of further development since have made nuclear power to an industrially mature and reliable source of energy and to a key component in the world's energy economy. In 31 countries, nuclear power plants are being operated for commercial electricity production provided by 442 reactors with a total capacity of 371 GW(e) (as of 2005) and the experience of over 12,000 years of operation. More than 44 GW of nuclear capacity went online since 1990. Currently under construction are 28 more nuclear stations with a capacity of 22.5 GW(e), and many countries are presently considering nuclear as a realistic option within their national energy policy. It is in 16 countries, where nuclear power provides more than 25 % of their electricity.

In Europe, nuclear power is one of the major energy sources providing more than a third of the electricity needs. Electricity is produced by 203 nuclear power plant units (145 in EU countries) with a total installed capacity of 172 GW(e) and 10 more units with 10 GW(e) under construction (as of 2006). The nuclear share, however, varies significantly between countries: while some countries completely refrain from nuclear, in other countries the nuclear shares range between ~3 % for the Netherlands to ~78 % for France.

2.2 Advantages and disadvantages of nuclear power

The major drawbacks of nuclear power often mentioned are basically related to the still unsolved problem of nuclear waste as a long-term threat,
the high capital cost (despite low fuel cost), limited uranium resources, and
since more recently the perception of a higher risk due to war/sabotage or
advancement of nuclear proliferation in certain countries. And it was in
particular the severe accidents of Chernobyl (1986), Kyshtym (1957),
Windscale (1957), Harrisburg (1979), Tokai-mura (1999), all of which
reached the upper accident levels of the IAEA nuclear event scale, which
weakened the public acceptance of nuclear technologies.

But despite the above accidents, nuclear power plant operation has
proven in the past to have reached a high level of safety, reliability and efficiency. With lessons learned from occurrences in the past during nuclear
plant operation, the IAEA has steadily improved its safety guidelines to a
best possible set of rules in order to minimize the risk of severe accidents.

Economic base load production of electricity makes nuclear energy
competitive to fossil fuelled power plants with the comparatively low fuel
cost to help stabilizing the price of electricity within the energy mix. It
helps saving fossil fuel resources and extending their availability, respectively. Nuclear energy as a source with no greenhouse gas emissions and
as a means to improve national energy security thus contributes to a "sustainable" development.

To keep the nuclear option alive in a longer-term perspective, the nuclear industries not only have to continue operating the existing plants at
that high level of safety and reliability, it also requires a steady development towards the next generation of NPP to meet the more and more insis-
tent concerns of the public about safety. Studies show that there is a high
potential for innovative nuclear techniques, systems and concepts which
can offer socially acceptable, environmentally benign and competitive solutions. There is also a large potential for nuclear power to play a major
and important role in the non-electric sector as a provider of process heat
in a great variety of industrial processes.

2.3 Three generations of nuclear power plants

Like any other established technology, nuclear power has passed through
different levels of development (see Fig. 2). The first generation of nuclear
reactors from the 1950s and 1960s was characterized by a relatively simple
and cheap technology for a rapid realization of electricity generation. Initially developed for naval applications and taking benefit of uranium en-

richment experience achieved in military programs, most of them were one-of-a-kind reactors. Since fissile material for the reactors was still deemed scarce, the light water reactor (LWR) with its low-enriched fuel was considered the most promising concept to be applied to commercial electricity production. The plutonium generated during LWR operation was early recognized as excellent nuclear fuel. Activities of reprocessing and radioactive waste treatment and disposal, however, remained on a low level at that time.

The second generation is formed by the plants constructed from the 1970s to the early 1990s. The deployment of nuclear reactors was advancing rapidly in that period. Like for the reactors of the first generation, operation was basically in an open fuel cycle with the uranium burnt once and then disposed. But activities were intensified on the reprocessing of nuclear fuel by extraction of the fissile material, the generated Pu and still unburnt U, using a mixture of Pu and U oxides to form MOX fuel for nuclear plants, a way towards closure of the fuel cycle. Since MOX fuel requires a slight modification of the operational parameters, only up to a third of the fuel is typically replaced by MOX. Its commercial use started in the 1980s and is in the meantime being employed in some 30 nuclear reactors (22 in Europe). About 2 % of today's new fuel is MOX.

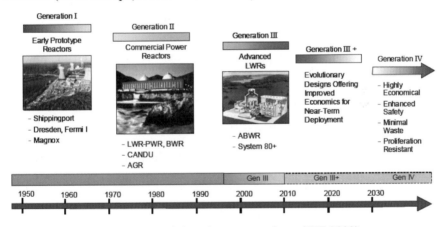

Fig. 2. Generations of commercial nuclear power plants (GIF 2002)

Up to today, most of the nuclear power plants constructed are of LWR-type, either pressurized water reactors or boiling water reactors. Light Water Reactors will most certainly continue to dominate for the next decades. New nuclear reactors which are ready for today's market are counted to the third generation. They are characterized by a simpler design with a higher level of passive (inherent) safety systems based on the physical principles of gravitation, natural circulation, evaporation, condensation rather than active components, thus reducing the need for operator intervention and further lowering the probability of a core melt accident. Design improvements are given in the fuel technology allowing higher burnups to reduce the amounts of fuel and waste. Such reactors are also designed for a longer expected lifetime of 60 yrs. Furthermore shorter construction times will reduce capital cost by some 30 %.

The first of such concepts is the "Advanced Boiling Water Reactor", ABWR, type. Four units of this type have been deployed already in Japan since 1996. But also various other concepts have been developed. The Gen-IV International Forum (see Chap. 4), a group of countries that cooperates in research on future nuclear reactors, has identified a total of 16 nuclear systems which are considered to have a better performance than the present systems and to be deployable in near term, i.e., by 2015 (GIF 2002). These are:

- Advanced Boiling Water Reactors
  - Advanced Boiling Water Reactor II, ABWR-II
  - European Simplified Boiling Water Reactor, ESBWR
  - High Conversion Boiling Water Reactor, HC-BWR
  - Siedewasser Reaktor-1000, SWR-1000
- Advanced Pressure Tube Reactors
  - Advanced CANDU Reactor 700, ACR-700
- Advanced Pressurized Water Reactors
  - Advanced Pressurized Water Reactor 600, AP-600
  - Advanced Pressurized Water Reactor 1000, AP-1000
  - Advanced Power Reactor 1400, APR-1400
  - Advanced Pressurized Water Reactor Plus, APWR+
  - European Pressurized Water Reactor, EPR
- Integral Primary System Reactors
  - Central Argentina de Elementos Modulares, CAREM
  - International Modular Reactor, IMR
  - International Reactor Innovative and Secure, IRIS
  - System-Integrated Modular Advanced Reactor, SMART
- Modular High Temperature Gas Cooled Reactors
  - Gas Turbine Modular High Temp. Reactor, GT-MHR
  - Pebble Bed Modular Reactor, PBMR

These systems both include evolutionary and innovative approaches. To the former group belongs, apart from the ABWR, the "System 80+", a US design of an Advanced Pressurized Water Reactor. The large-scale APR-1400 with 1,450 MW(e), which incorporates many features of the System

80+ design, will be marketed worldwide and, e.g., forms the basis of Korea's next generation reactor fleet soon to be constructed.

More innovative reactor systems are given with the AP-600 or the upscaled version AP-1000, a PWR with predominantly passive safety systems and also the ability to run on a full MOX core. It is the first design of the generation that may be called "Gen-III+" that received in 2005 final design certification from the US-NRC. The "Economic Simplified Boiling Water Reactor", ESBWR, with 1,550 MW(e), also uses a high degree of passive safety.

A European development is the "European Pressurized Reactor", EPR, with an electric power in the range of 1,550–1,750 MW(e) representing the new standard design in France and being ready for deployment. It is expected to reach a maximum burnup of 65 GWd/t and a thermal efficiency as high as 36 %. The first unit of this type is currently under construction in Finland with a second one to follow soon in France. Another European evolutionary design is the SWR-1000 BWR with a power of up to 1,290 MW(e), also equipped with various passive safety systems.

The Canadian development of the heavy-water-cooled CANDU reactor was besides LWR the only other type that has reached commercial level. Based on the present standard reactor, CANDU-6, of which first units started operation in China, Canada is developing, improved designs. CANDU-9 belongs to the larger units (925–1,300 MW(e)) and is flexible in its fuel design capable of using natural or low-enriched uranium, MOX, spent PWR fuel or – as a near-breeder – the fertile material thorium. Another Canadian concept being currently developed is the ACR-700 (Advanced CANDU Reactor) as a next generation reactor.

Much experience has also been gained from alternative concepts studied such as metal-cooled fast reactors or gas-cooled thermal reactors. Although never having reached commercial level, they have made it so far at least to the prototype level. Both types are extensively being developed by several countries.

Fast reactors characterized by their ability to produce fissile fuel material are using plutonium as fuel and natural or depleted uranium as fertile material, which is a much more efficient way of exploiting the uranium resources. Originally intended to generate more fuel than is consumed during operation, its present purpose is more the "incineration" of the plutonium stockpiles from dismantled nuclear weapons. Coolant exit temperatures around 550°C make fast reactors also appropriate to some extend for process heat applications.

The other promising alternative concept of high temperature gas-cooled reactors will be described in more detail in the following chapter.

3 High-temperature gas-cooled reactor

3.1 Technical characteristics of HTGRs

High temperature gas-cooled reactors are characterized by an all-ceramic core, a core structure made of graphite as moderator and reflector, helium gas as a single phase inert coolant, coated particle fuel and a low power density core. The use of refractory core materials combined with helium coolant allows high coolant temperatures up to 950°C or even higher and a high thermal efficiency results in a number of significant advantages. The low power density and large heat capacity of the graphitic core, the absence of coolant phase changes, and the prompt negative temperature coefficient represent inherent safety advantages.

The basic fuel containing unit is given in form of a tiny coated particle with a fuel kernel surrounded by a so-called TRISO coating which consists of a porous buffer layer plus an interlayer of SiC between the two layers of high-density isotropic PyC. The coating represents the primary barrier to the release of fission products which are generated in the kernel during operation. A reactor core for 400–600 MW(th) will contain between 10^9 and 10^{10} individual fuel particles. The particles are embedded in a graphite matrix to form the fuel elements.

The type of fuel element has developed in two directions (see Fig. 3):

1. The pebble bed concept pursued in Germany, Russia, China, South Africa, which consists of a 60 mm diameter sphere composed of a 50 mm diameter fuel zone with some 10^4 coated particles (or 1 g of U-235) uniformly dispersed in a graphitic matrix, surrounded by a fuel-free carbon outer zone;

2. The prismatic core concept pursued in the United Kingdom, the United States, Russia, based on a hexagonal graphite block with bore holes which are, in the US design, either coolant channels or filled with coated particles containing fuel compacts. In the Japanese design (not shown in the figure), the bore holes of a fuel assembly are filled with fuel rods ("pin-in-block") to contain the compacts (some 180,000 coated particles per rod), and the coolant flowing through the annular gap between rod and inner surface of the bore hole.

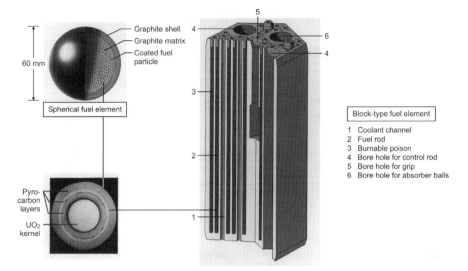

Fig. 3. Fuel elements for High Temperature Gas-Cooles Reactors

One of the attractive features of the HTGR is its flexibility in the use of fuel cycles. The thorium cycle with both separable and mixed fuel, the LEU cycle, and even cycles based on plutonium fissile particles are feasible. Fuel particles characteristic of those required for each cycle have been successfully tested in both prototype HTGRs and Materials Test Reactors.

3.2 Experience with HTGR operation

Graphite-moderated, gas-cooled reactors (GCR) are in operation since the 1950s with units sizes of up to 670 MW(e). The first commercial plant was Chinon in France, a CO_2-cooled Magnox-type reactor with 70 MW(e) starting in 1957. The next generation of GCR was the Advanced Gas-Cooled Reactor (AGR) operating at higher temperatures ($\sim 650°C$) and with higher efficiencies, the first commercial plant Dungeness B in the UK with 1,110 MW(e) in two units starting in 1983.

The final stage of GCR development is represented by the HTGR deserving its name from the high coolant outlet temperature in the range of 750–950 °C, which has the potential for high temperature process heat applications. There was even an experimental GCR designed for a very high coolant outlet temperature of 1320°C. The 3 MW(th) He-cooled UHTREX (Ultra High Temperature Reactor Experiment) built at Los Alamos Scientific Laboratory was mainly used for HEU UC_2 fuel testing. GCRs have accumulated so far more than 600 operation years. A comprehensive outline of the historical development of GCR in the world is given in (IAEA 1984).

Direct experience with HTGR plants up to now has been gathered in the countries United Kingdom, Germany, the USA, Japan, and China. A total of seven units have been constructed as given in the following Table 1.

The HTGR era started in the 1960s with the construction of experimental reactors, the 20 MW(th) Dragon in Winfrith, UK, which was a joint project of the OECD countries, the 46 MW(th) AVR reactor in Juelich, Germany, and the 115 MW(th) Peach Bottom Power Plant in the USA. All three mainly served as research instruments and were operated with high availability. In addition, the plants in Germany and the USA were also able to generate a significant amount of electricity.

During the successful operation of the experimental reactors, two HTGR prototype plants have been developed as a follow-on step: the 330 MW(e) Fort St. Vrain (FSV) reactor in the USA and the 300 MW(e) Thorium High Temperature Reactor (THTR) at Schmehausen in Germany. The former uses a prismatic core, while the second one uses a pebble bed core, both being connected to a conventional steam cycle. These demonstration units, however, were not very successful during their several years of operation, mainly suffered from technological problems (which could be solved though), and were eventually lacking the political and financial support at a time where the severe accidents of Three Mile Island in the USA and, in particular, of Chernobyl in the former Soviet Union changed worldwide the attitude towards nuclear. But both plants proved at least the feasibility of larger-sized HTGRs and the operation of respective system components, an experience of additional 17 years of operation which is beneficial for future designs.

The operating reactors

All five HTGRs which were operated in Germany, the USA, and the UK have stopped operation in the meantime. The only presently operated HTGRs are in Japan and China.

The Japanese High-Temperature Engineering Test Reactor, HTTR, is a 30 MW(th) experimental reactor which became critical in 1998. The active core with 2.9 m of total height and 2.3 m of equivalent diameter is composed of 150 hexagonal fuel assemblies. The average outlet temperature of the helium coolant is 850°C and 950°C, respectively. The heat is removed alternatively by a water cooler only or, in a parallel operation mode, 20 MW(th) by water cooler and 10 MW(th) via an IHX to a secondary helium circuit. The latter is for testing process heat application conditions.

Table 1. Main features of HTGRs which were or are being operated

	Peach Bottom (USA)	Dragon (UK)	AVR (Germany)	Fort St. Vrain (USA)	THTR (Germany)	HTTR (Japan)	HTR-10 (China)
Reactor type	Experimental	Experimental	Experimental	Prototype	Prototype	Experimental	Experimental
First criticality	1967	1964	1966	1974	1983	1998	2000
Out of operation (status)	1974 (safe encl.)	1975 (safe encl.)	1988 (defueled)	1988 (decomm.)	1988 (safe encl.)	(operation)	(operation)
Thermal power [MW]	115	20	46	842	750	30	10
Electric power [MW]	40	-	15	330	300	-	-
Power density [MW(th)/m³]	8.3	14	2.6	6.3	6	2.5	2
He inlet / outlet temp. [°C]	377 / 750	350 / 750	270 / 850, 950	405 / 784	270 / 750	385 / 850, 950	250, 350 / 700, 900
He pressure [MPa]	2.5	2.06	1.1	4.9	3.9	4.0	3.0
Fuel element type	Pin	Pin	Spherical	Prismatic	Spherical	Pin-in-block	Spherical
Fuel	Carbide	Oxide	Carb / Oxide	Carbide	Oxide	Oxide	Oxide
Enrichment	HEU	HEU, LEU	HEU, LEU	HEU	HEU	LEU	LEU
Coating	BISO	TRISO	BISO, TRISO	TRISO	BISO	TRISO	TRISO
Pressure vessel	Steel	Steel	Steel	PCRV	PCRV	Steel	Steel
Electricity production [MWh]	1,380	-	1,670	5,500	2,890	-	-

The Chinese HTR-10 with a pebble-bed core producing a power of 10 MW(th) is also basically a research tool. First criticality was achieved in 2000. The active core consists of some 27,000 fuel spheres with low-enriched (17 %) UO_2 TRISO particles. Coolant outlet temperatures are 700 and 900°C, respectively. Reactor core and steam generator are housed in two steel pressure vessels in a side-by-side arrangement. The HTR-10 is currently planned to be equipped with a gas turbine cycle (HTR-10GT). A respective basic design was completed in 2003. The main components are now under manufacture and their operation is expected to start soon.

The experience achieved with HTGR operation so far has shown a steady development in the improvement of components and fuel. Safety tests in the AVR and more recently in the HTTR and HTR-10 confirmed the predicted excellent physical behavior of the core under abnormal operating conditions. A test with sudden shutdown of the blowers during full power operation was simulated in the AVR in 1970 demonstrating for the first time the self-acting stabilization of an HTGR core (Kirch and Ivens 1990). Also numerous safety demonstration tests in the HTTR and HTR-10 recently have proven the safety features of small HTGRs.

3.3 Follow-up designs for future HTGRs

A whole variety of conceptual designs for follow-up HTGRs has been developed mainly differing in their core type (pebble/block), power conversion system (direct cycle/steam cycle/intermediate heat exchanger), but also in their application target (electricity/CHP/process heat/district heating) and with thermal powers ranging between 10 and 1,640 MW.

The trend towards large-sized plants, originally following the size of commercial LWRs, was gradually abandoned in the 1980s in favor of small modular-type HTGRs, which have been used since in many variants for future reactors. Although lacking the advantage of economy-of-scale, it is attractive because of its simplicity in design, lower capital cost, and serving the lower-power market, therefore appropriate candidates in deregulated industrial countries or developing countries. In particular, the tall and slim core geometry allows for the inherent temperature limitation of the fuel temperature due to a self-acting decay heat removal which works just with heat conduction, radiation and free convection such that the maximum fuel temperature under loss-of-forced-convection (LOFC) accident conditions, the equivalent of a LOCA in LWRs, remains below 1600°C, a level that can be sustained by the fuel without further damage.

The first concept in this development was the 200 MW(th) HTR-MODUL designed by the former German company INTERATOM with a tall (9.4 m) and slim (3 m diameter) core. An arrangement of the active core in an annular geometry permits to raise the thermal power output from 200 to 600 MW without changing the basic philosophy of the 1600°C limit.

Another feature introduced for various advanced designs is direct-cycle operation promising an increase of the power conversion efficiency. It was since long already subject of intensive studies in Germany, the United States, Russia, and Japan. A joint German-Swiss project called "High Temperature Reactor with Helium Turbine of Large Power", HHT, was active throughout the 1970s. It resulted in the construction of experimental facilities where components were successfully tested under nuclear conditions, and in a design of a large-scale HTGR with helium turbine and an electric power output of 1,640 MW at 40 % efficiency. The project ended in 1981 with a positive statement for the licensability of the basic concept.

All new modular HTGR concepts use steel vessels as the primary enclosure. The reactor building as protection against outer impacts is designed as a confinement, which is not necessarily gas-tight (unlike a containment). It is a controversial issue, since the containment is "traditionally" recognized as a necessary safety barrier. This concept makes the fuel element an eminently important component as the principal containment for radionuclides.

The production of high-quality steam at 530°C and 15 MPa in an HTGR allows the operation of a steam turbine prior to the "normal" steam application system for electricity production. The operation of such a process heat HTGR is flexible in the choice of the coolant temperature at steam generator inlet such that a chemical industrial complex could be optimized in terms of product spectrum and product volume. Efficiencies possible are 40–43 % for the electricity generating system with steam turbine, up to 48 % if a closed-cycle gas turbine is employed, and even up to 50 % for a combination of both. In the CHP mode, the efficiencies are estimated in the range of 80–90 %.

3.4 International activities on near-term HTGRs

Among the various nuclear reactors with a near-term deployment chance, there were two concepts of modular HTGRs.

Gas Turbine Modular High Temperature Reactor, GT-MHR

The GT-MHR is a joint US-Russia design with a prismatic and annular core design should go up to 600 MW(th) and with direct gas turbine cycle.

For commercial use, the GT-MHR reactor of 600 MW(th) power will use an LEU fuel cycle. A parallel GT-MHR design is developed for surplus weapons plutonium disposition. The GT-MHR project coordinating committee decided that before final design development starts, all efforts and funds should be concentrated on technology demonstration.

France is currently investigating the concept of an HTGR for electricity generation in an indirect cycle power conversion system. Based on principles of the HTR-MODUL concept and evolving from the US GT-MHR concept, the company AREVA-NP is developing the conceptual design of ANTARES. It has a block-type core generating a power of 600 MW(th) with a helium outlet temperature of 850°C.

Pebble Bed Modular Reactor, PBMR

In South Africa, the national utility ESKOM successful operating the two unit Koeberg PWR station, is pursuing the project of modular HTGRs for electricity production to meet the demand of its growing economy. The proposed demonstration power plant is a 400 MW(th) Pebble Bed Modular Reactor (PBMR) with a direct cycle gas turbine to convert the heat into 175 MW of electric power. Operation targets are a nominal lifetime of 40 yrs and an availability of 95 %.

The fuel design corresponds to the latest version of the German high quality spherical fuel element originally developed for the HTR-MODUL with 9.6 % enriched UO_2 TRISO coated particles and 9 g of (total) uranium per sphere. The PBMR uses the closed Brayton cycle with recuperator and inter-cooler arranged in a non integrated design with the reactor and the power conversion unit in two different vessels side by side. The design of the PBMR turbine plant is a horizontal, single shaft design.

The project to construct the PBMR Demonstration Plant on the Koeberg site near Cape Town has now been approved with a planned fuel load date of April 2010.

In China, based on the experience with the HTR-10, a demonstration plant of a pebble bed HTGR, the HTR-PM, has been proposed by a development team around the Institute for Nuclear and New Energy Technology (INET) of the Tsinghua University. The reactor is presently considered to consists of two 250 MW(th) units, and adopting a Rankine steam cycle for power conversion. Main goal of the project is to be competitive with the investment to come from the future utility company. Standard design is planned to be finished by 2006, end of construction is presently foreseen for 2012.

4 The fourth generation

4.1 Requirements to a new generation of nuclear plants

The next, fourth generation of nuclear reactors is expected to be introduced in about 25 years from now. The designs of such reactors, however, are being developed today based on requirements leading to a further progress of the nuclear technology by addressing the areas of safety and reliability, proliferation resistance and physical protection, economics, sustainability (GIF 2002).

Safety and reliability

Future nuclear power will require a further enhancement of the safety standards. A safer operation of nuclear reactors can be obtained by designs with a very low probability and degree of core damage, minimal consequences even after severe accidents and limitation of the consequences to the plant site, and by a further improvement of accident management such that the public will not be affected and off-site emergency response be near to unnecessary. It can be achieved by a robust design, a high level of inherent safety, and transparent safety features. In order to be competitive, also reliability and performance must be at a very high level achievable by considering both technical improvements and human performance.

Proliferation resistance and physical protection

Controlling and securing of nuclear material must be provided by modified design features and other measures to make it unattractive for potential diversion or theft or a sabotage act, and to make a dispersion of material more difficult. A robust design is also a means to protect against actions from war or sabotage.

Economics

The development towards a simpler reactor design, economic plant sizes and life cycles, improvement of the plant's efficiency will help achieving competitive cost. The economic risk can be reduced by employing modular reactor designs and use of improved or innovative construction techniques. Future nuclear reactors must also be flexible in meeting the needs for energy products others than electricity. The penetration of non-electricity markets by supplying process heat/steam for district heating, seawater desalination, or hydrogen generation will enhance efficiencies and thus competitiveness.

Sustainability

Sustainability as the ability to care about the needs of both the present and future generations is mainly looked at in terms of nuclear fuel supply and spent fuel management. Prolonged availability of resources can be achieved by breeding new fissile material or by recycling used fuel. Future radioactive waste management must include processes leading to significantly reduced quantities of waste to be disposed and its decay heat. Furthermore a significant reduction in lifetime and toxicity will simplify the requirement to a safe performance of the repositories.

4.2 The GIF initiative

"The Generation IV International Forum" (GIF) was founded in 2000 as a joint initiative of ten countries (Argentina, Brazil, Canada, France, Japan, Korea, South Africa, Switzerland, UK, USA), later joined by the EURATOM as well as China and Russia since recently, to develop the fourth generation of nuclear reactors by 2030, which is, apart from being safer, more reliable, more economic, more proliferation-resistant, also expected to penetrate non-electrical markets like the supply of heat or hydrogen on a large scale.

Selection of Generation-IV reactor systems

After the evaluation of some 100 potential Gen-IV designs that have been suggested, the GIF agreed in 2002 to continue enhanced studies on six selected nuclear reactor designs which were deemed to meet the above mentioned requirements and to be deployable in due time (GIF 2002):

- Gas-Cooled Fast Reactor System (GFR)
- Lead-Cooled Fast Reactor System (LFR)
- Molten Salt-Cooled Reactor System (MSR)
- Sodium-Cooled Fast Reactor (SFR)
- Supercritical-Water-Cooled Reactor System (SCWR)
- Very High Temperature Reactor (VHTR)

The principal characteristic features of these Gen-IV nuclear reactor systems are summarized in Table 2. All systems are intended to have a long service lifetime of 60 years. Except for the VHTR and one of the two

SCWR types with a one-through cycle, all systems are designed for a closed cycle allowing and actinide management by generating fissile material and use recycled fuel thus alleviating the radioactive waste issue.

In addition, the Gen-IV concepts are designed for higher coolant temperatures than most of the today's nuclear reactors. It enables the future reactors to generate electricity at a higher efficiency as well as heat or steam which can be transferred to various processes depending on the temperature level provided. All proposed Gen-IV systems are able to supply heat for the lower temperature ($\sim 120°C$) process of desalination of seawater.

In expectation of a future significantly increased demand for hydrogen, mass production of H_2 is a major goal for Gen-IV systems. Due to the need of high-temperature heat, it will be basically the helium-cooled reactors GFR and VHTR, the MSR and also one of the LFR designs appropriate to be linked with a H_2 production system.

The principal methods of nuclear H_2 generation considered are thermochemical cycles such as the sulphur-iodine cycle or the calcium-bromide cycle, and high temperature electrolysis (see Chap. 5).

The overall advancement in the development of Gen-IV nuclear reactor systems will be made in three distinct phases:

- a viability phase to investigate whether the basic concepts and key technologies and processes are feasible under the given conditions;
- a performance phase to verify and optimize processes, phenomena, material capabilities at an engineering level; and
- a demonstration phase to license, construct, and operate a prototype plant.

A parallel initiative has been started by the International Atomic Energy Agency, IAEA, called INPRO, the "International Project on Innovative Nuclear Reactors and Fuel Cycles". While the GIF, in which both IAEA and OECD/NEA are participating as observers, is serving essentially as a designers' initiative, INPRO includes also other countries and has a longer time horizon. It incorporates IAEA safeguard considerations more directly encompassing both designers and end-users and their requirements. A methodology for the assessment of innovative nuclear reactors and fuel cycles has been developed (IAEA 2003), which is currently being applied on a national base by the countries Argentina, India, the Russian Federation, and Korea.

Table 2. Main features of the six Gen-IV nuclear reactor concepts (GIF 2002)

	GFR	LFR	MSR	SFR	SCWR	VHTR
Reference thermal power [MW(th)]	600	125–3600	2250	900–3800	3900	600
Power density [MW(th)/m³]	100	~70	22	350, 150	100	6–10
Reference electric power [MW(e)]	288	50–150, 300–400, 1200	1000	500–1500	1700	300–360
Efficiency [%]	48	33–40	44–50	39	41–44	> 50
Neutron spectrum	fast	fast	epithermal - thermal	fast	a) thermal b) fast	thermal
Fuel cycle	closed	closed	closed	closed	a) open b) closed	open
Coolant	helium	lead, lead-bismuth	liquid fluoride salt	sodium	water	helium
Coolant outlet temperature [°C]	850	550–800	> 700	530–550	510	> 1000
Fuel	fertile U and 20 % Pu in composite ceramics	Metal alloy or nitride with fertile U and TRU	liquid mixture of Na-, Zr-, U-fluorides	U-Pu-minor actinides-Zr oxide or metal alloy	UO_2 with stainless steel or Ni alloy cladding	UO_2, ZrC-TRISO coated particles in pebble or block
Main purpose	electricity, actinide management, H_2 production	electricity, actinide management, H_2 production	electricity, actinide management, H_2 production	electricity, actinide management	electricity, actinide management	H_2 production, electricity, actinide management
Deployable by	2025	2025	2025	2015	2025	2015

VHTR – Most advanced among the Gen-IV systems

One of the most promising "Gen-IV" nuclear reactor concepts is the VHTR representing the nearest-term option of all reference plants selected and might be ready for deployment much sooner than 2030.

The characteristic features of the VHTR are a helium-cooled, graphite moderated, thermal neutron spectrum reactor core with a reference thermal power production of 600 MW. The coolant outlet temperatures of 900–1000°C or higher is ideally suited for a whole spectrum of high temperature process heat applications. The employment of a direct cycle gas turbine allows for a high efficiency of > 50 %.

The co-generation of hydrogen is done by connecting the VHTR to a H_2 production plant via an intermediate heat exchanger (see Fig. 4), which then represents an indirect cycle. Top candidate production methods are the sulfur-iodine (S-I) thermochemical cycle and high temperature electrolysis (HTE), considered presently by various countries.

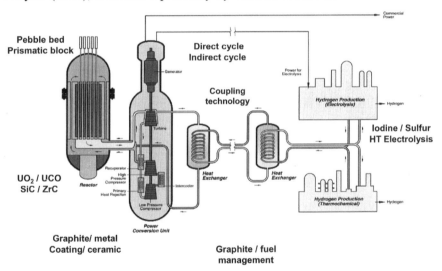

Fig. 4. Artist's depiction of the VHTR plus hydrogen production system (GIF 2002)

The technology of the VHTR takes benefit of the broad experience from respective research projects and from HTGR operation in the past, from the presently operated Japanese HTTR and the Chinese HTR-10, as well as from the comprehensive R&D efforts which were initiated in many countries to investigate HTGR systems in connection with nuclear hydrogen production.

4.4 The treatment of radioactive waste

In the area of nuclear waste treatment strategies, concepts are being developed towards a further closure of the fuel cycle, a requirement for environmentally benign nuclear energy. This, however, would be on the expense of the establishment of a complex spent fuel reprocessing technology.

Recycling of spent fuel, and partitioning and transmutation of actinides and long-lived fission product species, plus the immobilization of the remainder are the steps which eventually result in a minimization of radioactive waste in connection with much shorter periods of time in the order of hundreds rather than millions of years, during which the waste represents a jeopardy to public and environment.

Waste minimization and also the reduction of C-14 inventory in the graphite will become particular issues for the VHTR due to the typically extensive volumes of the low-power-density HTGRs. Very high temperature reactors may play an important role in reducing the toxicity of the waste through "deep burning" of plutonium and the long-lived actinides.

The aim of transmutation is to transform long-lived, highly radiotoxic actinides by reaction with neutrons into mostly short-lived and less toxic species. This process, however, does not change heat production of the waste nor does it reduce the waste. Current (and future) R&D is concentrating on efficient methods for the necessary isotope separation and the subsequent preparation of the "new" fuel, as well as the feasibility of transmutation either in specially designed nuclear reactors or in accelerated driven systems (ADS).

5 Penetration of non-electricity markets

5.1 Nuclear power for more than electricity production

A new, perhaps revolutionary nuclear reactor concept of the next generation will offer the chance to deliver besides the classical electricity also non-electrical products such as hydrogen or other fuels. In a future energy economy, hydrogen as a storable medium could adjust a variable demand for electricity by means of fuel cell power plants ("hydricity") and also serve as spinning reserve. Both together offer much more flexibility in optimizing energy structures. Prerequisites for such systems, however, would be competitive nuclear hydrogen production, a large-scale (underground) storage at low cost as well as economic fuel cell plants (Forsberg 2005).

There are many industrial sectors such as paper and pulp, food industry, automobile industry, textile manufacturing etc., which have a high demand for electricity and heat/steam at various levels of temperature and pressure, where nuclear energy specifically from HTGRs could play a major role in future (Schad et al. 1988; IAEA 2000, 2002). The data elaborated for different CHP applications in EU clearly show that unit sizes of some hundred MW are sufficient. The main challenge at present is to include nuclear combined heat and power applications into the general strategies and to establish transition technologies from present industrial practice or emerging new resources ("dirty fuels") in order to stabilize energy cost.

One example of CHP market penetration by nuclear power with near-term prospects is the provision of high temperature heat/steam and electricity in the chemical and petrochemical industries. Particularly in this sector, massive amounts of H_2 will be required in future for the conversion of heavy oils, tar sands, and other low-grade hydrocarbons (EC 2003).

An important example for the use of nuclear energy in the low temperature range is seawater desalination, a process with increasing importance due to growing drinking water shortage in many arid areas in the world. The nuclear driven process is technically and economically feasible (EURODESAL project (Nisan et al. 2003)), but is yet to be demonstrated on industrial scale. Its operation in the CHP mode will be necessary for reasons of competitiveness. This may also apply to other applications where electricity will be an essential by-product to improve economics and provide electricity for industrial sites.

Therefore, it is strongly recommended to investigate in more detail the CHP options and market requirements for nuclear CHP as it offers a considerable potential for fuel resource saving due to high overall efficiencies, improved economics and reduction of CO_2 emissions, already in the near and medium term. Simple non-electric applications like district heating and desalination may provide a psychological break-through, open new markets and contribute significantly to CO_2 reduction, too. Still CHP represents a challenging R&D target even if very robust technology and wide safety margins were applied.

5.2 Requirements for nuclear energy in industrial applications

Most industries need to rely on a secure and economic supply with energy to guarantee continuous and reliable operation of their process units. Ensuring supply security by diversification of the primary energy carriers and, at the same time, limit the effects of energy consumption on the environment will become more important goals in future.

A modular arrangement (2–6 units) will be necessary for redundancy, reliability and reserve capacity reasons which again reduce power size per modular unit. However, smaller power size allows for simplicity and robustness by higher safety margins even at higher operational temperatures.

The connection between nuclear and heat application plant is principally independent of the method of hydrogen production. The hot coolant transfers its heat to the chemical process via an intermediate heat exchanger (IHX). Main purpose of the intermediate circuit is to clearly separate the nuclear from the chemical island. In this way, the direct access of primary coolant to the chemical plant and, in the reverse direction, of product gases to the reactor building can be prevented. Thus it is possible to design the chemical side as a purely conventional facility and to have possible repair works being conducted under non-nuclear conditions.

Of particular significance is the consideration of conceivable accident scenarios in such combined nuclear and chemical facilities. Apart from their own specific categories of accidents, a qualitatively new class of events will have to be taken into account which is characterized by interacting influences. Arising problems to be covered by a decent overall safety concept are the question of safety of the nuclear plant in case of an explosion of a flammable gas cloud on the chemical side, or vice versa, the question of what influence the accident-induced release of radioactivity will have on the continuation of operation of the chemical plant. But there are also the comparatively more frequently expected cases of thermodynamic feedback in case of a loss of heat source (nuclear) or heat sink (chemical), respectively. For the specific example of the HTTR in connection with a steam-methane reforming device, the hazardous potential has been identified and evaluated (Verfondern and Nishihara 2004), and has been resulting in a respective proposal for a safety concept.

5.3 Nuclear process heat for the refinery industries

The processes of splitting hydrocarbons are presently widely applied production methods for hydrogen. The most important ones established on an industrial scale are steam reforming of natural gas, the extraction from heavy oils, and the gasification of coal. Biomass gasification is currently being tested on pilot plant scale.

Steam reforming of natural gas covering worldwide about half of the hydrogen demand, is one of the essential processes in the petrochemical and refining industries. This process was subjected to a long-term R&D program in Germany with the aim to utilize the required process heat from an HTGR. The Research Center Juelich has developed in cooperation with the respective industries the design of a nuclear process heat plant as well as the necessary heat exchanging components which according to their dimensions belong to the 125 MW power class. A particular 10 MW(th) component test loop was constructed and successfully operated over a total of 18,400 hours with about 7,000 hours at temperatures above 900°C. The components tested in terms of reliability and availability included two designs of an IHX, steam generator, decay heat removal cooler, hot gas ducts, and hot gas valves (Harth et al. 1990).

The steam reforming of methane was investigated in experiments conducted under the typical conditions of a nuclear reactor, i.e., in reformer tubes heated with helium of 900°C and 4 MPa with industrial-scale dimensions (15 m in length, 130 mm inner diameter). The first test facility was a single splitting tube (EVA), and the follow-on facility consisted of a tube bundle (EVA-II). A similar experimental program was recently conducted in Japan where the main focus was on the mutual thermo-dynamic interaction. Also EVA's counterpart, ADAM, a facility for the re-methanation of the synthesis gas generated in EVA, was constructed and operated, thus completing the system to a closed cycle and verifying the principle of a long-distance energy transportation system based on hydrogen as the energy carrier.

Within the frame of the project "Prototype Nuclear Process Heat", PNP, also the nuclear coal gasification processes for hydrogen production were investigated in Germany. These activities eventually resulted in the construction and operation of pilot plants for coal gasification under nuclear conditions. Catalytic and non-catalytic steam-coal gasification of hard coal was verified in a 1.2 MW facility using 950°C helium as energy source. The process of hydro-gasification of brown coal (lignite) was realized in a 1.5 MW plant operated for about 27,000 h with a total amount of 1,800 t of lignite being gasified.

5.4 Nuclear hydrogen production

Water and biomass are expected to become the main sources for hydrogen in the future with the necessary process heat for extracting the hydrogen to be provided by CO_2 emission free energy sources. With respect to hydrogen production on a large scale at a constant rate, nuclear energy may play an essential role.

Thermochemical cycles

Thermo-chemical cycles are composed of several reaction steps which in the sum are leading to a decomposition of water into hydrogen and oxygen. The supporting chemical substances are regenerated and recycled, and remain – ideally – completely inside the system. The only inputs are water and heat.

Numerous cycles have been proposed in the past and investigated in terms of their characteristics like reaction kinetics, thermo-dynamics, separation of substances, stability, processing flow scheme, and cost analysis. Only a few, however, were deemed to be sufficiently promising and worth further investigation. Among those whose partial reactions are being investigated in more detail also with respect to their coupling to an HTGR or a solar heat source, are the sulfur-iodine (S-I) process originally developed in the USA and later pursued and modified by various research groups.

The process scheme of the S-I cycle is composed of three principal steps as shown in Fig. 5. Step (1) represented by the medium part in the figure, corresponds to the so-called Bunsen reaction where at the presence of SO_2 and I_2 water is added. Products of this exothermal reaction are the two acids hydro-iodide and sulphuric acid which appear as HI and H_2SO_4-rich liquid phases. The acids are separated, purified, and concentrated. The final two steps are the decomposition of HI at a temperature of 400–500°C to liberate the hydrogen (left-hand side of the figure), and the decomposition of H_2SO_4 at high temperatures of 850–950°C to generate the oxygen.

This process was verified at the Japan Atomic Energy Agency (JAEA) and could be successfully demonstrated in a closed cycle in continuous operation over one week achieving a hydrogen production rate of 30 Nl/h. The next step which started in 2005 is the design and construction of a pilot plant operated under the simulated conditions of a nuclear reactor, i.e., (electrically heated) helium of 880°C at 3 MPa and with an expected yield of 30 Nm^3/h (Kubo et al. 2004). After 2011, it is planned to connect the S-I process to the HTTR for hydrogen production at a rate of 1,000 Nm^3/h.

Other processes considered worth of further investigation are the so-called Westinghouse process, a sulphuric acid hybrid (HyS) cycle where the low-temperature step runs in an electrolysis cell to produce the hydrogen, or the so-called UT-3 process based on a four-step cycle with calcium and bromine.

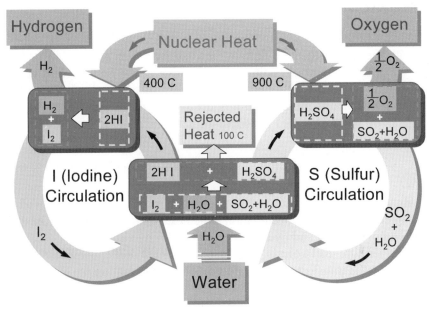

Fig. 5. Principal schematic of the sulfur-iodine (S-I) thermochemical cycle

High temperature electrolysis

The electrolytic decomposition of water is a widely applied technology on industrial scale accounting for approximately 4 % of the world's hydrogen production. It has, however, a comparatively low efficiency and is economic only, if cheap electricity is available. The electrolysis of water in the vapor phase at high temperatures of 800–1000°C has the advantage of a lower total energy input and, in particular, an electricity input reduced by about 30 % compared to "normal" electrolysis. R&D efforts in various countries are concentrating on the development and optimization of electrolysis cells, composition of cell stacks, and selection of appropriate materials. The development may benefit from efforts in the area of solid oxide fuel cells as the reverse process of high temperature electrolysis.

In Germany, the high temperature electrolysis process became known in the 1990s under "HOT ELLY" demonstrated in tubular cells in a 2 kW pilot plant. Japan's approach based on planar cells achieved hydrogen production rates of 3–6 Nl/h per m^2 of cell surface at a temperature of 850°C. The INL in the USA is presently conducting an experimental program to test solid oxide electrolysis cell stacks combined with materials research and detailed CFD modelling (Herring et al. 2005).

5.5 Worldwide activities on nuclear hydrogen production

Various countries have initiated ambitious programs within the GIF initiative with the main objective to bring nuclear hydrogen production to the energy market. Numerous institutions are active in the first stage of demonstrating the viability of nuclear hydrogen production to be followed by the stages of performance testing and demonstration of the pursued technologies. In the following, a couple of national research activities are shortly described.

France

The above mentioned French next-generation concept ANTARES can also be adapted to different cogeneration schemes. In the VHTR concept for electricity production and hydrogen cogeneration by either the S-I cycle or high temperature electrolysis, the reactor coolant outlet temperature is 1000°C. Heat is transferred in a plate-type IHX representing the only novel component which needs further R&D efforts. Process heat supply to the chemical process is finally at a temperature level of 925°C (Gauthier et al. 2004).

As a second Gen-IV concept, the Gas Fast Reactor is also under further consideration. Some viability and design selection studies will prepare the decision on the construction of a 50 MW(th) Experimental Technological Demonstration Reactor.

Japan

With the construction and operation of the 30 MW(th) HTTR, JAEA has early laid the basis for utilization of nuclear process heat for hydrogen production. After the successful demonstration of nuclear H_2 production with the HTTR by applying the S-I thermochemical cycle (at the moment scheduled for 2011), the follow-on HTGR plant, a VHTR, will be the GTHTR300C reactor (C = Cogeneration of electricity and hydrogen) (Kunitomi and Yan 2004). The direct cycle block-type HTGR with a thermal power of 600 MW provides a coolant outlet temperature of 950°C. In the IHX, a part of the thermal power, 168 MW, is transferred to the H_2 generation process (S-I) with the remaining power to be used for electricity generation of 202 MW(e). Assuming 50 % efficiency and a 90 % availability, the average amount of hydrogen generated is estimated to be 24,000 Nm^3/h.

Japan is also pursuing concepts based on Gen-IV fast reactors. The so-called "FR-MR" concept has been suggested to perform methane-steam reforming at lower temperatures around 550°C by employing membrane reformer technology. The FR-MR is designed to be a sodium cooled fast reactor with a thermal power of 240 MW(th). Nuclear heated membrane reforming can also be applied in the refining industries helping to save hydrocarbon feedstock (Hori et al. 2004).

Korea

In 2004, Korea has launched the "Nuclear Hydrogen Development and Demonstration" (NHDD) project with the ambitious goal to develop and demonstrate nuclear based hydrogen production technology by the year 2019 and to achieve commercial nuclear hydrogen production by the middle of the 2020s to cover 20 % of the total vehicle fuel demand corresponding to 3.3 million t/yr of hydrogen (Shin et al. 2006). The nuclear reactor is planned to be a VHTR with either a block type core to produce 600 MW(th) or a pebble bed core to produce 400 MW(th). The reactor pressure vessel will be based upon the Korean Advanced Power Reactor APR-1400 vessel which is the largest that can be manufactured today.

USA

In the United States, the goal of the "Nuclear Energy Research Initiative", NERI, is to demonstrate the commercial-scale production of hydrogen using nuclear energy by 2017. The Modular Helium Reactor (MHR) has been suggested as the Next Generation Nuclear Plant, NGNP, Gen IV reference concept for nuclear hydrogen generation on the basis of the S-I thermochemical cycle and alternatively high temperature electrolysis. The so-called H2-MHR based on the principle of the GT-MHR is helium cooled, graphite moderated, thermal neutron spectrum reactor directly coupled to a Brayton cycle power conversion system with a thermal power of 600 MW and designed for electricity production with an efficiency of 48–52 % (Richards et al. 2005).

Another NERI project is the so-called "Secure Transportable Autonomous Reactor Hydrogen", STAR-H2 project. It is designed as a heavy liquid metal cooled, mixed U-TRU-nitride fuelled fast reactor with a power of 400 MW(th) and with passive safety features. The primary coolant, lead, circulates by natural convection and transfers its heat to a molten salt coolant which then transfers heat to a hydrogen production system based on a thermochemical cycle. The heat rejected (~600°C) can be used for electricity production in a supercritical CO_2 Brayton cycle. Rejected heat here (<125°C) finally can be used in a desalination plant with a capacity of 8,000 m^3/d of potable water (Wade et al. 2003).

A new reactor concept, the "Advanced High Temperature Reactor", AHTR, has been suggested in (Forsberg 2004) incorporating essential features of the "classical" HTGR such as the coated particle fuel, coolant outlet temperatures between 700–1000°C, or passive safety systems. The reference design considers a large-size annular core of 2400 MW(th) cooled by a liquid fluoride salt. The large power size represents the main difference to a gas-cooled reactor whose decay heat removal capability limits its size to ~600 MW(th). Heat is transferred via an IHX to a secondary circuit which also uses a liquid salt coolant, before it is utilized for electricity or hydrogen generation.

European Union

Within the current Framework Programme 6 of the European Union, the engagement of research organizations, industry, and policy is given by the participation in nuclear and hydrogen activities. On the nuclear side, there is the RAPHAEL project, acronym for "Reactor for Process Heat Hydrogen and Electricity Generation" which started in April 2005 (Hittner et al. 2006). This EU Integrated Project with 33 partners from 10 countries is treating the pertinent aspects of material development, HTGR fuel technology, core physics, components, nuclear waste management, and coupling to hydrogen production technologies. Although there is no dedicated nuclear hydrogen project, a link is given to EU hydrogen projects to look at the coupling of nuclear (and solar) energy to H_2 producing systems and corresponding safety aspects.

6 Conclusions

Conclusions can be drawn as follows:

1. Nuclear power is a safe, reliable, clean, and economic energy source. The experience achieved and lessons learned from five decades of commercial nuclear power plant operation have resulted in a status of minimal risk of severe occurrences. It thus represents a powerful greenhouse gas emission-free option within the existing mix of energy sources to help meet the world's demand for energy in future and to reduce national dependencies on imports of fossil fuels.
2. The next generations of nuclear plants will be even safer, more reliable, more economic, and more proliferation-resistant, and will supply more than just electricity. This can be achieved by further raising

the degree of passive safety and developing concepts where catastrophic accidents can be excluded. Only if a nuclear concept has proven to meet the requirements of future reactors, the utilization of nuclear power will be generally accepted by the public.

3. Among the suggested types of nuclear reactors of the fourth generation, the VHTR represents a promising concept. It clearly shows the features of a catastrophe-free reactor and is most advanced in terms of R&D works. It will provide coolant exit temperatures of up to 1000°C, which can be utilized in the CHP mode for a broad range of process heat applications.

4. Particularly in the petrochemical and refining industries, there will be a growing demand for hydrogen due to the increasing share of "dirty fuels" such as heavy oils, oil shale, tar sands entering the market. Near-term and long-term options for a large-scale nuclear H_2 production are existing. Nuclear is also applicable to the production process of liquid fuels, compatible with the needs of the transportation sector. Technical and economical feasibility, however, remains to be demonstrated, since production processes have not yet been tested beyond pilot plant scale.

References

EC (2003) European Commission, Hydrogen Energy and Fuel Cells – A Vision of Our Future, High Level Group Summary Report

EIA (2006) Per Capita Total Carbon Dioxide Emissions from the Consumption of Energy, All Countries, 1980–2004, Energy Information Administration of the US-DOE, http://www.eia.doe.gov/emeu/international/ carbondioxide.html

Forsberg CW (2004) The Advanced High-Temperature Reactor: High-Temperature Fuel, Molten Salt Coolant, and Liquid-Metal Reactor Plant. In: 1st COE-INES Int Symp on Innovative Nuclear Energy Systems for Sustainable Development of the World, Oct 31–Nov 4, 2004, Paper 71

Forsberg CW (2005) What is the Initial Market for Hydrogen from Nuclear Energy?, Nuclear News, January 2005

Gauthier J-C, et al. (2004) ANTARES: The HTR/VHTR Project at Framatome ANP. In: Int Conf of High Temperature Reactor HTR, Sep 22–24, 2004, Beijing, China, Paper A10

GIF (2002) A Technology Roadmap for Generation IV Nuclear Energy Systems, US-DOE & Generation IV International Forum

Harth R, et al. (1990) Experience Gained from the EVA II and KVK Operation, Nucl Eng Des 121:173–182

Herring JS, et al. (2005) High-Temperature Electrolysis for Hydrogen Production Using Nuclear Energy. In: Int Conf on Nuclear Energy Systems for Future

Generation and Global Sustainability Global 2005, October 9–13, 2005, Tsukuba, Japan, Paper 501

Hittner D, et al. (2006) RAPHAEL a European Project for the Development of V/HTR Technology for Industrial Process Heat Supply and Cogeneration. In: 3rd Int Topical Meeting on High Temperature Reactor Technology, held at Johannesburg, South Africa, October 1–4, 2006

Hori M, et al. (2004) Synergistic Hydrogen Production by Nuclear-heated steam Reforming of Fossil Fuels. In: 1st COE-INES Int Symp on Innovative Nuclear Energy Systems for Sustainable Development of the World, Oct 31-Nov 4, 2004, Paper 43

IAEA (1984) Status of and Prospects for Gas-Cooled Reactors, IAEA Technical Report Series No. 235, International Atomic Energy Agency, Vienna, Austria

IAEA (2000) Status of Non-Electric Nuclear Heat Applications: Technology and Safety, Report IAEA-TECDOC-1184, International Atomic Energy Agency, Vienna, Austria

IAEA (2002) Market Potential for Non-Electric Applications of Nuclear Energy, IAEA Technical Report Series No. 410, International Atomic Energy Agency, Vienna, Austria

IAEA (2003) Guidance for the Evaluation of Innovative Nuclear Reactors and Fuel Cycles Report IAEA-TECDOC-1362, International Atomic Energy Agency, Vienna, Austria

Kirch N, Ivens G (1990) Results of AVR Experiments. In: VDI, AVR - Experimental High-Temperature Reactor, 21 Years of Successful Operation for a Future Energy Technology, Association of German Engineers (VDI), VDI-Verlag, Duesseldorf, Germany

Kubo S, et al. (2004) A Pilot Test Plant of the Thermochemical Water-Splitting Iodine-Sulfur Process, Nuc Eng Des 233:355–362

Kunitomi K, Yan X (2004) GTHTR300 for Hydrogen Cogeneration, HTTR Workshop on Hydrogen Production Technology, Oarai, Japan, July 5–6, 2004

Martin Bermejo J (2004) Fuel Cells and Hydrogen Research in the European Union, DOE Hydrogen and Fuel Cell Program Review, Philadelphia, USA, May 24, 2004

Nisan S, et al. (2003) Sea-Water Desalination with Nuclear and Other Energy Sources: The EURODESAL Project, Nucl Eng Des 221:251–275

Richards MB, et al. (2005) Conceptual Designs for MHR-Based Hydrogen Production Systems. In: Int Conf on Nuclear Energy Systems for Future Generation and Global Sustainability, Global 2005, October 9–13, 2005, Tsukuba, Japan, Paper 190

Schad M, et al. (1988) Project Study on Utilization of Process Heat from the HTGR in the Chemical and Related Industries, Report Lurgi GmbH, Frankfurt, Germany

Shin Y-J, et al. (2006) Nuclear Hydrogen Production Project in Korea, Third Information Exchange Meeting on the Nuclear Production of Hydrogen, held at Oarai, Japan, October 5–7, 2005, Proc OECD/NEA, pp 101–106

Verfondern K, Nishihara T (2004) Valuation of the Safety Concept of the Combined Nuclear/Chemical Complex for Hydrogen Production with HTTR, Report Juel-4135, Research Center Juelich, Germany

Wade D, et al. (2003) STAR-H2: Passive Load Follow/Passive Safety Coupling to the Water Cracking Balance of Plant. In: Symp on Nuclear Energy and the Hydrogen Economy, Cambridge, USA, Sept. 23–24, 2003

Transition Strategies for a Hydrogen Economy in Mexico

Sergio Dale Bazán-Perkins

División de Estudios de Posgrado, Facultad de Ingeniería, Universidad Nacional Autónoma de México, Coyoacán 04510, México DF.
E-mail: bazanperkins@hotmail.com

Abstract

The economic viability of the energy sector is fundamental for the development of Mexico. The main objective of the economic sectors is to reduce the cost of energy production. The application of the nuclear and renewable energy sources can be developed in Mexico with a limitless potential for the electricity production, heat, hydrogen and seawater desalination, without Greenhouse Gases (GHG's) emissions and under acceptable economic conditions. For the future development of this potential, it is required to make deep institutional changes by state promotion and with the public acceptance. However, it is at present when we most settle down the bases for a new energy politics by means of appropriate clear objectives. In order to attain such goals, the participation of the universities, institutions, public administration and companies is required. The proposed energy program until the year 2030 includes expansion mainly for the period 2007–2015 of the electrical sector using the renewable energy resources, hydro, geothermal, biomass and wind, for being more competitive for the country instead of a gas-based combined cycle plant. This paper analyses the benefits to develop the power sector mainly using High Temperature Gas-cooled Reactors (HTGR's), in 2015–2030, adding to renewable power technologies.

1 Introduction

Since the first oil embargo in the early 1970s, the concern for the security of the power provision has increased in the world. The developed countries chose to diversify their power supply. Also, the atmosphere acidification generated by fossil fuels and their consequences in environment and human health deterioration, incentive new technologies for the reduction of the emissions of NO_x, SO_x and NH_3 should be avoided (United Nations 1979).

Towards 21st century, the worldwide efforts to reduce the Greenhouse Gases (GHG's) effect, CO_2, CH_4, N_2O, HFCs, PFcs and SF_6, have become the main subject of the political international agendas. Beér (2000) expressed that the control for these emissions, is centered mainly in the technologies that allow diminishing those of CO_2.

The initiative to reach "zero emissions" is new and appears assumed by a growing number of nations to reach a sustainable society (Suzuki 2002). This initiative was presented in the University of the United Nations of Tokyo in 1994. The proposal included the recycling of industrial products not to create wastes and contamination, by means of the use of new technologies and the natural systems like model. By means of this political plan it is aimed to reach concrete objectives as "zero emission power plants", "zero loss of potable water", and "zero solid wastes", among many others (United Nations 1996).

On the other hand, Williams (2001) exposed the causes by which, at the beginning of 21st Century, the nations chose to promote the sustentability of their economies with the policy of "zero emissions". This is, the necessity of reduction of emissions of gases of effect hothouse, using alternative energy to the hydrocarbons, being aided by technological advances that allow covering the increase in energy demand.

One of the main economic reasons that has stimulated the use of new energy sources for the transports sector, is due to high prices of the petroleum that have been increasing from the year 2000 (EIA 2006). In 1999, Iceland astonished the world when announcing the creation, in a term of 30 years, of an economic sustainable source of low carbon dioxide emissions by using hydrogen. The hydrogen will produce it in large-scale, using the hydro and geothermal power stations (WI 2002).

Due the increasing interest of several nations to use hydrogen in their economies in the present Century, multinational organisms have been created to accelerate its implementation. Like the International Partnership for the Hydrogen Economy (IPHE). The European Hydrogen and Fuel Cell Technology Platform (EU TP H-FC) and Carbon Sequestration Leadership

Forum (CSLF). In the case of Mexico, the first steps to reach the future potential which is in hydrogen with own technologies, must develop in the civil associations, integrated by universities, research centers, private institutions and companies, like the Mexican Society of Hydrogen A.C. (SMH 1999) and the National Network of Hydrogen A.C. (RNH) according to Fernández-Zayas (2005).

Nevertheless, the research of Dunn (2002) also indicates that those initiatives of technological investigation and isolated promotions are not enough to reach a sustainable hydrogen economy. The governments represent the essential part to implement decisive public policies and the educational efforts that accelerate the transition toward a sustainable hydrogen economy. Barry and Abhijit (2004), conclude that for this transition, the main requirement it is to have a competitive economy. As well as, to have a long-term strategic energy plan, accepted by the consumers.

The present paper includes a series of strategies for the future development of the energy sector of Mexico. It outlines as objective the development of a sustainable economy in Mexico with collateral impact in lowering the carbon dioxide emissions significantly. In this sense, the production of electricity, cogeneration, seawater desalination and hydrogen, will be able to be obtained from the nuclear energy and the renewable ones. To reach these objectives towards the 2030, it is required to modify the program of expansion of electrical sector 2005–2014 and of other inadequate facilities. Due to economic inviability for the country of continuing installing gas based Combined Cycle power plant (CCPP). To reach these objectives will require of big technological, social and political transformations.

2 Economy of hydrogen with zero emissions

The development of an economy of hydrogen not necessarily implies the reduction of the polluting emissions; it depends on the technologies used to produce it. The countries that have big commercial deposits of coal, like the United States, India and China, promote the coal gasification, as energy source, as much for their CCPP and possible future hydrogen production, causing polluting emissions. The technologies of "zero emissions" to obtain hydrogen are mainly: of the biomass by means of pyrolysis, as well as of the water by electrolysis or the thermochemical processes, with no fossil energies.

The technologies to produce energy with "zero emissions" appear examined in detail by Williams (2001). He concludes that the nuclear and renewable power, allow reaching a sustainable economy. However, he puts in doubt that this objective can be reached by dioxide capture and geologic storage, when using the fossil fuels. Equally, Miller and Romney (2004) discussing the use of hydrogen as fuel in the transport sector, conclude that the transition to a sustainable economy happens if three conditions exist: 1) zero emissions. 2) high availability and accessibility of the matter of where the hydrogen will be obtained, and 3) the processes should possess, a system of continuous primary energy, in big quantities and of low cost. They conclude that the first two conditions are only possible when the hydrogen is obtained from the water. The third criterion only happens using the nuclear power.

3 Security of energy supply in Mexico

In the world, the interruptions of the electric power supply associated to the climatic change are more frequently and with greater intensity. The reason derives from the insufficient infrastructure in power plants that are not sensitive to the climate variability. For example, in the American Continent, these inefficient patterns cause periods of multiple power blackouts in: California (2000–2001), Brazil (2001–2002), Argentina (2004), Chile (1998–1999 and 2003–2005) and Uruguay (2005–2006).

Therefore, the dysfunction for the global climatic change has become a decisive factor in the selection of the technologies for the energy offer. When categorizing those for their sensibility to the climatology, it is found that the most sensitive correspond to most of the renewable energy, as the hydro, wind, waves and tides. The fossil fuels would be in an intermediate position. Then, the sources of energy of lesser vulnerability to climatic change are the nuclear, geothermal and biomass under special conditions. In order to solve this problem, they have considered nuclear power generation.

Hore-Lacy (2002) exposed the importance of the nuclear energy in countries that consider the transition towards a renewable energy system. He argues that in the power efficient systems, nuclear and renewable are complementary. The nuclear one contributes stability, viability and security to provide base-load electricity. This does not happen when the renewable energy which depends on the climatology where it is used. On the other hand, the nuclear technology does not count on the flexibility to cover the resounding variations with the electrical demand; the hydro-power plan can cover these variations with high efficiency.

In fact, in Mexico the development nuclear energy has been restricted, and we have limited use of the renewable energy sources. It is a strategy because when integrating them, the high costs of the renewable ones are compensated with the low costs of the nuclear one. With both technologies there could be produced advanced fuels to vehicular, hydrogen and the bio-oil (Bazán-Perkins 2005). Due to the high capacity factors of nuclear, geothermic and biomass energy, their employment will allow supporting the renewable energy whose operation depends of the variations of the climate, to produce electricity, desalination of the seawater and the hydrogen for the transport. With that exposed, it is considered that when maximizing the use of these three energy sources in Mexico, will be able to have security in the base-load energy, with low carbon dioxide emissions and competitive costs of electricity.

3.1 Energy-economic model

In Mexico, the prediction of the long-term national energy consumption is difficult, due to the uncertainty of the economic growth. In past decades, the high discrepancies among the predicted energy demand and the actual one; they were due to the recurrence of economic crisis, as those of 1976, 1982, 1985–1986, 1987 and 1994–1995. For this Century, the growth of the Gross Domestic Product (GDP) is smaller than the one that was projected. This way, SENER (2005) exposes that from the 2001 the growth of the national electric demand has decelerated significantly, causing that the generation capacity is bigger to required optimal level. All these presage is still a very imperfect undertaking, they have considered the prospective growth of the GDP, population, housing, prices of the fuels, inflation, exchange rate, energy saving, pluvial precipitation, among other more.

On the other hand, econometric models that they use a number of mathematical variables smaller than physical variables, has been more effective in recent years than in the past, for the predict of the growth of the national medium and long term electric demand. They were based on historical data, accumulated during more than 100 years. By means of econometric methodology, toward the year 2030 the demand of the Mexican electric system will be of about 537,000 GWh, considering the historical trajectory from 1900 to 2005. For the national transport sector, is considered that toward the 2030 the vehicular park. For the national transport sector, is considered that toward the 2030 the vehicular park will be 40.0 million vehicles, by means of historical data of 1970 to 2005. The effectiveness of these projections is evidenced by its high correlation coefficients, of 0.982 and 0.956, respectively.

The planning process for the expansion of the Mexican electric sector 2007 to 2030, considered for this paper includes the obtaining of the optimal mix of energy resources for electricity generation and their technologies, that guarantee the long term energy supply and at the smallest cost. This way their synergies are developed. The optimal mix of energy resources that minimizes the costs of production of a derived electric system of a single producing agent, like it is the case of Mexico, can be obtained by mathematical models of programming (Massé and Gibrat 1957; Carpentier and Merlin 1982; Dechamps 1983). Therefore, the proposal for the case of Mexico derives from conventional stochastic programming-based methods. The database includes the information projections, for each one of the projected years. In general, during the next decades the new energy plants that take advantage of the renewable and nuclear energy, will continue reducing their production costs. It is expected that these decrements are bigger for the technologies than use solar and tidal power (RWG 2004; Hirst 2006). For example, in Table 1, they are illustrated by technological and economic parameters of new power plants of the year 2005.

Table 1. Technical and economic parameters of new power units, in 2005 (Elaborated wth data of COPAR, 2006; Kutscher, 2006; RWG, 2004; NEI, 2006)

New power plant	Capacity power factor [%]	Total lifecycle levelized cost [¢US/kWh][a]
[c]Nuclear III Gen. (2×1400 MW)	93.0	3.4
Geothermal power (4×26.60 MW)	80.0	4.5
[c]Thermalelectric-biomass (1×1 MW)	80.0	4.0–9.0
Wind power (7×0.750 MW)	40.0	5.0–6.0
[b]Combined cycle gas (1×585 MW)	80.0	5.09
Thermalelectric Coal (2×350 MW)	85.0	5.4
[c]Tide power (1×1 MW)	40.0	6.0–9.0
Thermalelectric Oil (2×84 MW)	80.0	8.6
Hydro power (3×200 MW)	38.0	9.9
Thermalelectric-gas (1×85 MW)	25.0	12.3
[c]Solar power, PV (1×0.5 MW)	26.0	35.0

[a] Life time 30 to 40 years and capital cost 10 to 12%.
[b] Natural gas with subsidy.
[c] These technologies are not still integrated to the Mexican electric system.

Of these results, we can appreciate that the new technologies that take advantage of the nuclear, geothermal, biomass and wind energy would be the most competitive. Of these four, the biggest advantages are for the nuclear energy, when minimizing the electric systems production cost. Currently, the Mexican electric system doesn't include the employment the energy sources derived from biomass, tides and solar (Fig. 1). For the commercial production of electricity from nuclear energy, Mexico has a single plant located in Laguna Verde (Veracruz), integrated by two BWR-5, each one with capacity of 680 MW.

The linear programming model, for this study it contemplates the employment of new power generation technologies, using physical life of 30 to 40 years and capital cost 10 to 12 %. The equations of balance and restrictions of the system are defined by the maximum and minimum benchmarks of the variables of decision. In their group, they associate to restrictions of the power resources, restrictions of the plant factor (technological and operative), restriction of costs, load and peak reserves, balance of the offer, and demand of flow of power. The model is applied for every projected year, includes the parametric analysis of different scenarios. These variables have lineal restrictions, and they structure their participation in an electric system. In symbolic terms:

$$\text{Min } z = f(x_1,...Xn) = (PLLC_1)X_1 + (PLLC_2)X_2 ... + \Sigma (PLLC_n)X_n = PLLC_iX_i; \text{ from i, j} = 1 \rightarrow n$$

- Linear constraints

$$a_{11}x_1 + a_{12}x_1 + ...+ a_{1N}x_n \leq b_1$$
$$a_{21}x_1 + a_{22}x_2 + ...+ a_{2N}x_n \leq b_2$$
$$...$$
$$a_{m1}x_1 + a_{m2}x_2 + ...+ a_{mN}x_n \leq b_m$$

- And the variables may be located between the extreme values

$$x_1^{min} \leq x_1 \leq x_1^{max}$$
$$x_2^{min} \leq x_2 \leq x_2^{max}$$
$$...$$
$$x_N^{min} \leq x_N \leq x_N^{max}$$

$$x_1 \geq 0 + x_2 \geq 0 + x_3 \geq 0 \ x_N \geq 0$$

X_j: variables of decision, j = 1,2...,n.; n is a number of variables; m is a number of restrictions; a_{ij}, b_i and LEC, i = 1,2...,m, are well-known constants. $PLLC_1$, $PLLC_2$, ... $PLLC_n$ correspond to the Projecting Lifecycle Levelized Cost for each technology for electric generation. The $x_1...x_n$; $x_j \geq 0$, j = 1,2..., represent the installed capacity of each technology for electric generation. The restrictions, a_{1N}, $a_{2N}...a_{mN}$ are the technological coefficients that use the same type of power source. Then, $b_1...b_m$, are the vectors of terms to the right or constant associated to the demand, and are limited by the total capacity of each technology to supply electricity.

4 Renewable power program for 2007 to 2015

Mexico enters a new energy economy; its model of expansion of the sustained electrical sector mainly with the natural gas is not sustainable. The nation requires of a new energy program that breaks the scheme that lasts from the beginning of the 70's, where it exists an inadequate optimal mix of power sources for the electrical system (Eibenschutz et al. 1976). In effect, the employment of the fossil fuels is favored and ostensibly decreases the participation of the sources of energy non fossil (Fig. 1). All this, causes high production costs for the Mexican electric system, as well as the necessity of the subsidy to price of the electricity for the consumers. Also, it doesn't allow the diversification of energy sources, the economic growth and deteriorates the country's environment.

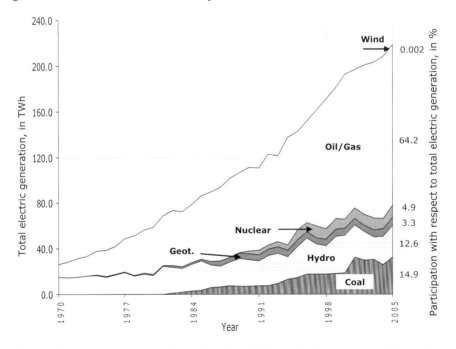

Fig. 1. Electric power generation in Mexico 1970 to 2005, is based mainly on hydrocarbons (Elaborated with data of CEFP 2001; SENER 2003–2005).

Along with the environmental reasons, the economic ones are: in Mexico, it has not been a suitable option to continue the increase of the use of fossil fuels for the electric power, due the high and volatile hydrocarbons price and their small commercial reserves in the country. The mineral coal deposits, they have been of low quality and high use costs. Its use for electrical sector expansion, implies an increase in import of fossil fuels and increased energy dependence. In fact, since the year 2000 hydrocarbon proven reserves of the country, are insufficient to sustain the 30 year-old operative life of new thermoelectric plants, using these fuels. For the 2005, in agreement with statistical data of BP (2005) and the INEGI (2006), proven natural gas reserves in Mexico are sufficient for about 8 years (14.9/1.75 Tpc). The proven crude oil reserves would last 10 years (14.6/1.46 barrels) (Fig. 2).

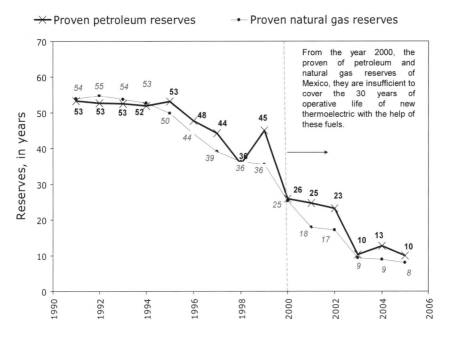

Fig 2. Proven crude oil and natural gas reserves in Mexico 1991-2006 (Elaborated with data of PEMEX, 2001-2005; BP 2005, INEGI, 2002-2006).

Since 1998, the expansion of the Mexican electrical sector was developed mainly using CCPP. This way, starting from 2004 most of electricity of Mexico it is sustained mainly by natural gas. Actually, it is not viable for the expansion of the electrical sector for the inadequacy of proven reserves. Also the high prices of the natural gas in the national market, the costs of production of CCPP in Mexico grew exponentially from the year 2002.

From the year 1990, the CCPP in Mexico lost their competitiveness with respect to the amortized nuclear plants in the United States. Similarly, in 1998, new geothermal power projects in Mexico were more competitive

than CCPP, situation that continues until the present time (Hiriart and Andaluz 2000; Bazán-Perkins 2004). Also, since 2003, the CCPP that operate
with subsidized natural gas prices in México, they have a bigger level total
cost of the electric production that new nuclear power plants of other countries (Bazán-Perkins 2005) (Fig. 3).

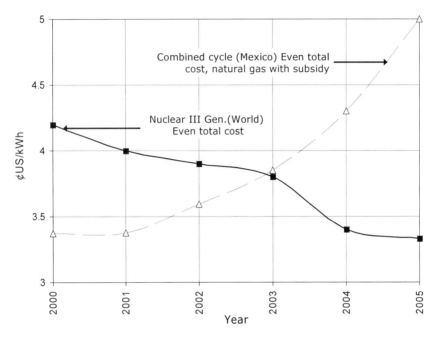

Fig. 3. Even total costs, gas-based combined cycle plant in Mexico and Generation III reactors of other countries, 1995–2005. Cost of Capital 10–12 % (Elaborated with data of Kayak et al., 1998; GE 2000; COPAR 2001–2006; WNA 2003;
UC 2004)

An additional comparison on the last one of competition of the CCPP
with respect to the renewable energy, like the geothermal one, biomass,
hydro and wind, is demonstrated during 2002. For this year, the level cost
of the electric production for CCPP is 3.8 ¢US/kWh (COPAR 2003). At
the same time, according to Mexican's Government (2002) the new projects of renewable energy obtain the following costs: geothermal 3 to 5
¢US/kWh, minihydraulics 3 to 20 ¢US/kWh, wind 3.5 to 4 ¢US/kWh,
biomass 4 to 6 ¢US/kWh. Being the solar, from 25 to 150 ¢US/kWh, the
one that obtain the higher costs.

In sum, the fall in proven hydrocarbon reserves in the country and its
high prices, imposed paradigm shift for model of the Mexican power sector. Conjuncture that represents enormous possibilities so that Mexico de-

velops to other technologies selections, for the energy production, based on the renewable, nuclear energy and hydrogen. When implementing their use, it would be reduced or annulated the greenhouse gas emissions, creating as well new industries and greater opportunities to the national technology (Fernández-Zayas 2001, 2003; Bazán-Perkins 2004, 2005).

In other words, the national energy program for electrical sector expansion, based on the installation of CCPP, represents commitments and different fastenings to the national interests. It avoids the economic growth, reducing social progress and environment is deteriorating. The most worrisome scene is in which it causes the fast exhaustion of the proven reserves of natural gas required for the economic sectors of the country (Bazán-Perkins 2006).

That is, in the measurement that are increased the participation of nuclear, geothermal and biomass energy, the electric power production costs of Mexico would be reduced and the security of the electrical system would be increased. Anther consideration is the use of hydropower, the data reported by COPAR (1997–2006); determine that level total costs of production of hydropower are higher to the CCPP. Nevertheless, from a macroeconomic perspective the hydroelectric power are more competitive, when avoiding the import of great volumes of natural gas (Table 1, Figs. 2 and 3).

Therefore, a change is required in the Mexican energy sector, congruent with the necessities of the country, the tendencies of the technological advances and the worldwide markets. In order to raise the competitiveness and innovation the Mexican energy sector in a short time, the efforts must be oriented to invest in the renewable energy and the nuclear one. In effect, the best strategy is to create for each nonfossil energy sources, its own economy. This excludes to install new thermoelectric energy obtained from coal, petroleum and natural gas.

During the period of 2007–2015, the proposal included to use renewable energy for the expansion of the Mexican electrical sector. These sources include hydro, geothermal, wind and biomass power. Inside these four energy sources, it is required mainly hydroelectric power (Fig. 4). At the same time, it would be developing the new human resources training, to develop plans and contracts for new nuclear power plants. The final decision, to modify the present power policy must be by means of the presidential initiative.

Besides, during this period, would be focused to the economic evaluation of the uranium deposits of the country. The most favorable regions in Mexico based on their potential conditions, are: In the first term, the Tlaxiaco Basin, state of Oaxaca and Guerrero (Mesozoic); Burgos Basin, Tamaulipas (Tertiary); sequence Trancas Formation, Querétaro (Mesozoic), the three in sedimentary environments of continental coastline. On a second hand, uranium deposits associated with volcanic rocks and hydro-thermal, like those of the uranium mining in the Villa Aldama, Chihuahua. In third term, non-conventional uranium deposits of San Juan de la Costa and Santo Domingo, Baja California, when being obtained as a by-product from phosphate rock (Bazán-Perkins 2005).

5 Nuclear and renewable power program for 2015 to 2030

After the period of 2015 to 2030, the Mexican electrical sector must reach a high level of economic viability, without polluting emissions. This is, to maximize the participation of the renewable energy in Mexico, to reach the energy autonomy and competitiveness; it would be by means of a mix of renewable and nuclear energy. To arrive at this objective, it is required that its expansion comes mainly from the nuclear one and the renewable ones. According with the plan, thermoelectric plants with fossil fuels would be substituted progressively in their entirety before the 2030. This proposal represents to reach the sustentability after the Mexican electrical industry, using the technological innovations.

The data in Fig. 4 shows the results of the optimal mix of generation sources of the Mexican electrical system in the year 2030, minimizing the production costs. In general, this mix would consist of a 57 % of nuclear power and 43 % of renewable energy (Table 2). Then, during the 2015 to 2030 the total capacity of the new nuclear power plants for the production of electricity is of the order of 36,000 MW. In this scheme, is considered the extension of the operative life of the nuclear power plant of Laguna Verde (Veracruz) to 60 years. The nuclear technology strategic would be integrated with small modular High Temperature Gas-cooled Reactors (HTGR's), as the one Pebble Bed Modular Reactor (PBMR) and Gas Turbine Modular Helium Reactor (GT-MHR). In addition to the HTGR's, are more attractive by inherent safety characteristics and efficiency operative (Fujimoto et al. 2004; Nakagawa et al. 2004).

According to DOE (2002), PBMR and GT-MHR, are within the 16 technologies of Third Generation that GIF (Generation IV International Forum) recommends for its international implementation of the 2002 to 2015. Tuohy (2006) exposes that new version of the PBMR's, it can be categorized as a first commercial Generation IV reactor, by its technological characteristics of sustentability. An analysis by Bazán-Perkins (2005), of the economic viability of nuclear power plants of III, III+ and IV Generation, concludes that the PBMR's could be a the main technological option to implement in Mexico.

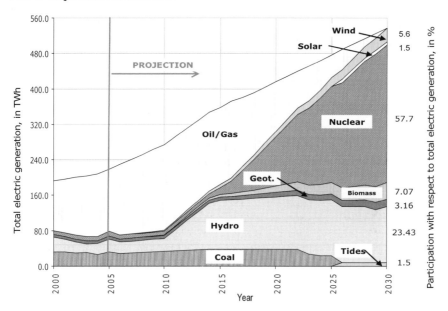

Fig. 4. Proposal for the expansion electric power generation in México 2007–2030

In fact, international interest in HTGR technology has been increasing in recent years due to high efficiency, cost effective electricity generation. Appropriate for the conditions in developing countries. Several international research and development as well as power projects for nuclear hydrogen and nuclear desalination applications are under way in China, France, Japan, South Korea, Russia, South Africa, USA, European Union and others (Methnani 2006).

In summary, the new technology of HTGR's represent a strategic advantage for Mexico. This due to its high versatility, modular construction and significant reducing costs, for production of electricity co-generation, seawater desalination and hydrogen. Still more, its viability is in the program to standardize its construction, according to the required specific necessities for the electrical, transportation, industry, agricultural and residential sectors. Conclusively, for the case of Mexico in where the production costs of the hydroelectric ones are elevated, the small reactors of modular construction could represent the future towards great operative and economic advantages. When settling next to the main regions of consumption, they would make the electrical office more efficient, to cause the saving of energy and water. All this is required to lower or to reduce the production costs of the electrical system due to the reduction of the hydroelectric ones.

Table 2. Optimal mix energy resources: minimizing the overall cost with "zero emissions" proposed for the electric sector of Mexico, at the 2030

Energy resources	Year 2030	
	Optimal power mix [MW]	Optimal power mix [GWh]
Coal	0.0	0.0
Oil	0.0	0.0
Gas	0.0	0.0
Nuclear	38,000	310,000
Geothermal	2,400	17,000
Biomass	5,400	38,000
Hydro	38,000	125,850
Wind	10,000	30,000
Solar	3,500	8,000
Tide	2,300	8,000
Total	99,600	536,850

The proposal for the expansion of the Mexican transport sector 2007–2030, that it bears to reach "zero emissions" toward the 2030 using the hydrogen, would require starting from the 2015 with the installation of a new nuclear capacity of 4,050 MW per year. Then, toward the 2030 with an installed capacity of 64,800 MW of HTGR's coupled with thermochemical plants to drive a sulfur-iodine (SI), to obtain a production of 21,600 t/day of hydrogen that would cover the operation from about 40.0 million vehicles to base combustion cells.

7 Northwest region, opportunity to impel the new clean energy technologies

The Northwest region of the country, embracing the states of Baja California, Sonora and Sinaloa, like an arid and uneven region, it could impel to their barely developed economy. When taking advantage of to their great potential of renewable energy geothermal, solar, wind and tides. Also, by means of the use of non-conventional uranium deposits of San Juan de la Costa and Santo Domingo in Baja California, where uranium would be recovered as a byproduct of phosphoric acid, in the exploitation of the phosphoric rock (Escofet and Castillo 1984; Castañeda 1986; Salas and Castillo 1988; Bazán-Barrón 1992).

In fact, the whole Mexican territory claims the uses of the nofossil energy. Therefore, by means of the external trade Mexico, will have the opportunity of accelerating the technological development of the clean energy, associated to the renewable ones and the nuclear one (PBMR and GT-MHR). In particular, can be potential exporter of the range of energy products required by the market of United States, great consumer of electricity, cogeneration, drinking water and in the future the hydrogen.

8 Conclusions

In the next years, Mexico it will have to make important decisions in their energy sector. Before the exhaustion of its proven reserves of hydrocarbon and their inviability from the year 2000, to use it as main energy source in the expansion of the electrical sector. The new plan should use the domestic energy resources, in where the nuclear and renewable energy and the hydrogen are acquiring greater importance.

Consequently, it is required of a national program of hydrogen, that it should be included in the National Plan of Development of Mexico, for its production, storage, transport, distribution and use. Within a developed chronogram of long term this includes promotion, formation of specialized human resources and financial supports. This program must be in revision periodically by the Senate and Deputies House. The program must include the participation of the international organisms that promote the sustainable economic development with very low carbon dioxide emissions.

References

Barry D and Abhijit B (2004) Energy Policy 34:781
Bazán-Barrón S (1992) GEOMIMET 177:39
Bazán-Barrón S (1994) El Universal Newspaper, 4 oct, Mexico
Bazán-Perkins S (2004a) Geos 24:SS3–37
Bazán-Perkins S (2004b) Geos 24:SE15–14
Bazán-Perkins S (2005) ISSN 1405-7743, VI
Bazán-Perkins S (2006) Edition EÓN, 41
Beér J (2000) Prog Energy Combust Sci 26:301
BP (2005) Statiscal Review of World Energy 2005
Carpentier J and Merlin A (1982) Energy Systems 4:11
Castañeda P (1986) Programa Universitario de Energía, UNAM, Mexico
CEFP (2001) Deputies House, H, CEFP/051/2001, Mexico
COPAR 1998–2006, CFE, Table A1, Mexico

Dechamps C (1983) Hemisphere Publishing Corp, USA, 2001
DOE (2002) GIF-002-00
Dunn S (2002) Int J Hydrogen Energy 27:235
EIA (2006) Annual energy outlook 2006, AEO2006, USA
Eibenschutz J, Guillen S, Fernández-Zayas J (1976) SePaNal, Energy Commission, Mexico
Escofet RA and Castillo FE (1984) PUE, UNAM, México
Fernández-Zayas J (2001) Cuaderno FICA, 1, Mexico
Fernández-Zayas J (2003) Sener, Morelos, Mexico
Fernández-Zayas J (2005) May 18, Reforma Newspaper, Mexico.
Fernández-Zayas J (2006) Mexican Academy of Science, AMC/08/06, Mexico
Fujimoto N, Fujikawa S, Hayashi H, Nakazawa T, Iyoku T, Kawasaki K (2004) IAEA-TECDOC 899:15
GE (2000) Advanced boiling water reactor general, plant description
Hiriart G and Andaluz J (2000) World Geothermal Congress 2000, Japan
Hirst N (2006) OECD/IEA, Tokyo, October 10
Hore-Lacy L (2002) Green Books
INEGI (2002–2006) Bank of economic information, Mexico
Kayak A, Ballinger R, Alvey T, Kang C, Owen P, Smith A, Wright M, Yao X, (1998) MIT ND
Kutscher C (2006) Aspen Climate Action Conference, NREL, October 12
Massé P, Gibrat R (1957) Management Science 3:149
Methnani M (2006) IAEA, TWG-GCR, HTGR, Activities
Mexican's Government (2002) SENER-Gtz, ISBN 970-9983-07-5
Miller A, Romney B, Duffey (2004) Energy 30:2690
Nakagawa S, Takamatsu K, Tachibana Y, Sakaba N, Iyoku T (2004) Nuclear Engineering and Design 233:301
NEI (2006) AGMA/ABMA Annual Meeting, March 3
PEMEX (2000–2005) Proceedings. Operative and Financial Results, Mexico
RWG (2004) Navigant Consulting Inc, reference 119888, December 15
Salas G, Castillo N (1988) Fondo de Cultura Económica, Mexico
SENER (2003-2005) Prospective of the Electric Sector 2003–2012, 2004–2013, 2005–2014, Mexico
SMH (1999) http://www.smh.org.mx/, last access 01.10.2006
Suzuki M (2002) United Nations University, Japan
Tuohy J (2006) American Nuclear Society
UC (2004) University of Chicago, August 2004
United Nations (1979) Protocol to the 1979, November 13, Geneva
United Nations (1996) ISBN: 4-906686-01-X
Williams R (2001) Nuclear Control Institute's 20th, Anniversary Conference, Washington, DC

Nuclear Energy Economical Viability

Gustavo Alonso, Jose R. Ramirez, Javier C. Palacios

Instituto Nacional de Investigaciones Nucleares, Apartado Postal 18-1027, México DF 11801, México. E-mail: galonso@nuclear.inin.mx; jrrs@nuclear.inin.mx; palacios@nuclear.inin.mx

Abstract

Recent construction technique developments and technology evolution have made the nuclear option a cost competitive option with other load base technologies such as coal and combined cycle facilities based on natural gas. Construction period, from first concrete to commercial operation, is around five years, as it has been confirmed by the most recent reactors built in Asia (e.g. Japan and China). At present, the cost for a new nuclear power plant has dropped overnight and is lower than in the past. The different reactors suppliers are offering new plants between 1200 and 1600 USD/kW with an output power between 1100 and 1600 MWe. In this work different scenarios of electricity generation using combined cycles by using natural gas and nuclear power stations are assessed from an economical point of view. The scenarios considered comprise three different discount rates, 5 %, 8 % and 10 %.

1 Introduction

Recent technology developments have achieved improvements in the performance of different nuclear reactor systems and also have made possible quantity reductions of equipment used inside the reactor buildings, thus resulting in smaller nuclear facilities. Along with that, new and more efficient construction techniques have reduced building costs for the latest nuclear plants. As an example, Figure 1 gives an idea about equipment reductions for the Advance Pressurized 1000 (AP1000) reactor from Westinghouse.

The new modular construction techniques permit many construction activities to proceed in parallel. These techniques reduce plant construction agenda, dropping the IDC (Interest During Construction) cost and reducing the risks associated with plant financing. For the so called Generation III reactors the site construction schedule is around 36 months; from first concrete to fuel loading. The most recent reactor development is the Generation III reactor. This category is classified into two types of reactors, namely: those that are an evolution of their predecessors, such as the Advance Boiling Water Reactor (ABWR) from General Electric, the European Pressurized Reactor (EPR) from AREVA and the Advanced CANDU Reactor 1000 (ACR1000) from AECL. The second type of reactor is that using passive safety systems. Among them for example: the Advance Pressurized 1000 (AP1000) from Westinghouse and the Economical Simplified Boiling Water Reactor (ESBWR) from General Electric.

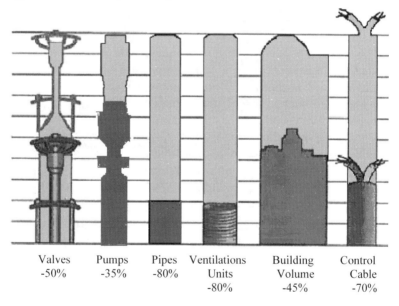

Valves	Pumps	Pipes	Ventilations	Building	Control
-50%	-35%	-80%	Units	Volume	Cable
			-80%	-45%	-70%

Fig. 1. Reductions in equipment for the AP1000 (courtesy of Westinghouse)

Among the fist class of reactors, the ABWR is the only one that has already been built and has operational experience. The first of its kind went into commercial operation in 1996, while the fourth unit came into commercial operation in March 2006, both in Japan. This kind of reactor has proven that it is possible to build them keeping the planned timetable and within the considered budget. Regarding the EPR reactor, its first unit is now under construction at Okiuloto, Finland and it is expected to go into commercial operation by 2009, the second unit will be built in Flamanville, France in 2007.

For the second type of reactors we must mention that the AP1000 is under consideration in USA by the following electricity companies: Duke Power, with two units, Nu-Start Energy with one unit, Progress Energy with 4 units, South Carolina Electric & Gas with two units, and Southern Nuclear Operating Company with 1 unit. It is expected that between 2007 and 2009, these companies will request a combined license to the Nuclear Regulatory Commission. Also in China, license applications for four units were recently requested. As for the case of ESBWR reactors, these are under consideration in the USA by the following utility companies: Dominion, Entergy and Nu-Start Energy with one unit each one of them.

The previous arguments show that the nuclear generation option is present and going strong. In what follows, this manuscript will perform an analysis of the competitiveness of the nuclear option from an economical point of view.

2 Levelized cost methodology

Base-load technology is the one that can operate the whole year the 24 hours per day, among these are coal, combined cycle using natural gas and nuclear. However, each technology used to produce electricity has specific characteristics among these are: construction time, electrical output, lifetime and different cost for investment, operation and maintenance. Due to these differences it is very difficult to perform a comparison between different technologies by only considering one of those characteristics.

One way to perform comparisons among different technologies is to use the levelized cost methodology (IAEA 1984), it allows to quantify the unitary cost of the electricity (the kWh) generated during the lifetime of the nuclear power plant; as it is a mean value, it allows the immediate comparison with the cost of other alternative technologies.

The total levelized generation cost is obtained by dividing the total electrical energy that the power plant will produce in its lifetime between the total cost generated by construction investment along with the interest rate and the cash flow during construction plus the operation and maintenance cost plus fuel cost, everything in present money worth. Therefore, this methodology eliminates uncertainty due technological differences.

Total levelized generation cost is given by the contribution of three concepts. The first one is the levelized investment cost, which comprises investment cost plus interest generated during construction time and indirect costs. Second one is the levelized cost of operation and maintenance, this one comprises direct and variable cost of operation and maintenance and the third one is the levelized cost of fuel. Addition of these three costs gives the total levelized generation cost.

Combined cycle using natural gas has been in the recent past the option mostly used to produce electricity worldwide because the low investment cost in comparison with other options, however the high volatility of gas prices makes wonder about this option. Fuel cost in this option is the major component, thereby in this study the nuclear option and the gas option will be compared under the levelized cost methodology.

3 Assumptions for the analysis

Nuclear power is considered as base-load technology to produce electricity along with the combined cycle plants based on natural gas and the coal plants. Thereby electricity generating cost-comparison can give some guidance about the economical viability of the nuclear option.

One of the main concerns for the investors in Nuclear Power is the construction time and the delays due to the licensing process. The Nuclear Regulatory Commission has modified the licensing process to decrease the possibility to stop the construction of a Nuclear Power Plant due to licensing conditions.

Two scenarios for the discount rate will be considered 5 and 10 %, three for the gas prices 4.44 (Gas 1), 5.20 (Gas 2) and 7 USD/mmBTU (Gas 3). Currently, AREVA, AECL, General Electric and Westinghouse have claimed in public documents that the overnight-cost for kWe can be between 1,200 and 1,600 USD/kWe (Ikehame and Sato 2001; Torgerson 2002; Redding 2003; Bruschi 2004; Twilley 2004).

Thereby, in this study we will consider three scenarios 1,200, 1,400 and 1,600 USD/kWe as nuclear low, medium and high scenarios, respectively. The output power for these reactors is between 1,100 and 1,600 MW, therefore we will use a medium value of 1,350 MW.

For the maintenance and operation costs we will use the ones reported by the OECD (OECD 2003). Table 1 shows all the characteristics of the electricity generating systems used along with their corresponding associated cost.

According to the USA experience, from 20 years ago, the construction time was double in some of their reactor projects. Thereby, there is uncertainty about the likeliness to meet the project schedule for a new nuclear reactor.

To perform a construction sensitivity analysis two scenarios will be considered. The first one will assume that the project will not be stopped but it will take more time to be built. It means that the cash flow will be expanded according to the construction period.

The new CFR10 part 52 assumes separate processes to apply for site permit (ESP), design certification (DC) and the combined license of construction and operation (COL). Figure 2 shows the process to deploy new nuclear reactors in USA.

This new process help to avoid construction stops due to public concerns. The internal testing analysis and acceptance criteria (ITAAC) stage comes after the construction is finished before the commercial operation. At this time, it can happen some delays due to public hearings. Thus we will assume that this delay can be of one, two or three years. This will be the second scenario.

The cash flow curve that we will use in this analysis is a normalized Boltzmann type curve as is shown in Fig. 3. This is a representative curve and it will vary slightly according to a particular vendor. Therefore, the results from this study will give only guidance about the levelized cost expectations.

Table 1. General characteristics of the Electricity-generating Plants

Plant Type	Gas	Coal	Nuclear
Lifetime [years]	25	40	40
Capacity Factor [%]	80	80	90
Proper Uses [%]	3.1	7.3	3.1
Power Output [Mwe]	560	700	1350
Thermal Efficiency [%]	52	37.24	34
Construction Time [years]	2	4	5
Overnight Cost [USD/kWe]	450	1000	1200
			1400
			1600
Fuel Cost	4.44[*]	1.78[*]	6.80[+]
[*][USD/mmBTU]	5.20[*]		
[+][USD/MWh]	7.00[*]		
O&M cost [USD/MWh]	2.77	4.75	7.83

One important issue in reactor operation is the nuclear power plant capacity factor. Currently, in USA, the average capacity factor is around 90 %. However, for a new power plant, the 90 % capacity factor could not be achieved. Thus, in this study a capacity factor sensitivity analysis is also performed.

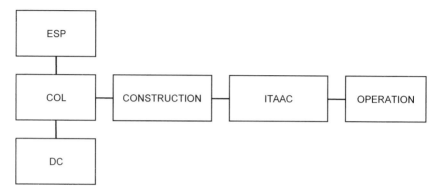

Fig. 2. Licensing process for New Nuclear Reactors in USA

Current Generation III reactors in operation are the ABWR in Japan. From ABWR experience, Kashiwazaki Kariwa 6 has an average capacity factor of 84.36 %, and it started on its first year with a capacity factor of 88.21 %. Kashiwazaki Kariwa 7 has an average capacity factor of 79.237 %, and it started on its first year with a capacity factor of 84.34 %.

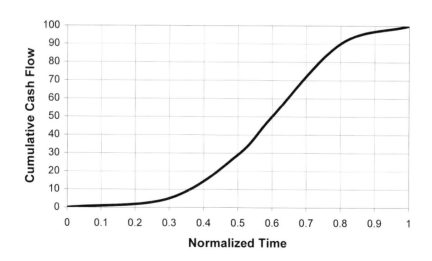

Fig. 3. Normalized cash flow curve

Therefore, our sensitivity scenario will comprise 80 %, 85 % and 90 % power plant capacity factors. It will assume that the reactors are built in a five years period.

4 Results

4.1 Total Levelized Generating Cost

The electricity generating cost for the three base-load technologies (combined cycle using natural gas, coal and nuclear) for three discount rates 5, 8 and 10 % is reported in Table 2.

To perform these calculations we use the data given in Table 1, and we apply the levelized cost methodology. Table 2 also shows the total investment in present worth money and the overnight cost plus investment.

4.2 Time Construction Sensitivity Analysis

The results for the first scenario considered when the cash flow is expanded through the time, 7 and 9 years, are shown in Table 3.

For the second scenario considered when the power plant is constructed but it suffers a delay of one, two or three years before going commercial operation. Table 4 shows the total levelized generating cost if there is a delay of 1, 2 or 3 years. Three different discount rates were considered 5 %, 8 % and 10 %.

Table 2. Levelized electricity generation cost

	Installed Capacity [MW]	Levelized Cost [USD/MWh]			Total Investment [Millions of USD]		
Discount Rate		5 %	8 %	10 %	5 %	8 %	10 %
Gas 1	560	37.86	39.52	40.74	282.42	292.83	299.84
Gas 2		43.01	44.67	45.89			
Gas 3		55.20	56.86	58.08			
Coal	700	32.83	38.16	42.18	882.19	942.80	984.16
Nuclear Low	1350	25.26	30.51	34.53	1933.71	2080.61	2182.96
Nuclear Medium		26.99	33.13	37.81	2256.00	2427.37	2546.78
Nuclear High		28.73	35.74	41.09	2578.28	2774.14	2910.61

Table 3. Levelized electricity generation cost by expanded cash flow

Time		Levelized Cost [USD/MWh]			Total Investment [Millions of USD]		
	Discount Rate	5 %	8 %	10 %	5 %	8 %	10 %
7 years	Nuclear Low	25.69	31.57	36.19	2014.14	2220.36	2367.37
	Nuclear Medium	27.50	34.35	39.75	2349.83	2590.42	2761.93
	Nuclear High	29.30	37.14	43.40	2685.52	2960.48	3156.50
9 years	Nuclear Low	26.14	32.70	38.02	2098.50	2371.14	2570.11
	Nuclear Medium	28.03	35.68	41.88	2448.25	2766.34	2998.46
	Nuclear High	29.91	38.65	45.74	2798.00	3161.53	3426.82

Table 4. Levelized generation cost assuming commercial operation delays

Delay		Levelized Cost [USD/MWh]			Total Investment [Millions of USD]		
	Discount Rate	5 %	8 %	10 %	5 %	8 %	10 %
1 year	Nuclear Low	25.78	31.77	36.49	2030.40	2247.06	2401.25
	Nuclear Medium	27.60	34.59	40.10	2368.80	2621.56	2801.46
	Nuclear High	29.42	37.41	43.71	2707.20	2996.07	3201.67
2 years	Nuclear Low	26.32	33.12	38.66	2131.92	2426.82	2641.38
	Nuclear Medium	28.24	36.17	42.63	2487.24	2831.29	3081.60
	Nuclear High	30.15	39.21	46.60	2842.56	3235.76	3521.83
3 years	Nuclear Low	26.90	34.58	41.04	2238.51	2620.96	2905.51
	Nuclear Medium	28.90	37.87	45.40	2611.60	3057.79	3389.76
	Nuclear High	30.91	41.16	49.77	2984.68	3494.62	3874.02

4.3 Capacity Factor Sensitivity Analysis

Results of the levelized cost and corresponding investment cost for the 80 % and 85 % capacity factors are shown in Table 5.

Table 5. Levelized electricity generation cost assuming different capacity factors

Capacity Factor		Levelized Cost [USD/MWh]		
	Discount Rate	5 %	8 %	10 %
80 %	Nuclear Low	27.28	33.20	37.71
	Nuclear Medium	29.24	36.14	41.40
	Nuclear High	31.19	39.07	45.09
85 %	Nuclear Low	26.21	31.78	36.03
	Nuclear Medium	28.05	34.54	39.50
	Nuclear High	29.88	37.31	42.97
90 %	Nuclear Low	25.26	30.51	32.42
	Nuclear Medium	26.99	33.13	37.81
	Nuclear High	28.73	35.74	41.09

5 Discussions

Because the investment for a nuclear project is high and it requires at least five year to be build, the levelized cost is very sensitive to the discount rate as can be seen in the results shown in the Table 2, where the differences for a discount rate from 5 to 10 % can be up to 45 % higher and the investment cost can be 333 millions of USD higher.

On the other hand, the combined cycle does not show a big difference when different discount rates are used. The difference is lower than 10 % when it goes from a discount rate of 5 % to 10 %.

However, the gas price is a key element in the electricity generating cost. The result shows that when gas price goes from 4.44 to 7 USD/mmBTU the levelized cost varies more than 40 %, no matter what the discount rate is.

One of the main constraints is to achieve the construction time goal. If the cash flow is expanded to a greater period of time the levelized cost will increase up to 11.3 % (from 41.09 USD/MWh to 45.74 USD/MWh for Nuclear High at 10 % discount rate), and the investment cost will increase in 516 million USD.

In the case of a 3 years delay to start commercial operation, the levelized cost will increase up to 21.1 % (it will be 49.77 USD/MWh for the nuclear high at a 10 % discount rate) and the investment will increase up to 33 % (963 million USD).

The sensitivity capacity factor analysis shows that going from 80 % to 90 % the levelized cost can decrease up to a 9 % (45.09 to 41.09 USD/MWh for the nuclear high at 10 % discount rate). It makes a difference of 4 USD/MWh.

6 Conclusions

In the base case analyzed, the levelized cost of the electricity by using nuclear power is lower than the ones given by using natural gas. Assuming that it takes more than 5 years to construct a nuclear power plant, the levelized cost shows that the nuclear option is competitive in comparison with the combined cycle based on gas, for gas prices up to 5.20 USD/mmBTU, and cheaper for higher gas prices.

In the case of the delays up to 3 years for a nuclear project, this option is still competitive with the gas option, and it is still cheaper if the gas price is greater or equal to 7 USD/mmBTU.

The difference by considering an 80 % capacity factor, instead a 90 % capacity factor, in a new nuclear power plant is that the levelized cost will increase by 4 USD/MW.

Therefore the nuclear power is a competitive way to produce electricity in all the scenarios considered, and it should be seriously taken into account in any electricity expansion program.

References

Bruschi HJ (2004) The Westinghouse AP1000 – Final design approved, Nuclear News, November 2004

IAEA (1984) Expansion Planning for Electrical Generating Systems, A Guidebook

Ikehame R, Sato G (2001) The World's first ABWR's: Kashiwazaki Kariwa-6 and –7, Nuclear News, December 2001

OECD (2003) Nuclear Electricity Generation: What Are the External Costs?, NEA4372.

Redding J (2003) Cost, schedule, and risk management – The building blocks of a U.S. nuclear project, Nuclear News, May 2003

Torgerson D (2002) The ACR-700 Raising the bar for reactor safety performance, economics and constructability, Nuclear News, October 2002

Twilley RC (2004) EPR development – An evolutionary design process, Nuclear News, April 2004

Natural Safety Storage of Radioactive Waste

Miguel Balcázar-García, Jesús Hernán Flores-Ruiz, Pablo Peña, Arturo López

Instituto Nacional de Investigaciones Nucleares, Apartado Postal 18-1027, México DF 11801, México. E-mail: mbg@nuclear.inin.mx

Abstract

The public acceptance of an increase program of nuclear energy requires an openly and straight forward discussion, in an understandable way of the main issues against nuclear energy as: nuclear accidents, proliferation of nuclear weapons and safety storage of nuclear waste. Regarding this last issue, there are doubts concerning stability of geological sites to storage nuclear waste as well as possible leakage and migration of radioactive waste from containers, which potentially could contaminate underground aquifers. Technical explanations about safety designs of those nuclear waste storages do not convince general public because of the thousand of year half-life of the radioactive generating material from uranium fuel.

Nature has contributed to present us a wonderful example concerning immobility of nuclear waste, not in a period of thousand of years, but thousands of million of years. This paper describes the discovery of several Nuclear Reactors (NRs) that operated two thousand of million years ago, in the Republic of Gabon, Africa.

The discovery of at least 17 NRs in Oklo uranium mine in Gabon was possible from the analysis of the nuclear waste that remaining undisturbed in that mine for nearly two thousand of million years. The reactors were very small of about ten centimeters in diameter and had a power of around 100 kilowatts and were intermittently operating during 150,000 years.

Xenon gases generated from the reactors were kept in the crystalline matrix of aluminium phosphate glasses. Recent analysis of the whole production and trapping process of xenon gasses permitted to know that reactors were intermittently operating in cycles of about 30 minutes and additionally aluminium phosphate glasses showed the capability of trapping xenon gasses in its crystalline matrix for nearly two thousand of millions years.

This paper presents the "design", functioning and safety storage of nuclear waste of NRs that were performed by interdisciplinary work by nature in Oklo, long before man appeared on earth.

1 Introduction

The population has grown up to a double in the last forty years, increasing the demand for food, water and energy. Therefore, the need of energy supply is not an isolated item, but rather is linked to sustainable development of the world. The eight millennium development goals for the year 2015 consider those three aspects (food, water and energy) as "Reduce by half the proportion of people who suffer from hunger; Reduce by half the proportion of people without sustainable access to safe drinking water and integrate the principles of sustainable development into country policies and programs; Reverse loss of environmental resources". The major percentage of energy production is by burning not renewable fossil resources as oil, coal and gas, raising the atmospheric concentration of carbon dioxide and producing the "greenhouse effect". During the last century global temperature rose $0.6°C$. To reverse this effect most governments have signed and ratified the Kyoto protocol aimed at combating global warming (Wikipedia 2006).

Therefore, energy production in a sustainable way by burning fossil fuel will be drastically limited to reduce greenhouse effect. Several countries have devoted substantial efforts in science and technology to improve the energy production by means of solar, wind, geothermal, biomass energy.

Up to now, the only ready to use massive energy production with zero carbon dioxide emission is nuclear energy. However, the public acceptance of an increase program of nuclear energy requires an openly and straight forward discussion, in an understandable way of the main issues against nuclear energy as: nuclear accidents, proliferation of nuclear weapons and safety storage of nuclear waste. There is sufficient technical information on the Three Miles accident in United States and the forbidden military test of the reactor at Chernobyl in the former Soviet Union, which has to be explained in a simple but reliable manner; at the same time nuclear proliferation has been drastically reduced in the last decade. Regarding nuclear waste, there are doubts concerning stability of geological sites to storage nuclear waste as well as possible leakage and migration of radioactive waste from containers, which potentially could contaminate underground aquifers. Technical explanations about safety designs of those nuclear waste storages do not convince general public because of the thousand of year half-life of the radioactive generating material from uranium fuel.

The purpose of this work is to describe how due to the natural safety undisturbed storage of nuclear waste during almost two thousand million years it was possible to discover the functioning of seventeen NRs in a uranium mine at Gabon, Africa.

2 Uranium: fuel of a nuclear reactor

Coal, oil and gas are the fuels of thermoelectric power facilities, as in the same manner uranium is the fuel of NRs. Uranium is manly composed by two naturally radioactive elements present in nature: U-238 and U-235; each radioactive element decreases its concentration according with a characteristic property called half-life, in such a way that at the end of its half life its concentration is half of that at the beginning.

The decreasing rate for U-235 is six times higher than U-238; therefore the natural elimination of U-235 during the live of earth has been higher than U-238. This implies that when the earth was formed 4,560 million year ago, the relative concentration of U-235 was 25 % with respect to U-238 and that 2,000 million year ago, when NRs started at Gabon, this relative concentration was 3.8 % and at present is 0.07202 %.

At present the uranium fuel in a nuclear power facility has to have a U-235 relative concentration of about 3 % for satisfactory functioning of the facility; this is why it is necessary to build up plants for uranium enrichment to raise the natural U-235 concentration from 0.07202 % to 3 %

The discovery of the natural reactors at the African Republic of Gabon starts in 1972 (Cowan 1976b), from the unexpected results on Oklo uranium mine obtained by H. Bouziques in his laboratory at Pierrelatte, France (See mine location in Fig. 1).

The high sensibility of the apparatus used by Bouziques, allowed him to detect that U-235 concentration was 0.7171 %, lower than the expected 0.07202 %. Miss-functioning of the equipment and other possible reasons for this disagreement were one by one abandoned; the only remaining explanation was that U-235 had been consumed in the mine by "burning it" in a natural NR.

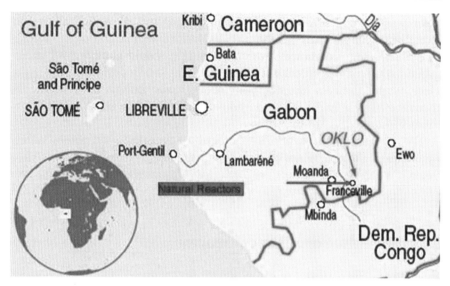

Fig. 1. Uranium ore mine location at Oklo, south-east of Gabon Republic, Africa

George Watherill and Mark Inghram from the Universities of California and Chicago, respectively had been previously proposed the existence of a Natural NR 20 year before Boutiques experimental discovery; however, it was Paul K. Kurada (Kurada 1954) from Arkansas University who described in detail what could have been the physical conditions for a natural reactor to function 2,000 million of year ago.

3 Interdisciplinary conditions for natural reactors at Oklo

The existence of natural reactors 2,000 million year ago, were possible due to the very high uranium concentration storage in Oklo mine of about 50 times higher than other part on earth; the functioning of natural reactors was due to the precise relative U-235 concentration of around 3 % and the accurate design of the reactors performed by nature which kept them functioning for about 150,000 thousand years.

The sketch process is given in Fig. 2. The three blocks at the bottom of the figure show the key sequence for having very high uranium concentration at the mine, the building up the natural reactors and the discovery of those reactors in our present era. The vertical arrows summarize the amazing work in physics, chemistry, geology, and biology performed by nature to achieve each one of the three blocks mentioned above.

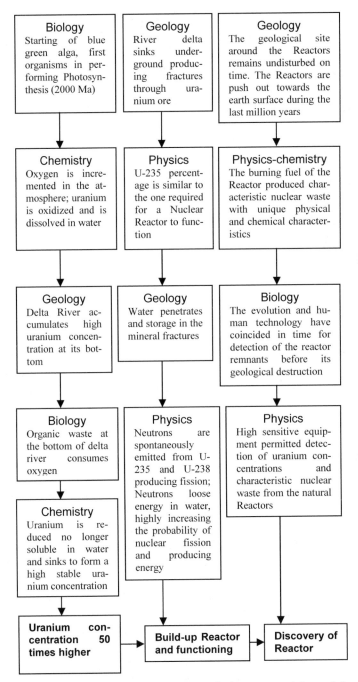

Fig. 2. Nature performed interdisciplinary work that gave origin and functioning of a NR.

Let us to describe each one of the vertical blocks of Fig. 2; starting from the first on the left hand side. The existence of a NR required high uranium concentrations. Two thousand year ago mineral uranium was spread at the south-east of Gabon Republic when "Biology" created the blue green algae, first organisms in performing photosynthesis and as result of that increasing the oxygen in atmosphere (see Fig. 3); the oxidation of uranium by "Chemistry" allowed to dissolve uranium by raining water, which is then transported down to a delta river; the high uranium concentration accumulated in the delta river was in presence of accumulated organic material which consumed oxygen, reduced the uranium, sinking it and kept trapped the uranium ore when the river disappeared; all this process took millions of years.

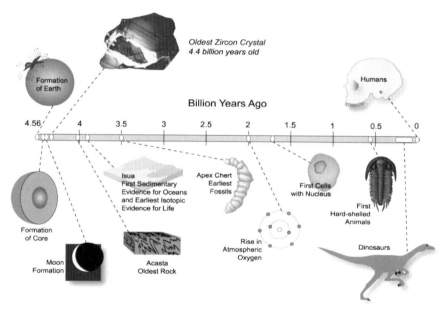

Fig. 3. Main events in earth history. Atmospheric Oxygen rose two thousand million years ago (http://www.geology.wisc.edu/zircon/Earliest Piece/Images/28.jpg)

The building up and reactor functioning followed a sequence outlined in second column of Fig. 2; during millions of years, "Geology" deeply sank the uranium ore fracturing it; at that time U-235 concentration was the required 3 %. Some parts of the uranium ore had spherical form of about ten centimeters in diameter, a correct shape "design" for a NR; natural reactors were triggered on when water penetrated along the fractures slowing down the spontaneous neutron emission from U-235 and U-238 and increasing the amount of fissions that liberated energy from the reactors (two thousand million years before Einstein calculated the total liberated energy). As a result of liberated energy the water rose its temperature to about 200°C without boiling due to the high pressure underground; water temperature went higher, starts to evaporate from the reactor core; then the neutrons are no longer slowed down by water escaping outside the reactor and turning it off; later more water came in the reactor turning it on again. This process kept intermittently running the reactor during 150,000 years.

The discovery of the reactor had again an interdisciplinary action by nature; in spite of the tremendous tectonic and volcanic activity of our planet during thousand of millions years the geological site at Oklo remained covered and undisturbed. In recent times the site is push towards the earth surface; the burning fuel, this is the nuclear waste, remained close were reactors were functioning; civilization and high sensitive devices were well developed to allowed detection of nuclear waste and to determine all the complex process.

The complexity of events shown in Fig. 2 carried out by nature gives an idea of the low probability for the reactors to function and to be discovered. The key of such discovery was the immobility of nuclear waste for thousand of million years without contamination leakage outside the reactors area. If humanity and civilization would have been fully developed in the Precambrian era, uranium fuel would have been ready to use without present enrichment processes. After the discovery of natural reactors at Oklo a science fiction idea was also proposed: a spaceship landed on earth disposed the used reactor, charged another one and returned to the cosmos again.

3.1 Immobility of nuclear waste at Oklo

Continents are formed by several pieces of tectonic plates that are constantly moved and give origin to earthquakes and volcanic activities. The tectonic plates move on the surface of earth at an average speed of 1.2 cm/year (Cowan 1976a). If we compare the earth size with an apple, the thickness of the continents is similar to the thickness of the skin apple; so fragile is our planet! Geologists have proposed that present continental configuration had its origin from a super continent named "Pangea" whose configuration was 225 million years ago (in the Triassic) as shown at the left in Fig. 4. At the right in Fig. 4 is represented the continental configuration 100 million years ago, Mexico and Europe were partially covered by water; Australia was together with the Antarctic continent.

It is surprising the geological stability of Oklo uranium mine at Gabon Republic, where natural reactors started 2,000 million year ago long before the continent separation. Gabon is one of the oldest and geological stable places on earth.

Fig. 4. 250 millions ago there was one continent "Pangea" (left). 100 millions years ago, Mexico and Europe were partially covered by water; Australia was together with the Antarctic continent (right). (Source: http://homepage.mac.com/uriarte/historia.html)

The discovery and operation characteristics of NRs at Oklo were possible due to the immobility of the nuclear waste at the reactors site for thousands of million years in spite of the tremendous volcanic and tectonic activity of the planet.

Neodymium (Nd) is and element that gave evidence of the existence of natural NRs. Neodymium is composed by "several neodymiums" (neodymium isotopes), all of them with the same chemical properties but with different masses and physical characteristics. Neodymiums are naturally encountered in the environment, and are also produced by "burning" uranium in a NR. However, the relative mass composition of both Nd is different.

Figure 5 shows the percentage mass composition of Nd: naturally encountered in environment (black colour), the one produced in a NR (light grey colour) and the one found at Oklo (dark grey colour). Nd-142 is the only one that is not produced in a NR. Because the Nd-142 concentration found at Oklo (dark grey) was small, it was expected that the rest of Nd (dark grey) had even lower concentration than Nd-142, in the same proportion that the natural content of Nd-142 (black) had with the rest of its similar Nd. This disagreement clearly showed that the Nd found at Oklo did not belong to the natural environment.

The hypothesis was then, that Nd found at Oklo (dark grey) was produced in a NR (light grey).

The straightforward percentage comparison of Nd produced in a NR (light grey) with the Nd found at Ocklo (dark grey) did not coincide. But, Nd formation in a nuclear is the result of radioactive elements from "uranium burning" which are spontaneously and successively transformed in other radioactive elements until the final product is neodymium; the time process of these transformations is very well defined by nuclear physicists; therefore after this process correction (white), the comparison with the Nd percentage found at Oklo did agree. Additionally the necessary time correction involved, gave the age or time when natural reactors were functioning at Oklo, between 1700 and 1900 millions years ago!

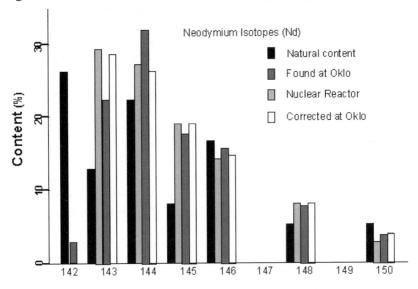

Fig. 5. Percentage comparison of Neodymium abundance from different sources at Oklo uranium mine. (Source: http://en.wikipedia.org/wiki/Oklo_Fossil_Reactors)

3.2 Operation characteristics of nuclear reactors at Oklo

Recently Meshik et al. (2004) were able to determine the operation characteristics of natural reactors at Oklo from the analysis of the non mobility of its nuclear waste. To his surprise, they found xenon gases, trapped in aluminum phosphate glasses, in samples few millimeters apart from the Oklo mine, where natural reactor number 13 was located. They were capable of detecting very small amounts of xenon masses using a high precision instrument. Similar to Nd, xenon (Xe) is composed by several masses of the same chemical characteristics but different physical ones.

One of the xenon gases trapped in the crystal array of aluminum phosphate was Xe-130; this is a non radioactive element, not directly originated from uranium fuel in the NR; its origin is the following: uranium fuel consumption in the reactor produced Iodine-129, which, was immediately trapped in growing crystals of aluminum phosphate glasses, then neutrons from the reactor bombarded Iodine-129 converting it in radioactive Iodine-130 which in turn changed to Xenon-130. The whole process can be summarized as:

Uranium consumption → ^{129}I production + neutron bombardment → 130I conversion → 130Xe production.

The physical properties of radioactive materials and the knowledge of crystal-growing properties were the key in defining the process mentioned above. Aluminum phosphate crystals rapidly grow in hydro environments rich in minerals at temperatures of 270–330°C. Water was present at those temperatures in the surrounding areas of the reactors due to the high pressure underground at the reactor site.

The operational sequence of the NR was then determined not only from Xe-130 analysis but from the rest of the "xenons" trapped in the aluminum crystal matrix (Meshik et al. 2004): Water from the earth surface permeated underground through the fractures into the reactor core, slowing down the energy of neutrons emitted by the uranium fuel and increasing the amount of energy produced in the core of the reactor. As the energy increases the reactor triggers on; water rises its temperature in liquid state well above 300°C without boiling due to the high pressure underground. Then the reactor keeps operating around 30 minutes. Iodine-129 is produced during reactor operation, among other radioactive elements. Water temperature rises and starts to evaporate. Without the water barrier, neutrons increase their velocity and escapes out of the core of the reactor turning it off, leaving behind small drops of water with Iodine-129 dissolved inside.

The cycle repeats over again, that is to say; more water, abundant with phosphorus and aluminium minerals, slowly penetrates into the pressured and hot environment around the reactor. Subsequently aluminium phosphate glasses begin to grow, trapping in its crystalline structure Iodine-129. After certain time, water fills in most fractures in the reactor core and the reactor is set in operation again. At that moment neutrons from the reactor impinge the Iodine-129 nuclei, trapped in the crystalline structure, transforming them into radioactive Iodine-130, which in turn spontaneously transform to the stable element Xe-130, the latter nuclei being trapped in the crystalline matrix of aluminum phosphate glasses for nearly 2,000 million years.

Plutonium-239 is one of the characteristic radioactive wastes from the reactors, which no longer exist at Oklo; nevertheless, Plutonium-239 was spontaneously transformed into other radioactive and stable elements that remained in the vicinity of the reactor. A study carried out on reactor 10 at Oklo by Bros et al. (1993) found that Plutonium-239 descendants migrated only 30 cm outside the reactor site in 2,000 million years!

There have been many research papers concerning the discovery, functioning and properties of the 17 natural reactors at Oklo. The key success of those studies is the geological stability that kept intact the resulting nuclear waste. There are many stable geological sites on our planet that have been identified by geologists. Because of one of the geologically stable sites is at Australia, it was possible to date the oldest zircons on earth from several samples containing small pieces of zircons, collected at Jack hills, west part of Australia. The age of those zircons was of 4,400 million years (Valley et al. 2002), in other words 2,400 million years older than the natural reactors at Oklo, just before the moon formation, as it can be seen in the time scale shown in Fig. 3.

4 Conclusions

The geological stability of Oklo uranium mine through thousand of million years allowed scientist to discover and understand the operation characteristics of seventeen natural reactors developed by nature. The fact that at Oklo, nature kept nuclear waste concealed and without any leakage of radioactive material gives mankind an option for a definite solution to the nuclear disposal problem.

References

Bros R, Turpin L, Gauthier-Lafaye F, Holliger PH, Stille P (1993) Ocurrence of Naturally Enriched [235]U: Implications for Plutonium behaviour in Natural Environments, Geochemical et Cosmochimica Acta 57:1351–1356

Cowan GA (1976a) The Break up of Pangaea. Sci Am 223:30–41

Cowan GA (1976b) A natural fission reactor. Sci Am 235: 36–47

Kurada PK (1954) On the Nuclear Physical Stability of the Uranium Minerals. Jour Chem Phys 25:781–782

Meshik AP, Hohe CM, Pravdivtseva OV (2004) Record of Cycling Operation of the Natural Nuclear Reactor in the Oklo/Okelobondo Area in Gabon, Physical Review Letters 93:182302-1–182302-4

Valley JW, Peck WH, King EM, Wilde SA (2002) A Cool Early Earth. Geology, 30:351–354

Wikipedia (2006) http://en.wikipedia.org/wiki/Global_warming

The Reinassance of Nuclear Power

Luis C Longoria-Gandara, Jaime Klapp, Gustavo Alonso, Salvador
Galindo

I Instituto Nacional de Investigaciones Nucleares, Apartado Postal 18-
1027, México DF 11801, México. E-mail: longoria@nuclear.inin.mx,
klapp@nuclear.inin.mx, galonso@nuclear.inin.mx,
sgu@nuclear.inin.mx

Abstract

Nuclear power set off commercially in the 50's, with very high hopes in
Russia, the United Kingdom and the United States among others. Under
the oil crisis, France started a very aggressive program to generate about
70 % of its electrical production from nuclear power. However, in the
eighties, as a result of the Three Mile Island accident, several safety concerns were raised, resulting in considerable construction and operation delays. Many planned reactors were left behind and after the Chernobyl accident, reactor construction was stopped in the western world. However in
the meantime, the Industry made improvements in safety measurements
and performance of its nuclear power facilities. The capacity factor of nuclear plants grew to be around 90 %. The operation licensees of many units
were renewed for another 20 years after their first 40 years of operation.
Today there are new expectations for nuclear power not only because its
economical competitiveness but its potential to produce electricity without
greenhouse gas emissions consequently making feasible to achieve Kyoto
protocol goals to alleviate global warming. This paper shows the potential
of nuclear power as a clean, safe and innovative energy.

1 Introduction

Nuclear power started commercially in the 50's, with such optimism that someone once said that nucleoelectricity assured forever the world energy supply, and that it would be so cheap that no one would ever care any longer to install meters. Soon Nuclear power reactors were constructed in the United Kingdom, the United States, Russia and other countries. It seemed that some countries decided to adopt nuclear power as the answer for their energy needs. However, accidents such as the one in Three Mile Island and Chernobyl virtually stopped the construction of new reactors in the West, and produced a shocking image of nuclear power, to the public opinion

After the Chernobyl accident in 1986, only countries in the Far East continued with the construction of new nuclear plants, however the improvement of operation of the existing plants and the development of new concepts for better reactors never stopped. Countries like France adopted the nuclear option as its main source of electricity. This happened after the 70's oil crisis and France has continued along this path ever since. The United States together with Canada and some other countries in Europe and Latin America adopted the strategy of "wait and see".

Nuclear power electricity accounts for one sixth of the total electricity produced in the world. This is the same proportion as electricity produced from hydroelectric facilities. In recent years the West has increased the amount of nuclear electricity through increasing the existing reactor efficiency and upgrading mayor plant components. With 80 % of all the electricity produced in the country, France is still in proportion the largest nuclear power producer, however in terms of power generated, the United States is the largest producer in the world operating more than 100 reactors. The design of nuclear reactors in the West greatly differs from the concepts developed by the ex Soviet Union, however, nowadays there is more worldwide sharing of nuclear technology and many concepts have been standardized.

In recent years as the world's energy demand has increased and oil and gas prices have dramatically grown. For the time being no new mayor alternative energy sources have been found. This situation has forced many countries to have a second look at the nuclear power option.

In 2005 Finland, was the first country from the West (excluding France) to start the construction of a new 1,600 MWe plant. By then, France had already started the development of a new site for a nuclear plant and the UK, in the summer of 2006, decided to retake the nuclear option.

The United States took a diverse approach, on the one hand, by steadily increasing its nuclear capacity as a result of upgrading existing plants and through extending the operation lifetime of their active nuclear plants. It must be mentioned that the United States Nuclear Regulatory Commission (USNRC) has been extending operating plant licences to 20 more years in addition to the present 40 years licences. On the other hand, applications for licensing construction sites for new reactors in the United States have been already approved and more are expected. In 2005 the United States introduced a new bylaw in its energy policy aimed at giving economic incentives to the first new nuclear plants constructed.

2 Electricity Economics

Certain fossil fuel plants are economically competitive when compared to nuclear power facilities. Some combined cycle Gas plants have low capital costs, can be ordered almost "off the shelf", and can be bought in very short periods of time, however their operating costs are higher than a nuclear plant and a large percentage of these costs are related to fuel costs. Gas prices have been steeply increased in recent years and with a higher volatility. The operating cost of a coal plant is also high but coal prices have remained steady for the last few years, however new coal plants will require systems for capturing carbon dioxide which will increase the coal electricity prices substantially.

Nuclear electricity is much less dependant on fuel prices, and although uranium prices have been increasing in recent years, the operation cost of a nuclear plant is less than that of a Gas plant and only 20 % of its operation expenditures are fuel costs. At current fuel prices, nuclear electricity is cheaper than gas and about the same as coal generated electricity, however nuclear electricity prices have included external costs such as dealing with waste, if carbon dioxide emission bonus are included, nuclear electricity is clearly the most economic option.

Recent studies suggest that nuclear electricity cost about 0.06 USD per kilowatt hour whereas gas generated electricity is about 0.07 USD per kilowatt hour. Coal generated electricity costs are around 0.05 USD per kilowatt hour. The mayor hurdle to the construction of nuclear plants from the economics point of view is the high capital cost, however lowering construction expenses, reducing construction times and lowering operating costs will reduce nuclear electricity considerably. The cost of a 1,200 MWe nuclear power plant is estimated to be about 2,500 million USD.

3 Nuclear electricity and the environment

One of the single most important issues regarding the environment is global warming. This is considered the cause of mayor world weather changes, emissions of carbon dioxide is a mayor contributor to Greenhouse gases to the atmosphere. Worldwide governments have, in principle, agreed to reduce carbon emissions. However, reducing these emissions while supplying the energy needed to sustain economic development is an enormous challenge. At present time the burning of oil contributes 43 %, coal 37 %, and gas 20 % to carbon emissions, which are estimated to be about 7 billon tons of carbon per year. Roughly every 2 billon tons of carbon corresponds to one part per million of carbon dioxide and in 50 years the concentration of carbon in the atmosphere will reach 1,200 billion tons.

Nuclear electricity can play a mayor role in reducing carbon emissions to the atmosphere, it is estimated that the use of nuclear power in a year save at least the CO_2 emission of 1,400 million tons to the atmosphere. Table 1 shows the energy equivalent among the main fuels used to produce electricity, also it shows the amount of CO_2 by every 1,000 MWh generated, it includes efficiency factors by source. Also, Fig. 1 shows the CO_2 emissions from 1950 to 2002, according to the Carbon Dioxide Analysis Centre from Oak Ridge National Laboratory. Furthermore, the greenhouse gases are also the NO_x and the SO_2, Table 2 shows, as an example, the amount of greenhouse gases avoided by the use of nuclear power. The example corresponds to the United States of America, which has 103 nuclear power reactors that contribute with around 20 % of the electricity generated per year in that country. Figures 2 and 3 show the greenhouse gas emissions during electricity generation and from the other stages needed to produce and transport the fuel to the site. In particular Fig. 2 is given for base load capacity; it means the sources that can provide energy during the 24 hours of the day, the 365 days of the year. Figure 3 is given for renewable sources. From the same figures it is also clear that Nuclear Power is the one that produces the lowest amount of greenhouse gases, if it is considered all the stages to generate electricity and in addition, it does not produce any greenhouses gases during the generation stage. Nuclear Power can be compared with Hydro and Wind Power for their null Greenhouse gases emissions.

From the previous paragraphs it can be concluded that Nuclear Power is a clean energy that can help to lessen the climate change.

Table 1. Energy source emissions during energy generation. Note that nuclear energy has nil emission

Source	Energy Content	Amount to produce 1 GWe /year	CO_2 released Per 1 GWh
U-235 (1g)	8.2×10^{10} J	3.6 t	0
Oil (1 barrel)	6.3×10^{9} J	15 M barrels	798 t
Coal (1 ton)	2.9×10^{10} J	3 M t	940 t

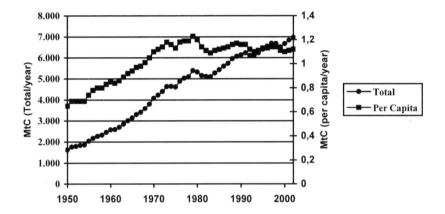

Fig. 1. Worldwide CO_2 emissions released by the use of fossil fuels, 1950–2002

Table 2. Greenhouse gases not released by the use of nuclear power in USA

Year	NO_x [t]	SO_2 [t]	CO_2 [Mt]
1995	2.03	4.19	670.6
1996	1.89	4.16	645.3
1997	1.76	3.97	602.4
1998	1.76	4.08	646.4
1999	1.73	4.13	685.3
2000	1.54	3.60	677.2
2001	1.43	3.41	664.0
2002	1.39	3.38	694.8
2003	1.24	3.36	679.8
2004	1.12	3.43	696.6
Total	15.88	37.71	6657.95

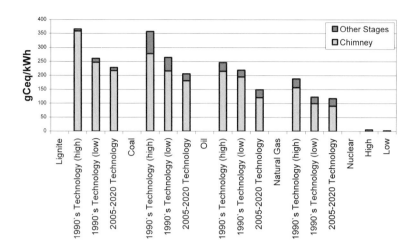

Fig. 2. Greenhouse gas emissions from base load electricity sources

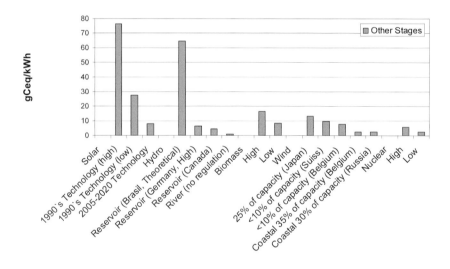

Fig. 3. Greenhouse gas emissions from renewable sources

4 Radioactive wastes

One of the mayor obstacles, that nuclear power faces, is the waste management issue. Although technically this problem can be solved through natural and engineering barriers, governments and the public are still not convinced. The radioactive waste comes from the different applications of nuclear energy and can be by itself radioactive or can be mixed with radioactive substances. The radioactive waste according to its radioactivity level can be divided into low, intermediate and high level waste. It is very important to mention that the radioactive waste is always confined, whereas in other industries there is not a strict control of the waste that is produced. Table 3 shows the amounts of waste by human activity.

There are technical solutions to deal with the radioactive waste, it can be stored or disposed. In the first case, the storage of radioactive waste is a solution for low level waste because the waste is collected in a suitable site for a period of time long enough to avoid any leakage or outside contamination. Because that type of waste has isotopes with small activity and short half life, it can be stored in repositories close to the surface using adequate physical barriers. This type of solution can last up to 300 years; currently there are around 70 facilities in the world.

Table 3. Waste produced by human activities

Waste Type	Amount [Tons]
Agriculture and Domestic	50 millions
Industrial	3.5 millions
Toxic and Dangerous	0.35 millions
Radioactive	100

The high level waste is the depleted nuclear fuel used to generated electricity. It has high activity and long half lives, currently all of them are stored and confined in repositories that offers intermediate solutions as pools inside the reactor building and dry-well storages. To be able to dispose the high level waste it is necessary to guarantee that the repository will last at least 10,000 years confining it. It means that the repository will be in a stable geological site in a very deep underground. To reduce the amount and volume of high level waste, the spent nuclear fuel has to be reprocessed, thereby concentrating the long lasting radio nuclides. This technology adopted in England and France has the advantage of reducing waste volume and reusing the unburned uranium as well as producing plutonium fuel.

Up to now there is no a single disposal repository for radioactive waste. It has been due to policy and public opinion problems. There is technology to deal with the management of radioactive waste, as an example of this technology is the Yucca Mountain site in the United States which plans to start receiving waste by the year 2015. Figure 4 shows this type of technology.

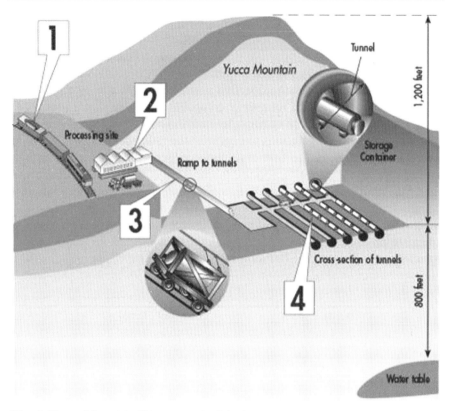

Fig. 4. Yucca Mountain Site, conceptual design

The process to deal with the radioactive waste in Yucca Mountain is as follows:

1. Containers transportation in special casks
2. Removal of special casks
3. Automatic tunnel introduction
4. Storage in parallel tunnels, at 366 m from the land surface

It can be seen in Fig. 4, that the ground water is 244 m under the storage tunnels, this guarantees that there will be no water contamination. As it is presented this is a viable solution. However, as it was mentioned before, the problems arise from the public opposition. Yucca Mountain Site, has established its feasibility for 10,000 years and now as consequence of public pressure it needs to show its feasibility for 100,000 years.

5 Nuclear future

Reactor types are usually classified by generations, the generation I being the reactors built till the early 60's, generation II reactors till early 90's and generation III till today. Generation IV reactors are in the drawing board now, these will come into operation in about 30 years from now and will have advantages such as increased safety, more efficient and economic, and less suitable for nuclear proliferation. These reactors can also include technology to utilise the reactor heat to desalinate water as well as to produce hydrogen. The United States along with other six countries have currently a program called the GIF initiative, "The Generation IV International Forum", which is working on six different concepts of new reactors. The International Atomic Energy Agency has also formed a group to study the design of new reactors. There are reactors considered to be generation III plus, such as the design of the IRIS (International Reactor Innovative and Secure) reactor from Westinghouse which incorporates some features of generation IV reactors but that can be build with today's technology. These types of reactors will be available for construction by the year 2020.

The current inventory of uranium in the world guaranties the supply of nuclear fuel for many years to come, and if new designs reactors can burn thorium or use the spent fuel from the uranium cycle, the amount of fuel will last hundreds of years. For the next few decades however, the reactors been constructed will be primarily generation III and III+ reactors. Figure 5 shows the evolution of the reactor types classified by generations. Against a background of rising demand for energy, mainly in developing countries, and increasing interest in reducing greenhouse gas emissions from fossil-fired plants, the Generation IV nuclear reactors are coming not only as a solution but mainly as a challenge. Nevertheless, a great number of issues concerning new technologies and innovative R&D activities, including fuel and management, have to be addressed and solved to ensure full participation of nuclear power in the future competitive energy market. Attention has been focused on small and medium reactors and their various combinations of relatively simple design, fuel economy, reduced sitting cost and long life cores, with enhanced safety and proliferation resistance features. On these arguments, it is not difficult to foresee Generation IV Fast Reactors as a very good alternative to face the nuclear challenge (dos Santos 2002).

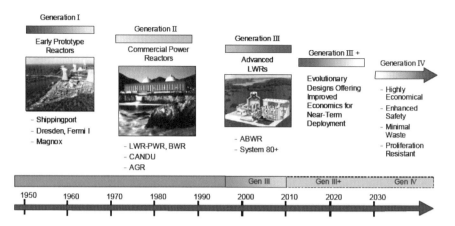

Fig. 5. Evolution of the reactor types classified by generations

6 Conclusions

The commercial nuclear power industry started in the 50's in several countries, including Russia, the United Kingdom and the United States. For a number of years many nuclear power plants were built, but after some nuclear accidents, in particular those in Chernobyl and the Three Mile Island, the growth of the nuclear industry almost stopped.

From the analysis presented in this paper, it can be concluded that currently the nuclear option is the best alternative to generate base load electricity, taken into account the main global concerns about environmental pollution, safety of supply, and economic competitiveness. However, there are some public and political aspect that must be overcome, as public risk perception and acceptance of nuclear power, together with the risk of nuclear proliferation in some countries. With regard to the nuclear waste, there are technical solutions that must be implemented but mainly the reprocessing technology of spent fuel has medium and long term significant benefits. The current inventory of nuclear fuel in the world guaranties the nuclear energy generation for many years to come, and if new design reactors can burn thorium or use the spent fuel from the uranium cycle, the amount of fuel will last hundreds of years. For the next few decades however, the reactors been constructed will be primarily generation III and III plus.

The goal of the Generation IV initiative is to work on an international basis to identify, assess, and develop sustainable nuclear energy technologies that can be licensed, constructed, and operated in a manner that will provide a competitively priced supply of energy while satisfactorily addressing nuclear safety, waste, proliferation resistance, and public perception concerns of the countries in which they are deployed.

There are also the coming fusion reactors that for a number of years are promising an efficient, clean and safe nuclear alternative. Unfortunately this is still far into the future.

References

dos Santos W (2002) A Introduction to Fast Reactors of Generation IV, Nuclear Energy Institute, CNEN

Part III

Alternative Energy Resources

Renewable Energy in Mexico: Current Status and Future Prospects

Jorge M. Huacuz

Non-Conventional Energy Unit, Electrical Research Institute (IIE), Av. Reforma 113, Cuernavaca, México, C.P. 62490, Phone/Fax: +52 777 362 1806. E-mail: jhuacuz@iie.org.mx

Abstract

This paper reviews the current situation and the future prospects for the application of renewable energy in Mexico. It shows that, in spite of the abundance of renewable energy resources, generation of electricity and other non-electric applications are minimal. Opportunities to use renewables as part of the Mexican energy mix are many, and could bring a number of benefits, social, economic, political and environmental, among others. Barriers to do so are also many and are outlined here. It is concluded that Mexico is lagging behind other countries of similar economic capacity with respect to the development of its renewable energy resources, and that a concerted action among different sectors of the economy is necessary to alter the present situation. Otherwise, the opportunity ahead will be lost and Mexico will remain a net importer of new energy technologies.

1 Introduction

The use of renewable energy (solar, wind, biomass, ocean, geothermal and others) is a major objective in many countries. Technologies for the conversion of renewable energy resources into electricity, heat and cold have undergone substantial progress over the past two decades: systems efficiency and reliability have improved, while life-cycle costs have decreased and markets have expanded. For instance, the world-wide capacity installed in wind generators now exceeds 50,000 megawatts (MW), from a mere 1,500 MW fifteen years ago (IEA 2005); global sales of photovoltaic (PV) generators in 1995 amounted to almost 1,093 MW in a market that practically did not exist some 20 years ago (IEA 2006). Renewables offer a number benefits more than energy. They include new jobs, a cleaner environment, energy security and others. In view of this, most OECD countries and a number of developing nations have set targets for renewables in their energy plans. Many of them have also fostered the development of local renewable energy industries, and are aggressively entering this new technology market. In contrast, Mexico has remained dormant in this field. Mexico trails the rest of the OECD countries in terms of practical applications of renewables, as well as on the necessary programs and infrastructure to foster their development. At the same time, the gap between Mexico and other nations of similar development, such as India, Brazil and China, is growing; even smaller economies, such as Costa Rica, are more advanced than Mexico in this field.

In this paper the present situation in Mexico regarding the use of renewable energy is described. Future prospects in several fields of application, including electric and non-electric uses, are also discussed. This review focuses on three main aspects: availability of renewable energy resources; local technology stocks and current applications; and the required enabling legal and regulatory framework. Human capital, sources of financing and the institutional setting, three most important elements for the development of any national project, are published elsewhere (Huacuz 2005).

Mexico is endowed with plentiful renewable energy resources. However, with the exception of geothermal energy and large hydropower, they remain virtually untapped. Reasons for this are many and include: oil and associated fossil fuels are also plentiful; per-capita electricity consumption is still low; environmental concerns have not been high enough in the country's agenda; the cost of renewables is perceived to be higher than that of conventional energy; a set of direct and hidden subsidies distort the Mexican energy market. Nevertheless, as the economy grows and the standard of living increases; and as local environmental awareness grows, the need for more and cleaner energy will certainly grow. Hence, Mexico needs to draw a strategy to tap renewable energy resources. In this regard, the present petroleum reserve can be used to promote the transition to a cleaner and more sustainable energy base. It can also buy time for the local industry to enter the new energy business, not only as users, but also as producers, and even exporters, of new energy technologies. Revenues from oil exports can also help to create the infrastructure (human, technical and institutional) necessary to support the transition. But as time goes by and no firm action is taken, this opportunity may be lost.

2 The present situation

Interest in the application of new renewables has been traditionally low, but a number of changes in the landscape have occurred in the past few years: the number and importance of stakeholders promoting and supporting renewables has increased; the institutional and regulatory frameworks have slightly improved; and a variety of projects are at different stages of development. Nonetheless, progress has been slower than desirable for a number of reasons which will be addressed later in this paper. At the present time, renewables represent around 7.1 % of the overall energy supply in Mexico. A bit less than one third of this contribution comes from firewood, used in rural areas mostly for cooking and space heating. Salt production in open pans, such as the Guerrero Negro salt works in the Baja California peninsula and other low technology uses of solar energy such as open-air drying of agricultural products, are substantial but not accounted for in the energy balance. In the electric sector, slightly over 20 % of the total electricity for public service comes from large hydroelectric power plants (16.09 %), geothermal (4.25 %), and a tiny portion from wind and solar photovoltaics (SENER 2006).

The Energy Sector Program 2001–2006 calls for an increased use of the nation's renewable energy resources, with a target of 1000 MW of new capacity for electricity generation from renewables, additional to CFE (Federal Electricity Commission)'s planned renewable energy projects (SENER 2001). Unfortunately, this goal did not materialize for a number of structural reasons to be discussed later in this paper.

2.1 The renewable energy resource base

Renewable energy resources are abundant in Mexico. However, availability of site-specific and detailed information on the new renewables (i.e. large-scale hydro and high-temperature geothermal excluded) is very limited, and usually not good enough to support commercial green power projects. Available information shows the following:

- Synoptic solar radiation and wind maps are available (Hernández 1991). Solar irradiance is excellent throughout the country and has an average density around 5kWh/m^2-day.

- Several regions with good wind potential have been identified; preliminary figures indicate around 5,000 MW that could be now economically tapped, but according to some experts, further exploration could add up to 15,000 MW in new inventories (Borja et al. 1998).
- The full potential of small hydro resource is unknown; some estimates indicate that at least 3,550 MW could be harnessed at several river basins (CONAE 2002).
- The potential of biomass has not been fully assessed: recent estimates indicate that the theoretical potential for large-scale electricity generation with bioenergy in this country is substantial. The technical limit for power generation is estimated at around 23,500 MW, equivalent to about 50 % of the total generating capacity currently installed in Mexico. However, after taking into account a number of non-economic factors that could limit project development, the current practical limit for bioelectricity can be set at around 1,790 MW (IIE 2005). In calculating this figure, only the biogas route from sanitary landfills was considered to recover energy from urban solid waste. The use of other processes such as combustion and gasification could substantially increase this potential.
- Ocean energy (tides, waves, currents, etc.) could represent a substantial resource for power generation along the over 11,000 kilometers of Mexican coastline; however, practically no work has been done to assess this resource in its entirety.

Activities to overcome this situation are still limited. They include the implementation by the Electrical Research Institute (IIE) of Mexico of a reference anemometric network with stations at several sites around the country, as part of a project financed by the Global Environment Facility (GEF) through the United Nations Development Program (UNDP). Data from these stations are publicly available on the Web and are meant to help removing the important barrier represented by the lack of reliable information for project development (IIE 2007). A number of private companies interested in the development of commercial wind projects have also implemented anemometric stations of their own at several sites around the country.

Evaluation of other renewable energy resources is currently done either on a site-specific basis or through indirect techniques, such as mathematical modelling for full regions or the whole country, since fieldwork for this purpose is too expensive and resource-consuming. A geographical information system for renewables (SIGER) has been under development at IIE for the past few years. It has the objective of providing good and timely information to facilitate the development of commercial renewable energy projects and to support planning and decision-making processes. Maps of the various energy resources, built from a combination of indirect methods and field data where available, can be accessed from the Web (GENC 2007). An interactive version to facilitate pre-investment analysis is under development and will be uploaded on the Internet in due time.

2.2 Project inventory as of 08/2006

Local technology stocks for green power are almost nil. Most developments undertaken to date are quite modest (hampered mostly by a limited vision of the potential for business, weak links between industry and the technology development centres, and scarce research, development and demonstration funds) and remain at an embryonic stage. Few examples of commercial projects using indigenous or successfully appropriated technology could be cited. Studies show that over 200 Mexican companies have the right infrastructure to produce parts and components for the wind industry (Mejía-Neri 1999) but only one has taken this opportunity thus far. Most large-scale green power projects with new renewables built or planned to date in Mexico are based on imported technology. Things are similar at the small-scale, where the local photovoltaic (PV) market has grown to over one MW per year, basically for remote power supply, using imported PV modules. Nevertheless, the emergence of a modest manufacturing capacity for balance of system components can be observed. This section summarizes projects already built or under development as of this writing.

Electric projects

The so-called "green electricity" is in its infancy in Mexico. Few projects have been carried out to date, notably those for rural electrification in the recent past and current grid-tied applications including wind farms in the development process. A quick summary follows:

Solar PV. Off-grid photovoltaic (PV) systems and hybrids thereof have been implemented in Mexico for over fifteen years, mostly to bring electricity-based services to remote rural communities. Applications include basic lighting and entertainment in rural houses (more than 90,000 installations); more than 13,000 PV-powered rural telephones; and electricity for hundreds of rural schools, medical dispensaries and communal buildings in over 2,500 rural communities (Huacuz and Agredano 1998). Over 200 small PV-powered water pumps have been also implemented through a FIRCO (a trust fund of the Mexican Government for agriculture infrastructure)-GEF-World Bank project (WB 2002). New programs to serve native Indian communities currently without access to the electrical grid are under preparation. Electricity supply in off-shore oil rigs, cathodic protection of oil and gas ducts, signalling and telecommunications, eco-hotels, natural preserves and forest surveillance posts, are also growing applications of PV in Mexico. The total PV power currently installed in Mexico for off-grid applications is estimated at 18.65 MW (PVPS 2005).

Grid connected applications have been experimented over the last ten years in the modality of net-metering for domestic applications. Six grid-connected PV systems, between 1.5 and 2 kW each, have been in operation for the last six years in the cities of Mexicali, Hermosillo and Monterrey in Northern Mexico. As of this writing, a larger domestic PV system of almost 6 kW is being installed in the city of La Paz, Baja California Sur.

Through an initiative of the Baja California State Government, the first PV neighbourhood is being built in the city of Mexicali. The first phase of this project consists of close to 220 new houses each equipped with one grid-connected PV system of 1 kW in capacity. The purpose of this project is to test PV as an option to low electricity consumption in low income housing as a means to relief the domestic economy from the burden of high electricity bills during the summer time. The second phase of this project will carry a similar amount of installations.

The first 3-phase grid-connected installation was carried out in Mexico City in December 2005. With 32 kW in capacity, this is the largest grid-connected PV system in the country. It is also the first application in the commercial sector, where the PV panels serve as roof for a health food store. Monitoring of this installation will provide information to derive technical norms for similar applications elsewhere in the country (González et al. 2006).

Solar thermal. A large solar concentrating collector, coupled to a multi-megawatt natural gas-fired combined cycle power plant, to be built in northwest Mexico, has been on the drawing board for a number of years. Parabolic trough technology is being considered to supply heat for a 20 MW fraction of the total power plant capacity of over 400 MW. Once built, the project will benefit from a grant by the GEF-WB to cover the marginal cost of the solar field. Pre-investment studies have already been carried out and a call for bids has been issued to build the facility. However, the project has been delayed until today for administrative reasons.

Biomass. Electricity generation with biogas from sanitary landfills was first demonstrated in Mexico with a pilot installation of 20 kW, built in 1991 by IIE in cooperation with the national electric utilities CFE and LFC, at the Santa Cruz site in eastern Mexico City (IIE 1991). The first commercial plant in Mexico to generate electricity with biogas from a sanitary landfill was built in northern Mexico in 2003. This is a 7.4 MW facility erected in the municipality of Salinas Victoria as a public-private partnership to supply a fraction of the electricity needed for street lighting, water pumping and electric transport for the metropolitan Monterrey area in the state of Nuevo León. The project was supported by a grant from the GEF-World Bank which helped to remove some of the financial constraints that in the past had prevented construction of this type of plants (WB 2003).

The city of Monterrey in northern Mexico has been also a pioneer in the use of the biomass contents in residual municipal water for the production of biogas at its water treatment plants. The "Dulces Nombres" plant currently treats 5 m^3/s of residual waters. This facility includes a sludge processing plant that produces biogas destined to supply 60 % of the fuel for a 9.2 MW electricity generating station; the remaining 40 % of the fuel comes from natural gas. The Monterrey IV plant treats 2.5 m^3/s of residual water. The biogas produced here has been used to power a 1.6 MW cogeneration plant for electricity and processes heat. Current operational conditions of these plants are unknown.

The town of Tizayuca, just a few kilometers to the north of Mexico City, is well known for its dairy farms. For years, manure production has been a major environmental problem and a headache for milk producers who have to spend important sums of money for the proper disposal of this polluting material. In the period 2003–2004 a project to generate 10 MW of electric power from biogas produced by anaerobic treatment of manure was developed by a consortium of national and international stakeholders. Once built, this will be a hybrid biogas-natural gas facility with total planned generating capacity of 70 MW. The project stalled apparently due to financial difficulties of the consortium promoting it.

Biogas is now being produced by anaerobic treatment of fruit and vegetable residues at a food cannery and juice making plant in the town of Ecatepec, near Mexico City. The gas is used to generate 1 MW of electric power and process heat for internal use at the plant.

Wind. In spite of the large wind potential available in Mexico, only around 2.5 MW of grid-connected wind generators have been built so far. This includes the La Venta I pilot wind farm with 7 wind machines of 225 kW each, built by CFE over 10 years ago in La Ventosa, Oaxaca, and one 600 kW wind machine in Guerrero Negro, Baja California Sur, also built by CFE. The APASCO cement company installed a 700 kW wind machine in the premises of its plant near the town of Saltillo in the state of Coahuila. Some years ago this machine caught fire for unknown reasons and is now out of operation. La Venta II is the first large-scale wind farm being built in Mexico, also by CFE. It is currently under construction in the region of La Ventosa Oaxaca and when finished, by the end of 2006, it will have a total capacity of 83 MW. Over half a dozen small wind generators, around 10 kW in capacity each, have been installed to power mini-grids in rural communities throughout the country. They operate in a hybrid mode combined with PV and diesel generators to supply 24 hours electricity to the user.

Small hydro. It is estimated that there are about 45 MW in small hydroelectric plants privately owned, scattered throughout Mexico, but information on their specific location, individual capacities, operational status and ownership is scarce. In the government sector, CFE and LFC jointly own 58 small hydroelectric facilities, with individual capacities below 5 MW. All of them are old plants, ranging from 40 to over 100 years of age, but only 22 are reported to be operational today and interconnected to the national electric grid in 9 different states. The total capacity of these plants is close to 38 MW, while the capacity of the non-operational plants is close to 37 MW. Three recent projects totaling 52 MW in capacity have been developed in the states of Guerrero and Jalisco. Already existing water dams have been retrofitted for electricity generation, which represents an economic premium for this type of projects. Another four projects with a total capacity of 28 MW are under construction in the states of Puebla, Oaxaca, Guerrero and Sonora.

Non-electric applications

As already mentioned, a large number of low-technology traditional applications of renewable energy, such as drying of agricultural products and salt production by evaporation of sea water, can be found in Mexico. Other applications with more advanced technology focus mainly on the production of low and medium temperature heat using flat-plate and parabolic trough solar collectors.

Solar heating. Over 400,000 m^2 of flat plate solar collectors are estimated to have been installed in the country for applications such as swimming pool heating, domestic water heating, and institutional supply of hot water. Local production of flat plate solar collectors dates back to the first half of last Century; however, imported products are increasingly found in the national market. Typical applications of flat plate collectors include swimming pool heating as well as domestic and institutional water heating.

Parabolic trough applications for industrial process heat are in its infancy in Mexico, in spite of the fact that a large portion of fossil fuels spent in this country are used to produce low to medium temperature process heat, used in the food, beverage, pharmaceutical and textile industries. A couple of modules designed and built at IIE have been applied to substitute LPG at an industrial laundry in the city of Torreon and another at a car assembly facility in the city of Cuernavaca. Studies are underway to assess the market potential for this particular technology.

Solar cooling. Cold production using solar energy has been of interest among researchers and renewable energy promoters, for at least three decades. Because of its tropical climate Mexico is in need of cold chains to prevent spoiling of perishable goods produced in remote areas where grid electricity is not available. Several national academic institutions have developed prototypes of solar-powered cold and ice making machines. These efforts, however, have not been complemented by programmatic actions to deploy this technology in the field. A few pilot and demonstration installations of this type have been carried out in remote communities using imported technology. Experience with these installations has been far from satisfactory and hence no further progress in the application of the technology has been observed.

Solar-powered air conditioners have undergone a similar fate. The Sonntlan project, a cooperative effort between the Mexican and German governments back in the 1980's, implemented the largest solar cooling system ever in this country. This was a centralized absorption chiller powered by heat-pipe solar collectors, built to cool individual apartments in a multi-story building in the city of Mexicali. In the same city, and almost at the same time, the IIE implemented a smaller solar cooling system to test this option as a means to reduce electricity use in the domestic sector. Both projects operated for a number of years, after which they were terminated. No similar applications have been reported in subsequent years.

Biofuels. The trend towards biofuels is very recent in this country. The only commercial application known to the author as of this writing is a factory to produce biodiesel from raw animal grease, recently put in operation in the city of Monterrey. Other biodiesel and bioethanol initiatives are in the making as of this writing, inspired by the possible approval by Congress of a biofuels bill currently under revision.

3 Market supporting elements

Sustainable markets demand a number of soft technology support elements, such as technical norms, certification procedures, best practices manuals for project replication, guidelines for project development, follow up and monitoring and so forth. Availability of these elements in Mexico, applicable to renewable energy projects, is still very limited. The few ones available apply only to off-grid PV projects and constitute a good precedent. However, much remains to be done in terms of developing, adopting and adapting technical standards, guidelines, norms and specifications, to facilitate the correct identification, development and implementation of renewable energy projects.

Mexico is a fairly well developed economy and, as such, a variety of capabilities necessary to support the deployment of renewables are already in place. However, many of these capabilities need to be expanded and/or updated in the context of the new energy-environment dimensions, such as the Clean Development Mechanism and the Carbon Economy. Knowledge about the business opportunities with renewables must grow among consulting and engineering companies, while capabilities for project identification, development and implementation need to be developed. In general, the commercial banking system has little or no mechanisms to finance renewable energy investments.

The legal framework for energy matters is contained in Articles 27 and 28 of the Mexican Constitution. It is comprehensive and clear, but fails to recognize renewables as national assets (in contrast to petroleum and big hydropower). Hence, no provisions are found in the secondary laws to foster their application. The Public Electricity Service Law (PESL) sets the operational rules for the electric power sector. It was reformed in 1992/93 to facilitate private investment in non-public service applications, listed here in Table 1. The PESL mandates the electric utilities to purchase the least cost power options for public service supply, levelized over the plant's estimated useful lifetime; needless to say, few renewables could meet this criterion at the present time, but self-supply of electricity is a good niche of opportunity for renewables. *Ad hoc* regulatory instruments are being developed and implemented by the National Energy Regulatory Commission (CRE) to foster this niche market. In one of them, CFE is entitled to swap electricity under preferential terms and conditions for the self-supplier (CRE 2001).

A new initiative by SENER-GEF-WB seeks to establish a green fund to foster green power projects (GEF 2003). The United Nations Development Program (UNDP) is implementing jointly with IIE the GEF-supported project "Plan of action for removing barriers to the full-scale commercial implementation of wind power in Mexico", which includes the creation of the Regional Wind Energy Technology Centre in the state of Oaxaca (IIE 2007).

Table 1. Concepts excluded from the definition of "Public Service" in the Mexican Electricity Service Law

Concept	Description
Self Supply	Electricity produced with the sole purpose of satisfying the generator's own needs. No sell to third parties allowed. Under the current law, electricity users can form self-generating companies with third parties
Co-generation or Small Generation	Generating capacity under 30 MW for sell to CFE, or under 1 MW for the supply of remote rural communities
Independent production	Electricity generation with no capacity limits, for the only purpose of selling to CFE
Export	Generation of electricity for export, either from co-generation, independent power production, or small generation
Imports	Electricity imported by individuals or formally established entities, for the sole purpose of self supply
Emergency	Generation of electricity in case of emergency caused by the interruption of the public electrical service

4 Future prospects

Renewable energy resources offer Mexico potential benefits, more than energy. They are environmentally sound and can help to solve hard recurring local problems; for instance, urban solid waste can be used as fuel, thus reducing the problem of final disposal, or reclaiming already deforested land for the production of energy crops. On the social dimension, renewables are oftentimes the only reasonable possibility of providing electricity-based services to remote communities, improving quality of life and facilitating local economic development through productive projects. In the urban and industrial sectors, renewables can constitute a "democratising force" to move away from centralized forms of energy supply: individuals and businesses can generate their own electricity and hence financially contribute to the creation of energy infrastructure. Large green power projects can attract fresh private capital to build new generating capacity. Green power technologies are well within the existing capabilities of the Mexican industry and represent a good opportunity for participation in this new power technology market. This will mean new jobs, new forms of the energy business, and reactivation of stagnant industries.

The use of renewables can help to extend the lifetime of the national oil and natural gas reserves, lowering at the same time the carbon intensity index of the economy. This will put Mexico in a good position to honour international environmental obligations and to benefit from the economic mechanisms deriving from the Kyoto Protocol, which in turn may pay off in political benefits both at home and abroad. The question is how to move forward in a strategic manner to make all these potential benefits come true.

4.1 What can be expected for the short term?

Little systematic work has been carried out to forecast the penetration of renewables in Mexico and to assess the impacts from their large-scale deployment. Pioneering work can be found in the literature (Islas et al. 2001, 2002, 2003). Four medium-term scenarios for the future penetration of renewables can be identified, the first three dealing with electricity only. The first two of them derive from official documents, including the 2002 prospective for the electric sector and the Energy Sector Program 2001–2006. They could be combined into a single baseline of 1,686 MW for green power. The enhanced-penetration scenario was built by the author and is contingent upon the introduction of changes in the regulatory framework (for wind), effective environmental regulations (for biomass), programmatic elements (for small hydro), and lower technology costs (for solar thermal and photovoltaics), in order to reach a green power capacity of 2,720 MW in ten years. The high penetration scenario includes non-electric applications and was developed by a panel of national experts convened by the National Solar Energy Association (ANES 2000). It is assumed that renewables will have a contribution of between 5 % and 10 % to the total energy consumption in Mexico by the year 2010. This scenario assumes an aggressive activity in several fronts, and sets the upper bound for green power at 5,000 MW.

No renewable energy program has been implemented in Mexico up to the present time. Nonetheless, a number of projects in both the public and private sectors are underway or are being planned, but it remains to see whether any of the scenarios presented in the previous paragraphs have any meaning. The Energy Regulatory Commission (CRE) is the best source of information for electricity generation projects currently underway or approved for the short term. According to CRE, permits for over 1,360 MW have been awarded for green power projects, including 1,252 MW for wind farms, 80 MW for small hydro, and 31 MW for biomass (CRE 2006). Electricity projects below 500 kW are not required to get generating permits from CRE. Hence, no official record on the upcoming capacity for small-scale distributed generation is available. However, it could be expected that a few MW of PV installations could be added through initiatives such as the GEF-WB funded rural electrification project and domestic or commercial grid-connected projects.

Non-electric projects are even more difficult to forecast, as no registration or tracking system is available for such applications, except in the case of flat-plate solar collector for water heating. However, it is expected that the recently enacted regulation for the application of solar water heaters in Mexico City will increase the contribution of solar energy for thermal applications. Contribution of biofuels to the Mexican energy supply after the expected approval of the bioenergy bill is also expected to grow.

4.2 Drivers to move forward

The energy value of renewables may not be enough to justify their incorporation in the supply mix of an oil-dominated economy such as Mexico's. And as long as the economic competitiveness of many of the associated technologies is still marginal, exogenous drivers are needed to break the inertia. Three of them are discussed in this section; others, such as the impacts on the economy, including job creation, diversification of productive activities, and the beneficial aspects on health from avoided pollution from fossil fuels, are left out of the scope of this work, due to the present lack of quantitative data.

Protection of the global environment

Global environmental protection has been a good driver for the use of renewables in other economies; however, Mexico's contribution to global CO_2 emissions from fuel combustion has been traditionally low as compared to that of other OECD economies (IEA 2002). Hence, global environmental protection might not be a strong argument for the mainstream incorporation of renewables in this country. Nevertheless, in solidarity with other nations, Mexico ratified the UNFCCC in 1993, and consequently later on, in 1997, implemented a National Action Program for Climate Change. Furthermore, Mexico ratified the Kyoto Protocol in the year 2000 (Fuentes 2002), which is currently the main legally binding instrument for most industrialized nations to protect the environment, but with no quantitative obligations for Mexico and other emerging nations.

Nonetheless, Mexico can benefit economically from curbing CO_2 emissions within the framework of the Clean Development Mechanism. On the other hand, Mexico, along with a few other emerging nations, signatories of the Kyoto Protocol, are good candidates for mandatory emissions reduction once the Protocol enters its second phase in 2012. In either case, the use of renewables is an attractive option for Mexico.

The emerging legal framework

Two bills for the promotion of renewable energy currently await final approval in Congress. One bill, known as LAFRE for its name in Spanish, conceived to foster the use of renewables in general, was introduced to the House of Representatives in December of 2005. Among other things, if approved, this bill would create a trust fund of about 120 million US dollars per year to: provide incentives to the commercial application of renewables; foster research, development and innovation; and facilitate renewable energy resource assessment and mapping. Additionally, LAFRE sets a 2012 target for green power (large-scale hydro excluded) of no less than 8 % of the total electricity generation in the country at that time. The second bill exclusively deals with biofuels. It was submitted to the Senate almost at the same time as LAFRE. This bill emphasizes production of ethanol from agricultural crops and seeks to open new business opportunities for farmers. Among other things, it proposes the creation of a national biofuels commission and suggests funding to be approved as per yearly project portfolio. Once approved, these two bills may become strong drivers for the application of renewables in this country. Net-metering schemes, now under consideration in official energy circles, could also open interesting opportunities for the application of grid-connected photovoltaics and other distributed generation options in the domestic and commercial sectors.

New business opportunities

Energy supply in modern Mexico, be it either electricity or fuels, has always been in the domain of the Public Sector. The rest of the Mexican population has played the sole role as consumers, or at best franchisers in the case of gasoline distribution. With the advent of the new energy technology, a number of possibilities emerge for new business in the energy field. For instance, private entrepreneurs could let small photovoltaic systems for leasing to homeowners, who could pay a monthly fee from the savings brought about by using solar electricity instead of grid electricity, if costs allow. Or else, farmers could grow oil-producing plants as feedstock for the production of biodiesel. A number of such possibilities could be thought of, which could represent an important driver for the deployment of renewables in an economy in bad need for jobs.

4.3 The approach by sectors

Different sectors of the economy can have different drivers for the introduction of renewables. The need for additional bulk generating capacity would be the driver in the power sector. Water supply, treatment and reuse require substantial amounts of electricity. Costs and energy security in this sector could be a good driver for the introduction of hydroelectric turbines in the downhill portion of aqueducts, or the use of windmills for pumping water uphill. Environmental, political and economic problems could impel municipalities to use currently despised urban solid waste for electricity generation. Production of energy can become a new agricultural activity, which means creating jobs, increasing local economic turnover and cutting harmful emissions to the environment. Technology is available to produce electricity from energy plantations, cattle manures, agriculture waste and forest residues; biofuels to power agricultural machinery can also be locally produced. Hence, local economic and human development would be the main drivers for the production of green power in the agricultural sector. Human development and environmental factors have already driven the implementation of renewables in the rural sector in Mexico (Huacuz and Martinez 1995). The use of renewables for productive applications can help to improve the economic conditions of the rural population.

Drivers for the introduction of renewables in the industrial sector would be basically economic. Self supply with intermittent sources is not attractive for continuous industrial operation, but could help to reduce the total electricity consumption, and thus help companies avoid buying electricity in the higher tariff blocks. The new interconnection contract for renewable energy sources mentioned above favours this possibility and polls conducted among Mexican industries show that the ten largest industrial companies in Mexico have expressed their will to pay a monthly premium for electricity produced from renewables (De Buen 2001). High power quality and process heat requirements represent areas where renewables could play an important role in the industrial sector.

Finally, technologies such as grid-connected photovoltaics could have an important contribution to resolve complex issues, such as summer peak power demand from air conditioning in desert regions. Traditionally, the government of Mexico has granted direct subsidies to electric tariffs for domestic consumers living in places with high summer ambient temperatures. Removing subsidies has turned politically unattractive and, hence, growing sums of money are allocated for this purpose. Schemes can be envisioned on how to remove subsidies and avoid hidden generation, transmission and distribution costs by means of renewable-based distributed generation systems.

5 Barriers to the implementation if renewables in Mexico

The large-scale introduction of renewables in Mexico will not be easy. A number of barriers of different kinds have to be removed for this to happen (Huacuz 2001). As already mentioned, the current legal framework does not favour the adoption of new renewables by the electric power sector. On the other hand, distributed generation may be perceived as risky within a centrally structured utility (due to possible loss of political control over the electricity business; negative impacts on the integrity, safety and quality of the grid, etc.), or the "bigger is better" paradigm, followed by many power engineers, can inhibit needed decisions. From the planning point of view, availability of fossil fuels challenges the wisdom of developing local renewable energy sources. Nevertheless, the national electric utilities are taking the first steps towards a more diversified technology portfolio where renewables are given an increasing role.

The current political support for renewables from the federal administration and some state governments is unprecedented in Mexico. However, regulatory and institutional changes to level the playing field cannot go beyond the limits set by the current Law. Further legal and regulatory changes will require approval of Congress, where some political barriers need to be removed. For instance, land ownership issues may represent important political barriers for wind farms and energy plantations projects.

Much needs to be done to foster a healthy domestic industry for the local supply of renewable energy technologies on a competitive basis. Among other things, broader exchange and closer practical cooperation between the manufacturing industry and the R&D establishment, as well as human resource development are critical to achieve this purpose. The considerable theoretical and empirical knowledge that exists in Mexico's R&D institutions needs to be transformed into practical applications and national training programs.

Further actions need to be carried out to upgrade data bases, improve prediction models and expand the geographical coverage of monitoring stations, so that development of commercial projects is facilitated. International standards, technical specifications, and recommended engineering practices need to be assimilated for the same purpose by local organizations. Pilot and demonstration projects in several technologies need to be carried out, and lessons from international demonstration projects need to be assimilated, to build local capacities. There is a generalized lack of knowledge about the potential market for green power in Mexico and its economic area of influence.

An effective coordination among different agencies at the three levels of governments, and among other national and international stakeholders, is necessary to fully realise the benefits from the large-scale implementation of renewables. Lack of interagency coordination represents an important barrier towards a more dynamic renewable energy activity in Mexico.

6 Conclusion

Mexico is at cross-roads regarding renewable energy. On the one hand, this country has all the necessary elements to become a major user and developer of the technology. Because of its geographical position and foreign policy instruments in place, such as NAFTA and the Puebla-Panama Plan (between Mexico and the Central American countries), Mexico could also become an important exporter of this technology to its neighbouring markets. On the other hand, however, the big oil paradigm is too deeply imprinted in the minds of many people, energy officials and industry leaders included. Hence, unless a quick and conscious effort is made to alter this situation, the opportunities ahead (economic, social, political, environmental and otherwise) can be lost, as the window of opportunity to close the gap is too narrow. The effort to be made includes the implementation of aggressive renewable energy policies and a variety of technical and non-technical changes in the energy market. Barriers that could inhibit progress need to be identified and strategies to remove them in the short to medium terms must be developed; new capabilities and infrastructure (human, technical and physical) to identify and tap niches of opportunity where green power is technically and economically viable must be created, so that enough experience is gained within the country in this new field of the energy business; finally, mechanisms to assure a level enough playing field for renewables to compete with other alternatives, under equitable and transparent rules of game, must be introduced. Consumer awareness has to be raised, new capacities of the public, private and social entities have to be built, technology intermediation centers have to be strengthened or created, and new financing services have to be established. Above all, a new energy culture must be created.

References

ANES (2000) Strategies for the Development of Renewable Energy in Mexico (in Spanish). National Solar Energy Association (ANES). September 2000

Borja MA, et al. (1998) Estado del Arte y Tendencias de la Energía Eoloeléctrica. Universidad Nacional Autónoma de México, Programa Universitario de Energía, Instituto de Investigaciones Eléctricas, Cuernavaca, México

CONAE (2002) Renewable Energy Resources in Mexico (in Spanish). In: Proceedings of a Meeting to Foster Renewable Energy Investment in Mexico. February 2002. Studies by IIE indicate that around 10 % of this figure can be tapped from irrigation channels

CRE (2001) Contrato de Interconexión para Fuentes Renovables de Energía, Resolution N° RES/140!2001. Energy Regulatory Commission. http://www.cre.gob.mx/registro/resoluciones/

CRE (2006) http://www.cre.gob.mx/registro/resoluciones.html

De Buen O (2001) CONAE's proposal for a program to promote the generation of electricity from renewable energy (in Spanish) Comisión Nacional para el Ahorro de Energía, September 2001

Fuentes CC (2002) Renewable Energy for Sustainable Development: The Kyoto Protocol. International Experience and the Mexican Case. (in Spanish). Master's Thesis. Universidad Nacional Autónoma de México (UNAM), Jan 2002

GEF (2003) http://www-esd.worldbank.org/gef/fullProjects.cfm

GENC (2007) Gerencia de Energías No Convencionales http://genc.iie.org.mx/genc/index2.html

González R et al (2006) First 3-Phase Gris-connected Photovoltaic System in Mexico. In: 3rd International Conference on Electrical and Electronics Engineering (ICEEE 2006). Veracruz, Mexico, September 6–8 2006

Hernández HE, Tejeda MA, Reyes TS (1991) Atlas Solar de la República Mexicana, Textos Universitarios, Universidad Veracruzana, Mexico, pp 45–47

Huacuz JM (2001) Renewable Energy in Mexico: Barriers and Strategies. Renewable Energy Focus. The International Renewable Energy Magazine Jan/Feb pp 18–19

Huacuz JM (2005) The Road to Green Power in Mexico: Reflections on the Prospects for the Large-scale and Sustainable Implementation of Renewable Energy. Energy Policy 33:2087–2099

Huacuz JM, Agredano J (1998) Beyond the Grid: Photovoltaic Electrification in Rural Mexico, Prog. Photovolt. Res. Appl. 6, 1998

Huacuz JM, Martinez AM (1995) Renewable energy rural electrification. Sustainability aspects of the Mexican programme in practice, Nat. Res. Forum, 19, No. 3, August 1995

IEA (2002) CO_2 Emissions from Fuel Combustion 1971–2000, International Energy Agency, 2002 Edition

IEA (2005) IEA Wind Energy Annual Report. Executive Committee for the Implementing Agreement for Cooperation in the Research, Development, and Deployment of Wind Energy Systems

IEA (2006) IEA Trends in Photovoltaic Applications. Survey report of selected IEA countries between 1992 and 2005. IEA-PVPS Report T1-15: 2006

IIE (1991) Planta Piloto para la Generación de Electricidad con Biogás del Relleno de Santa Cruz Meyehualco. IIE/10/14/3128/I-04/P Instituto de Investigaciones Eléctricas, Cuernavaca, Morelos, Nov 1991

IIE (2005) Preliminary Assessment of the Potential for Electricity Generation in Mexico with Renewable Energy other than Wind. IIE Report for the World Bank

IIE (2007) Plan de acción para eliminar barreras en el desarrollo de la generación eoloeléctrica en México   http://planeolico.iie.org.mx/iiepnud.htm

Islas J, Manzini F, Martínez M (2001) "Reduction of greenhouse gases using renewable energies in Mexico 2025" Int. J Hydrogen Energy 26:145–9; Ibid (2002) "Renewable energies in electricity generation for reduction of greenhouse gases in Mexico 2025" Ambio 31(1):35–9; Ibid (2003) "Cost-benefit analysis of energy scenarios for the Mexican power sector" Energy 28:979–992

Mejía-Neri F et al (1999) Data base of companies and institutions with technical capabilities to conform the basis for a Mexican wind energy industry (in Spanish) Technical report IIE/01/14/10819/I003/A4/F/V2. Electrical Research Institute, December 1999

PVPS (2005) PVPS Annual Report 2005. International Energy Agency

González R et al (2006) First 3-Phase Gris-connected Photovoltaic System in Mexico. In: 3rd International Conference on Electrical and Electronics Engineering (ICEEE 2006). Veracruz, Mexico, September 6–8 (2006)

SENER (2001) "Programa Sectorial de Energía 2001-2006". Chapter 6 Electricity. Secretaría de Energía

SENER (2006) Balance Nacional de Energía 2005. Secretaría de Energía

WB (2002) World Bank Projects section of the Bank's Web site http://www.esd.worldbank.org/gef/

WB (2003) The World Bank. Case Study of Landfill Gas To Energy Project in Monterrey Mexico. Final Report. http://www.esmap.org, http://www.bancomundial.org.ar/lfg/default.htm

The Development of Thermal Solar Cooling Systems

Roberto Best-Brown

Centro de Investigación en Energía, Universidad Nacional Autónoma de México (UNAM) Privada Xochicalco s/n, 62580, Temixco, Morelos, México. E-mail: rbb@cie.unam.mx

1 Introduction

The sustainability of the energy sector in México, as in many countries of Latin America will depend on decisions that have to be taken now in order to expand the use of clean and renewable energy sources. The use of solar, wind, biomass, as well as other renewable energy sources, will provide México with the diversification of energy resources that will help to reduce the dependence on fossil fuels and decrease CO_2 and other emissions. The increased role of renewable energy sources in Europe and specifically in Germany will have an impact on México, where the development of clean energies needs collaboration between the two countries to accelerate its use.

Refrigeration is available in the industrialized countries through the availability of electricity but is not readily available in the major part of the world. An alternative solution for this problem is solar energy, available in most areas and it represents a good source of thermal energy. In addition, combination of solar energy with absorption, adsorption, desiccant, and others technologies less studied for refrigeration are being currently investigated and improved around the world. Many arrangements or cycles are possible: solar collectors can be used to provide energy for absorption cooling, desiccant cooling, and Rankine-vapour compression cycles. Solar hybrid cooling systems are also possible. The concept of cooling is appealing because the cooling load is roughly in phase with solar energy availability.

One of the most interesting areas of use of clean energies is the use of renewable energy for cooling and refrigeration. In a report produced by the International Institute of Refrigeration, "Sustainable Development: Achievements and Challenges in the Refrigeration Sector", the Refrigeration Institute explores promising refrigeration technologies and applications using non-vapour-compression technology that will undoubtedly also play important roles in ensuring sustainable development (IIR 2006):

- Absorption and adsorption cooling systems, which quite often are fuel-fired, are a practical means of providing both commercial and industrial cooling without imposing a major drain on a developing electric infrastructure and therefore a major drain on the limited developmental capital available to most developing countries. Absorption based air conditioning, in the form of large absorption chillers for major commercial-building or industrial applications, is the most widespread application of these technologies today. Low energy efficiency is still the major drawback of this technology. Further development and simplifications are needed in order to enable this technology to be more widely applied.
- Solar refrigeration is technology that should be given priority when choosing sustainable development options in developing countries. The growing demand for ice for the conservation and transportation of perishable products, the development of cold storage for food storage, the freezing of fresh and cooked products, space air conditioning, among other refrigeration applications, are only a sample of the potential applications of this technology. Priority actions are: The establishment of the infrastructure required for the production of solar refrigeration units and the setting up of educational programmes and training in the operation and maintenance of solar plants as well as in the design and instrumentation aspects.
- Desiccant technology includes a broad spectrum of systems providing cooling, dehumidification, and ventilation in order to control the quality of the indoor environment in the industrial and commercial sectors. But many production and technical issues still have to be addressed.
- Trigeneration (combined power production, heating and cooling) has considerable benefits from an energy standpoint. It makes it possible to totally or partially utilize the heat rejected to ambient as waste heat generated during electrical power production and use part of it in refrigerating applications. The development of high performance absorption plants will enhance the benefits of trigeneration plants.

In this work, the status of the development of solar cooling and refrigeration is analysed from the perspective of its development in México.

2 The energy sector in Mexico

The energy consumption in México increased 5.3 % from 2003 to 2004 (IEA 2005), where electricity consumption increased 4.5 % in the same period. This is due to an increase in population and social development (quality of life). Fossil fuels account for 90 % of energy production, with a large increase in natural gas use for electricity production. Energy/GNP decreased by 0.9 %. Energy consumption/inhabitant increased by 3.7 %. All new electricity power plants and large energy investments are directed mainly to combined cycles with natural gas (CCGT), and in less extent to hydro and geothermal energy.

The government energy policy is that México has to grow at the least energy cost for the growing population: this leads to a growing dependence on the already scarce fossil fuels and a short-sighted view of the urgent need to develop indigenous renewable resources now for the near future.

3 Renewable energy in Mexico

A brief review of renewable energy (RE) in the world from the perspective of México shows a great difference between the potential and actual use of RE and the potential in other countries. One of the most developed RE's is wind power. By 2005, 60 GW had been installed worldwide; Germany, Denmark and Spain are the countries where wind energy has increased more rapidly. Spain, with 216 MW installed in 1996, has moved to 10,000 MW in 2005. Costs of power production are in the range of 4 to 6 cents/kWh. In México there are good wind resources in Oaxaca (Itsmo de Tehuantepec), Quintana Roo, Veracruz, Baja California, Zacatecas, San Luis Potosí and Hidalgo. The only existing wind farms are in Oaxaca (1.5 MW) and Baja California (600 kW). An 83 MW wind farm is being developed in La Venta zone in Oaxaca, and 588 MW are to be installed by 2014. An additional 956 MW have been approved by the Regulatory Energy Commission (CRE) as self supply private projects. A potential of 5,000 MW exists in those areas. Research work in wind power has been carried out for more than 20 years at the Instituto de Investigaciones Eléctricas (IIE). There is only one successful private company, which manufactures small systems with their own technology and large systems in a joint venture with international companies.

4 Solar thermal energy in Mexico today

Around 30 registered, all relatively small, manufacturing companies exist in México. A total area of 633,000 m^2 has been installed in 2004, with an annual increase of 10 %. Research activity is important in research centres, most of it however not in joint form or financed by industry. Important activities in standards and promotion are developed by the Mexican Solar Association (ANES) and the Energy Savings Commission (CONAE), through COFER. An increase in energy costs, especially LPG and natural gas, together with environmental restrictions, will accelerate the use of solar collectors for water heating in large cities as México City. Official Mexican Norms (NOM, for its acronym in Spanish) to determine the thermal performance and functionality of solar heaters is in force. A norm for solar heaters, with the goal of setting up criteria for the use of solar energy in new establishments and remodelling in México City, that requires hot water for productive uses. This NOM now in force establishes that at least 30 % of the annual energy consumption needs to come from solar heating systems (SENER 2005).

4.1 PV in Mexico

There is no commercial PV production, although a small plant existed in CINVESTAV of the National Polytechnic Institute in the seventies. A large number of small systems have been installed in rural areas for lighting and some water pumping. Installed capacity in México 2004 was 1.09 MW for a total of 16.1 MW, compared with the 7 GW to be installed worldwide by the end of 2006. A large FIRCO project from the Agricultural Ministry has installed a large number of water pumping systems with World Bank support. Good research in thin film and other technologies for low cost cells is carried out at the National University (UNAM).

4.2 Small hydro in Mexico

76.3 MW of power was installed by electric utilities and private companies in 1999. The potential is calculated at 3,000 MW. Only in Veracruz and Puebla states 1,100 sites have been identified for a capacity of 400 MW.

Around 184 MW of new installations have been approved by the Energy Regulatory Commission (CRE).

4.3 Biomass in Mexico

Wood and sugar cane bagasse accounted for 3.6 % of primary energy production in 1998. About 75 % of rural homes use wood as their main energy source. 135 MW of biomass-based power plants have been approved by the CRE, mainly running on energy from sugar cane bagasse. There is a lack of information on other biomass resources and a need for a strong collaboration in new research areas, such as biofuels. A bio-energy network has been established recently in México and small projects concerning biofuels, bio-diesel fast growing trees and biomass gasification are being developed.

5 Challenges for renewables

The government's energy policy at the moment considers only economic dimensions (minimum cost) over environmental or social aspects; until recently there was no clear policy for renewables. The *Programa Sectorial de Energía* (2001–2006) was announced by the government in November 2001 to increase the use of RE and to promote energy efficiency and savings. The action lines envisaged included design and implementation of programmes for the use of RE, promoting power generation with RE, intensification of RE research and technological development, and the formation of human resources for the use of RE. This program was more a wishful thinking than a reality, as most of their actions were not achieved. An important step could be the approval of the Law Initiative for the Use of Renewable Sources of Energy (Ley para el Aprovechamiento de las Fuentes Renovables de Energía, LAFRE). This law contains a range of instruments of this type that if approved by the Senate, would contribute to RE development. The LAFRE Initiative has, among its more powerful instruments, the creation of a trust fund that would grant temporal incentives to projects that generate, through renewable energy sources, electricity for public service. It is also relevant to mention that SENER (SENER 2006), GEF and the World Bank are jointly developing a scheme for the implementation of a Green Fund. This Fund would provide incentives to RE independent producers that deliver power to CFE for public service.

6 Perspectives for renewables

A barrier in México for RE is that energy prices are low or subsidized. Investment costs for renewables are high. Most of the population has a low income. Few private companies are promoting renewables and there is a lack of local technology development for most renewables. The number of research centres where renewable technology can be developed and/or tested is small. There is a lack of human resources trained in renewable technology. Utilities need to be restructured. Increasing the availability of energy to a growing population with higher social and environmental expectations needs all forms of energy production, and not only fossil fuels. Globalization should enhance and facilitate technological transfer for renewables. Many technologies show a technical and economic maturity. There is a tendency to lower the costs of renewable energy technologies. México has a R&D capacity to develop technology. Industrial capacity to produce renewable equipment must be enhanced.

7 The solar network

In 1998, the National Council of Science and Technology (CONACyT) created the Research Network on Solar Energy. This Solar Network aims to confront and resolve joint research challenges, employing staff, capabilities and infrastructure of participant groups in joint projects. The objectives of this network were:

- to fortify the R&D of solar energy in México and to channel the results to the productive sector;
- the formation of human resources in this area;
- to spread the knowledge of solar energy as a sustainable and trustworthy energy resource;
- to integrate groups and sectors related with solar energy;
- to develop capacity building programmes;
- to guide, co-ordinate and orientate research efforts in proposed multidisciplinary research areas.

No research projects have yet been approved.

8 National programmes in solar energy for buildings

Various programmes are currently being developed for savings and efficient use of energy in the government's building-related sector. These programmes are implemented through the instrumentation of standards, such as lighting for non-residential buildings, energy-efficient systems for lighting on traffic ways and building exteriors, energy efficiency on the use of compact fluorescent lamps, and energy efficiency in non-residential buildings, in particular the standard for the building's "envelope". CONAE is the agency responsible for the co-ordination of these programmes.

9 R&D Programme

The main research areas include the development and demonstration of solar technologies. The number of researchers, engineers and technicians that are involved in R&D topics related to solar energy is calculated at 400, spread over more than 40 research institutions. The research areas are solar architecture, heating and cooling of buildings, solar materials, solar chemistry, flat plate collectors, concentrating collectors, refrigeration and air conditioning, water purification, drying, thermal systems, evaluation of the solar resource, basic research in solar cell materials, electro-chemical cells, photovoltaic applications and instrumentation.

10 Solar cooling and refrigeration

The use of solar energy for cooling is an obvious advantage as the cooling demand increases when the incident solar radiation is at its peak. The energy utilized to produce cooling is mainly electricity and the power demand in the summer for cooling increases every year, as the quality of live in developing countries increases as well as the increase in population triggers the use of air-conditioning.

10.1 Energy consumption in cooling

In many regions of the world room air-conditioning is responsible for the dominant part of electricity consumption of buildings. Electrically driven chillers cause high electricity peak loads of electricity grids, even if systems are used that reached a relatively high standard concerning energy consumption. This is becoming a growing problem in regions with cooling dominated climates. Figure 1 shows the total newly installed electric capacity due to room air conditioners (RAC units) since 1998 (assuming a replacement of 10 % per year and an average electric capacity of 1.2 kW per unit). In recent years an increasing number of cases occurred in which summer electricity shortages were created due to air-conditioning appliances. In some regions or municipalities building regulations were set up in order to limit the application of active air conditioning systems, unless they are operated using renewable energies.

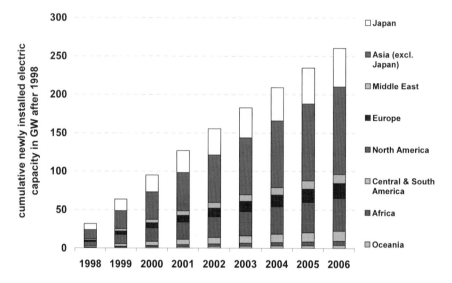

Fig. 1. Acumulative newly installed electric capacity due to RAC units since 1998 (IEA 2005)

A small window system which is sold in every discount store in México consumes at least 1.2 kW of electric power that is produced by fossil fuel combustion. With a COP of around 3, these systems although efficient are affecting the balance in peak hours and increasing the load and the need for power construction and are either idle in winter or only or partially used.

11 Solar cooling technologies

To use solar energy for cooling has two to be focused in two ways, using PV systems to operate a DC or an AC compressor in a mechanical vapour compression system or a thermal operating system consisting of either absorption, adsorption system or a desiccant system. Figure 2 shows a PV refrigerator system that is commercially available. The use of PV systems with air washers and evaporative cooling is also available, although only a very small number of systems have been installed. Thermoelectric systems operated with PV are also a possibility but efficiencies as well as that the temperatures reached are not as low as required for ice making.

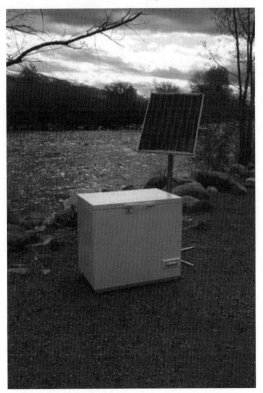

Fig. 2. Sun Danzer Refrigerator, PV solar refrigerator without batteries (Photo from Sun Danzer, www.sundanzer.com)

Absorption systems have been available for solar operation since the end of the 1970 by a series of American companies ARKLA, CARRIER, although in the small capacity applications only a Japanese company, Yazaki, offered systems between 7 to 35 kW were commercialized.

Figure 3 shows an installation of a solar air conditioning system in Mexicali, Baja California, México that consisted of a 7 kW Yazaki cooling system driven by 40 m^2 of double glazed flat plate collectors of Mexican design. Today only 35 kW systems are available. After the cost of fossil fuels decreased in the later 80's, solar cooling was too costly to compete with conventional systems, although many research centres continued the research in order to reduce costs.

Fig. 3. Yazaki 7 kW absorption air conditioning system, with a 1000 L storage tank auxiliary gas heating system and cooling in Mexicali, Baja California, Mexico

Although a large potential market exists for this technology, existing solar cooling systems are not competitive with electricity-driven or gas-fired air-conditioning systems because of their high first costs. Lowering the cost of components and improving their performance could reduce the cost of solar cooling systems. Improvements such as reduced collector area because of improved system performance and reduced collector cost will lower the cost of solar components.

Several solar driven refrigeration have been proposed and are under development such as sorption systems including liquid/vapour and solid/vapour absorption, adsorption, vapour compression systems, photovoltaic/vapour compression, adsorption; but most of them are not yet economically justified. The main technical problem of solar refrigeration is that the system is highly dependent upon environmental factors such as cooling water temperature, air temperature, solar radiation, wind speed and others. On the other hand, its energy conversion efficiency is low, and from an economic point of view, solar cooling and refrigeration are not competitive with the conventional systems.

11.1 Potential of solar cooling technologies

In order to evaluate the potential of the different solar cooling systems a classification has been made (Best and Ortega 1999), it uses two main concepts: solar thermal technologies and technologies for cold production.

The solar technologies relevant are:

- Flat plate collectors,
- Evacuated tube collectors,
- Stationary non-imaging concentrating collectors,
- Dish type-concentrating collectors,
- Linear focusing concentrators,
- Solar ponds,
- Photovoltaic and
- Thermoelectric systems.

The cooling technologies are:

- Continuous absorption,
- Intermittent absorption,
- Solid/gas absorption
- Diffusion,
- Adsorption and
- Desiccant systems.

The photovoltaic/vapour-compression systems and the photovoltaic/thermoelectric have predominated in the application of small refrigerators for medical use in isolated areas like vaccine conservation where high system cost is justified.

Solar thermal systems such as flat plate collectors and lithium bromide/water absorption cooling systems are in the stage of pre-production and commercial introduction, also for small capacities. Five companies are not fabricating systems due to a lack of a market for their products. The global efficiency of solar refrigeration systems for below freezing applications oscillated between 7 % and 20 % and differs because of insolation conditions. Even though electricity prices continue to increase, the solar refrigeration systems will have to reduce their costs by a factor of 3 to 5 times the actual costs in order to become competitive with the traditional vapour compression systems. It is difficult to predict the date when these solar technologies reach maturity.

The problem is not only technical but it incorporates economical, social and environmental aspects. In general it is expected that a technology will have to wait 15 years to pass from the stage of commercial introduction to the one of maturity. It does not always happen that a technology that reaches a development stage continues to the next.

Apart from flat plate collectors LiBr-H_2O, the photovoltaic/vapour compression and photovoltaic/thermoelectric the other technologies that have the major potential to reach maturity are:

• For uses between 4 and 25°C evacuated tube with ammonia/water absorption systems.
• For uses between -10 and 4°C dish type concentrating collectors with solid/gas absorption.
• For uses between -20 and -10°C parabolic trough concentrators with ammonia/water absorption.

11.2 Recent solar cooling installations in Mexico

Figure 4 shows a CPC collector used in an intermittent solar refrigerator for ice production, the system uses ammonia as refrigerant and a lithium nitrate/ammonia solution as absorbent. It was designed at the Centro de Investigación en Energía of UNAM, México, to produce 10 kg of ice. The cost of the system was estimated for a mass production to be around 2000 Euros.

Figure 5 shows an adsorption cooling system that has been installed in Cuernavaca, México that produces 15,000 litres per hour of chilled water using an adsorption system to cool a textile factory. It utilizes 440 m^2 of advanced flat plate collectors with water temperatures between 75 and 90°C to operate.

12 Conclusions

Solar cooling has a great potential to reduce fossil fuel consumption in the near future, and the environmental drawbacks of conventional air conditioning and refrigeration. An implementing agreement front the International Energy Agency IEA, Solar Heating and Cooling has started a new project, Task 38 (IEA 2005) to advance in this technical solar application, the needs have been identified for standardizes systems, small capacity systems and advanced operation and control.

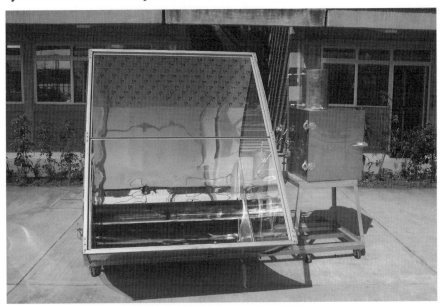

Fig. 4. Intermittent solar refrigerator for ice production (Photo CIE-UNAM)

Fig. 5. Adsorption cooling unit at Guetermann-Polygal – Cuernavaca-Mexico (Photo Modulo Solar, México)

Acknowledgements

The author would like to thank the organisers for financial support in order to attend the meeting and to M.E.S. Naghelli-Ortega for editing this work.

References

Best R, Ortega N (1999) Solar refrigeration and cooling. Renewable Energy 16:685–690

IEA (2005) Task 38, Solar air-conditioning and refrigeration, solar and heating programme, International Energy Agency, http://www.iea-shc.org

IIR (2006) Sustainable development: Achievements and challenges in the refrigeration sector, International Institute of Refrigeration, http://www.iifiir.org

SENER (2005) Balance nacional de energía 2004, http://www.energia.gob.mx

SENER (2006) Renewable energies for sustainable development in Mexico. SENER-GTZ, http://www.energia.gob.mx

Converting Solar Radiation to Electric Power in Mexico

Antonio Jimenez Gonzalez, Aaron Sanchez-Juarez, Arturo Fernandez, Xavier Mathew, PJ Sebastian

Departamento de Materiales Solares, Centro de Investigación en Energía.
Universidad Nacional Autónoma de México, Temixco, Morelos, México.
E-mail: ajg@cie.unam.mx

1 Introduction

The generation of electric power for supplying the populations of large cities is effected at an elevated cost, given that this cost does not include solely the price of generating the electric power itself, i.e., production, storage, labor, maintenance of installations, distribution, etc., but, in addition, bears the associated costs of damage to the environment due to the production of residual pollutants, and the consequent damage to the health of humans, and to the flora and fauna.

The use of fossil fuels in the generation of electric power contributes to environmental pollution by way of the emission of greenhouse gases such as CO, CO_2, and ozone, etc. These non-renewable sources, for example, oil, are also beginning to be exhausted. In Mexico, oil production has reached its peak and the cost of crude oil production has begun to increase.

In Mexico's case, in 2004, the composition of electric power generation by power was: 79 % fossil fuels (natural gas, oil, diesel and coal), 16 % renewable energy (hydroelectric, geothermic and eolic) and 5 % nuclear (SENER 2005). This consumption structure, based principally on hydrocarbons, translated into CO_2 emissions that represented, in 2003, 49 % of the total emissions generated in the country. Finally, according to SENER, the country has 11 years of proven oil reserves at current production rates. Given this tendency, the cost of producing crude has begun to rise. According to Hubbert's model, on a global scale, the level of oil production has begun to diminish. On the other hand, however, in recent years, the cost per barrel of crude has risen until reaching its maximum of 72 dollars. Another example of electrical energy generation is that undertaken using nuclear energy, which certainly has a very high capacity for generating electrical energy. Nevertheless, this form of energy generation has several important disadvantages: the great danger involved in working with nuclear reactors, which, in the event of an accident, could produce considerable harm to human health and ecosystems, etc., as was the case at Chernobyl, due to massive radioactive emissions. Moreover, the generation of residual radioactive materials represents an unresolved problem, and one which is being left to future generations, given that these wastes have a radioactive half life of more than 200 years.

Considering the existing problems associated with the generation of electrical energy in a country like Mexico, where production costs are high and associated environmental deterioration considerable, the alternative of using resources that are non-polluting, readily available, abundant, economical and easily manageable for the generation of electrical energy presents an attractive proposal. Indeed, there exist many different energy alternatives, such as wind power (eolic), sea wave power, geothermic, solar power, and the energy that may be extracted from the biomass, etc. This group of energy sources is today attaining considerable importance worldwide, and it is predicted that these alternative energy sources may replace, in a large scale, those sources currently responsible for ensuring the global supply of electrical energy.

Photovoltaic energy is among these alternative energy resources, and in this work, we highlight it, as a source of electrical energy that is both environmentally friendly and highly promising as a substitute supply for a large percentage of the electrical energy currently produced by more traditional means. The generation of this type of energy has its base in the photovoltaic effect that takes place in a p-n semiconductor junction, which transforms radiative energy (for example solar energy) into electrical energy (Lorenzo 1994). In the last two decades, both in industrial and academic spheres, a strong technological development has been observed in the area of photovoltaic conversion, transforming the status of this technology from an exotic and expensive energy generation source to that of a practical source of alternative energy having a wide range of potential applications. Thus, photovoltaic technology represents one of the most promising transformations of solar energy for future wide-scale use.

Solar energy is a renewable resource that is widely available in Mexico. The country enjoys an abundant and high quality solar resource throughout its territory, with an average solar irradiation of 5 kW/m^2/day (as shown in Fig. 1), which is equivalent to an average annual radiation of 1,825

kW/m^2/year. This represents a vast quantity of radiative energy that may be converted into electric power with a high degree of efficiency.

Taking into account the nation's annual electrical energy demand, which, for 2000, was 37,000 MWe, this equates to a total of 3.2×10^{11} kWh/year. If it was desired to meet this demand for electrical energy using commercial photovoltaic modules operating at just 10 % efficiency, an area equivalent to 3,507 km^2 (a square of 59 km sides) would have to be covered. If 15 % efficiency photovoltaic modules were used, that area would be reduced to one of 2,350 km^2 (a square of 48 km sides). As a reference, consider that Mexico City has a diameter of approximately 50 km. Ideal locations, with enhanced insulation indices and cloudless skies can be found in the States of Sonora and Chihuahua for the installation of photovoltaic plants. This brief calculation gives us an idea of the exploitation capacity of the solar resource available in Mexico, which is free and does not pollute. Another alternative would be to generate electric power in distributed form along the length of the country, in photovoltaic modules of the order of 1 kWh. The reader conversant with photovoltaic technology will surely be able to envisage many more possibilities for taking advantage of our solar resource. Actually, the PV systems installed in Mexico reach almost the 10 MW-p expecting for a market growth of 45 % in the next 5 years (De Buen 2006).

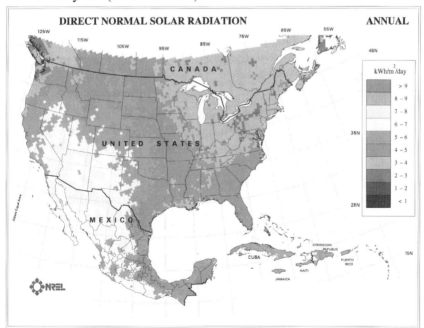

Fig. 1. Direct normal solar radiation map produced by NREL

2 Basic principles of photovoltaic conversion

A solar cell is a device that transforms the solar energy radiation in electric power. Fig. 2 shows the schematic structure of a solar cell; it consists of a junction of two semiconductor layers, one is *n*-type and the other one is *p*-type. This configuration has two electric contacts, one on the top and another at the bottom The thickness of the whole device can vary from less than a micron for the case of very absorbent semiconductors (the amorphous silicon), up to some hundred of micron for the case of semiconductors with low absorption coefficient (the crystalline silicon).

The solar radiation is absorbed by the electrons in the *n* layer and also by the *p* layer, which are excited from its initial state (the valence band) toward an excited state (the conduction band). During the excitation, the electron leaves a vacant site (or hole) in the valence band, which has a positive electric charge, and creates an electron in the conduction band with a negative electrical charge. Therefore, the electromagnetic irradiation of a solar cell generates electric charges, positive and negative, which are separated by an internal electric field formed due to the nature of the cell. The internal electric field is formed by the union of the *n* (negative side) and *p* (positive side) layers. The photogenerated charges travel under a concentration gradient toward the union where they are separated by the electric field. The electrical field sends electrons to the *n* layer and the holes to *p* layer; this effect creates a potential difference between the top and the bottom layers. The electric phenomenon that appears when a solar cell is illuminated is called photovoltaic effect and the generated voltage by the potential difference between both layers, when they are illuminated, is called photovoltage.

If an external electric circuit is connected to both layers, the accumulated electrons will flow through the circuit, returning them to the initial position. The flux of electrons constitutes the photovoltaic current through the circuit. Under open circuit conditions, the photovoltaic effect generates a different potential V_{oc} between the top and bottom part of the structure. Under short circuit conditions, the process generates an electric current I_{sc}.

Figure 3 shows a typical electrical behavior I vs. V. An I vs. V plot is required to find out the efficiency of the solar cell. This curve shows the most important electrical parameters: V_{oc}, I_{cc}, and the maximum power generated at some irradiation level. The solar cell efficiency η is defined as:

$$\eta = \frac{Im \times Vm}{Ps},$$ (1)

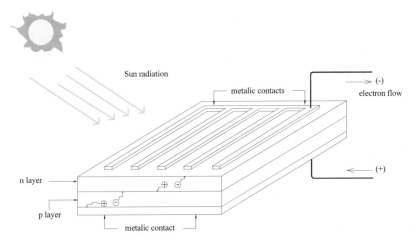

Fig. 2. Schematic drawing of how the solar cell operates

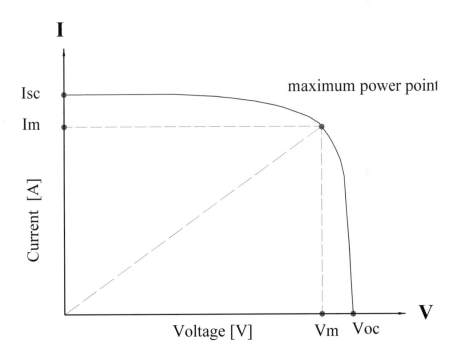

Fig. 3. Curve I *vs.* V shows the open circuit voltage V_{oc}, the short circuit current I_{sc} and maximum power P_M extracted from a solar cell

where Ps is the power of the incident light on the solar cell surface. Ps is defined by the product of the solar irradiance times the solar cell area. Figure 3 shows a photograph of a photovoltaic system installed at the Centro de Investigación en Energía of UNAM, Mexico. The PV system can supply approximately a third part the electric demand of our Center and it is planned to be connected to the grid.

Fig. 4. Photovoltaic plant of 10 kWe-p

3 Photovoltaic technology: state of art

Since solar cells generate direct current electricity they can be electrically connected (in series or parallel strings) and encapsulated for building photovoltaic modules (PV modules) or panels with appropriate voltage and current values for a particular application. Depending on the technology, silicon crystalline or thin film solar cells, the most common technology for encapsulating the solar cells strings is the lamination between a tempered glass and Tedlar or two glasses (Lorenzo 1994). EVA (ethyl vinyl acetate) is the best material for lamination purposes. The encapsulation insulates the cells electrically, provides mechanical strength for handling and protects them against the environment (humidity, dust, water or snow). A PV module has two output terminals (positive and negative) at the back side of the panel for the electrical connections. An aluminum frame as mechanical structure is used for handling and installation purposes.

The number of solar cells connected in series in a PV module depends on its applications. It is common that PV modules are built using 36 cells connected in series to storage the produced energy in 12 V batteries. However, this kind of module can be used for other applications such as water pumping and grid connected systems. The PV module is then used alone or connected in an electrical circuit with other similar modules to build the PV array. Thus, the energetic behavior of the PV technology is based on the PV modules as the building blocks of photovoltaic generators.

Table 1 shows the price evolution of crystalline silicon solar cells based on their market availability for terrestrial applications. It can be seen that silicon semiconductors have dominated the solar cells market due to its optimal physical properties and the world availability of the raw material. At research level, the efficiency of single crystal silicon solar cells has reached 24 % (Green et al. 2005), while for commercial purposes the module efficiency is around 11 % to 15 % depending on the manufacture process and the solar cells packing density in the PV module. The highest efficiency, almost 32 %, has been obtained in thin film solar cells based on GaAs, however, their production cost and price are too high, and hence they are only used for space applications.

Table 1. Price evolution of crystalline silicon solar cells and its conversion efficiency

Progress on the PV technology					
Price reduction [USD $/Watt-p]					
1958	1965-70	1980's	1990's	Present time	2010
≅1000	≅200	≅20	≅5-7	≅4-6	≅1-2?

Conversion efficiency in commercial crystalline silicon PV modules

5 to 15 %

Record conversion efficiencies for experimental solar cells

Single crystal silicon solar cells	Thin film solar cell based on GaAs
24 %	32 %

Table 1 shows also how the technology price in USD/W-p has been going down with the improvement of the solar cells fabrication process and to the demand increment. At present time, the price ranges between US$ 4.00 to US$ 6.00 /W-p. It is expected that in the near future this price can go down close to US$ 2.00 /W-p, due to the improvements in the photo-voltaic technology producing cheaper materials or increasing the production in the actual technology due to the demand increase.

Crystalline silicon solar cells dominate the market because the manufacturing process has reached its maturity compared to other technologies such as amorphous silicon thin film solar cell or thin film solar cells based on CdTe or CuIn Se_2. The maturity of the technology is reflected in the energetic stability and mean life of a module. At the present, the average life of PV modules should exceed 20 years under normal conditions, depending mainly on the durability of its encapsulation quality. The warranty given by the manufacturers of silicon PV modules is for a period longer than 20 years.

Despite of their longevity and reliability, crystalline silicon solar cells are heavy devices with significant fabrications costs. Because of its low absorption coefficient, they need around of 250 μm thickness to absorb all active wavelengths from sunlight and to produce reasonable efficiencies. This means that more material is needed to absorb solar radiation. In order to reduce the material amount to be used in a silicon solar cell, a great deal of research has been done with the purpose of reducing the thickness of silicon wafers. However, the efficiency of these thin solar cells is still lower than thicker silicon solar cells.

Since the earliest 70's, when terrestrial applications of crystalline silicon technology began to emerge, there had been a parallel effort to develop semiconductors other than silicon, to elaborate thin film polycrystalline devices with a better light absorption, to consume less material, to get a better power to weight ratio for space applications and, if it were possible, to apply technological processes that would produce cheaper materials. An example of this effort was the cuprous sulphide/cadmium sulphide heterojunction that can be elaborated using a simple manufacturing process with an intrinsic low cost. However, the poor stability of these solar cells did not allow its commercial development.

Thin films of hydrogenated amorphous silicon represent a good choice as high absorption materials. With a thickness of 0.5 μm, the material can absorb completely the solar spectrum. In spite of a-Si:H can be doped with both n- and p-type, the *p-n* junction does not represent a good choice to become a solar cell; however, the *p-i-n* junction shows the photovoltaic effect. With a modest efficiency (less than 6 %), hydrogenated amorphous silicon (a-Si:H) PV panels based on the *p-i-n* junction appeared in the market in the earliest 80's for small appliances such as calculators and watches. At that time, a-Si:H solar cells suffered from a lack of stability due to the well known Staebler-Wronski effect. However, new developments in dual and triple junction devices, with better performance than a single p-i-n junction and stabilized efficiency around 10 %, have put the a-Si:H PV panel technology into the market of solar cells as a good choice to produce power electricity for any application.

Another solar cell technology is represented by the p-n junction formed by thin films layers of n-type CdS and p-type CdTe. Development efforts on this junction have improved the performance of thin film CdTe solar cells until producing PV modules, nowadays available in the market, with a 10 % efficiency.

Copper indium diselenide ($CuInSe_2$) semiconductor materials have a higher absorption coefficient compared to the majority of other semiconductor compounds. A p-type ($CuInSe_2$) material can be used to build a p-n junction with an n-type CdS layer. Solar cells made with these materials have shown conversion efficiencies around 18 %, and because its hardness, this solar cell can be used for space applications. There is a strong effort to put this technology in the market aiming at a module efficiency above 15 %, however an improvement of the junction during the fabrication process must be achieved.

Another semiconductor compound that had received a great deal of attention is gallium arsenide (GaAs). It has very good optoelectronical properties for solar cells applications. Since the material can be doped with both n- and p-type and it is possible to elaborate alloys with other elements, so that this compound has always represented the best choice for high efficiency solar cells. In fact, the best solar cell reported with the world record of high efficiency has been made of a tandem structure based on GaAs alloys. Although, its fabrication cost is too high compared to the-other technologies, the GaAs solar cells are the best choice for space applications due to its high power to weight ratio.

Summarizing, at present time, the PV technology is based on crystalline, polycrystalline and amorphous semiconductor materials. They can be manufactured as thin films or thick materials. Examples of thin films are amorphous silicon, CdTe, and $CuInSe_2$ which, with a thickness of 1 μm, can absorb the entire sunlight spectrum. Thick semiconductor materials such as crystalline silicon have been used to build solar cells. Polycrystalline silicon has been proposed to reduce the fabrication costs of single crystalline silicon; however, its efficiency is lower than those for the single crystal. All the solar cells made with crystal silicon technology have been named "first generation" solar cells. Meanwhile, the thin films solar cells described above are called "second generation" solar cells.

4 Solar cells manufacturing technology

4.1 Monocrystalline silicon

PV modules with the largest efficiency at the market are made from monocrystalline silicon. Most of the solar cells used in the PV modules have been fabricated using thin wafers obtained by cutting large cylindrical moncrystalline ingots prepared by the well established technology called Czochralski process (Cz). The starting material is silicon of semiconductor grade that has been obtained by a purification process of metallurgical silicon which is obtained also by a purification process of sand (SiO$_2$). The doping with boron doped is carried out during the ingot growth process to produce p-type silicon. Ingots are sliced to produce thin wafer almost 300 µm thick, with a typical diameter of 10 to 15 cm. Boron-doped silicon wafers are treated at high temperatures in a phosphorus atmosphere in order to produce a diffusion of phosphorus, in this way, around 1 µm from the surface of boron-doped silicon is doped with phosphorus to produce one side of the wafer as a n-type silicon semiconductor. Thus, the p-n junction required for the solar cell is fabricated. Metal contacts are put on the p- and n- sides of the wafer by several techniques being the screen printing the most common process. Cells are interconnected and laminated in a package known as a PV module (Green 2001). PV Modules are designed to charge batteries at 12 V of nominal voltage, so that each module contains 36 solar cells soldiered together in series. Every module has two terminals, positive and negative, inside of a junction box attached to the module rear. Such modules have proved to be extremely reliable in the field and manufacturers offer a least 20 years warranty on the module power output, which are the longest warranties offered on any commercial product. Monocrystalline silicon solar cells used in PV modules have conversion efficiency between 12 to 16 %. However, the module efficiency is slightly lower than that of the constituent cell due to the packing factor.

4.2 Polycrystalline silicon

The mono crystalline crystal structure is not the only way to produce ingots of silicon. The large grained polycrystalline form of silicon is obtained by the casting or directional solidification technique. The solidification is carried out in presence of boron for doping the polycrystalline silicon ingot. Wafers are obtained by slicing the ingots, and the solar cell is elaborated following almost the same steps used for the monocrystalline silicon devices. The conversion efficiency of the polycrystalline silicon solar cells is lower than that obtained in monocrystalline silicon, ranging in the interval of 11–15 %. Polycrystalline PV modules are fabricated in the same way that monocrystalline modules, using 36 solar cells connected in series. Since the polycrystalline wafers are almost perfect squares, the packing factor of this kind of modules is bigger than for monocrystalline silicon. Thus, the conversion efficiency of polycrystalline silicon modules available in the market varies from 11 to 13 %.

4.3 Amorphous silicon

Amorphous materials are solids formed by atoms distributed in the space without any order. Because the lack of atomic order, amorphous silicon (a-Si) is a semiconductor with different optoelectronic properties to those for crystalline silicon. Several techniques are used to prepare thin films of a-Si. Among of them, the plasma enhanced chemical vapor deposition (PECVD) is a technique that produces a-Si thin films with adequate optoelectronic properties to be used in a solar cell structure. This technique is based on the decomposition of a starting gas material into its constituents by a cold plasma generated inside a vacuum chamber at 13.56 MHz.

Since the starting material to produce a-Si by PECVD is the silane gas (SiH_4), the deposited material is an alloy of silicon with hydrogen. This alloy is known as hydrogenated amorphous silicon (a-Si:H). Under normal preparation conditions, the a-Si:H thin films are of intrinsic nature with a low dark electrical conductivity; however, the light electrical conductivity is about 5 orders of magnitude bigger than the dark electrical conductivity. This material can be doped with boron and phosphorus to produce p- and n-type a-Si:H, respectively. The solar cells are based on p-i-n junctions (Wronski and Carlson 2001). With a typical thickness of around 1.0 μm, this a-Si:H solar cell can produce electricity with conversion efficiency of around 6 %. This single junction has stability problems due to the well known Staebler-Wronski effect. New developments in dual and triple junction devices, with better performance than for a single p-i-n junction, with a stabilized efficiency of around 10 % and with a manufacturer's warranty of 20 years, are nowadays available in the market.

4.4 CdTe

Among the thin film photovoltaic competitors CdTe, CIGS, polycrystalline and amorphous Si, CdTe has some special advantages such as optimum band gap of 1.5eV for the efficient photo conversion, high optical absorption coefficient and a variety of well-established thin film fabrication techniques. The work on CdTe/CdS thin film solar cell started in 1970s (Panicker et al. 1978), however, a real progress in research and development achieved in 1980s with the development of various techniques such as electro deposition (ED), screen printing (SP), physical vapor deposition (PVD) and close spaced sublimation (CSS). The pioneering work of various groups (Panicker et al. 1978; McCandless et al. 1999) in the 80s and 90s contributed significantly to the advancement of CdTe photovoltaic technology, and efficiencies greater than 15 % have been achieved. The present status of the thin film CdTe/CdS solar cell in superstrate configuration is more than 16.5 % efficiency for devices on transparent conducting oxide (TCO) coated glass substrates (NREL 2001). In an attempt to develop alternate technologies to overcome the heavy weight and damage-prone nature of glass based devices, attempts were made to develop CdTe/CdS substrate configuration devices on flexible metallic and polymer substrates (Matulionis et al. 2001). The reported efficiencies of the CdTe/CdS devices on flexible substrates are 7 % efficiency for devices on metallic substrates (Matulionis et al. 2001) and 11 % efficiency for devices on polymer substrates (Romeo et al. 2002, 2004). The commercialization of the CdTe photovoltaic technology started in 2001 and the CdTe photovoltaic modules with output power ranging from 45 to 65 W are available in the market. The efficiency of the commercial CdTe solar module is about 8 % for 55 W modules (First solar 2006).

In the conventional glass based superstrate devices, the first step is the deposition of TCO film. The most common TCO films are either fluorine doped SnO_2 or indium doped SnO_2. The next step is the deposition of the heterojunction partner, an n-type semiconductor. The well known hetero junction partner to the CdTe absorber layer is the CdS. The CdS thin films can be deposited by a variety of methods such as electro deposition, chemical bath deposition, thermal evaporation, spray pyrolisis and sputtering. In glass based superstrate devices, the best cells are prepared with the chemical bath deposited CdS film. Major effort is to obtain very thin and pin-hole free CdS films. Thinner film can reduce the series resistance and absorption in the film which is not contributing to the photocurrent. In practical devices, the thickness of CdS film is in the range of 80 to 100 nm. The thickness can be further reduced by incorporating a high resistive transparent layer between CdS and the TCO.

One of the greatest advantages of CdTe photovoltaics is that the absorber material CdTe can be prepared by a variety of techniques such as electro deposition (ED), spraying, close spaced sublimation (CSS), vapor transport deposition (VTD), screen printing, electron beam evaporation, laser ablation, thermal evaporation, metal organic chemical vapor deposition and sputtering. The ED is simple and economical while VTD is suitable in large scale manufacturing and CSS can produce larger grain polycrystalline films. To date the highest efficiency CdTe photovoltaic devices on glass substrates were prepared by CSS.

4.5 Cu-In-Ga-Se

Ternary compounds of groups I-III-VI$_2$ in the periodic table have great potential for the construction of solar cells. Its application for photovoltaic structures constitutes an important aspect from a technological point of view. The most important ternary and quaternary compounds for this type of cells are the compounds Cu-In-Ga-Se-S which have a wide of band gap from 1.0 to 1.5 eV. For example, CuInS$_2$ posses an ideal band energy gap that produce acceptable efficiencies. This compound has band gap of approximately 1.5 eV and high coefficient of optic absorption (10^{-5} cm^{-1}) (Rajaman et al. 1983; Meese et al. 1975). CuInS$_2$ can be prepared with conductivity type p or n, depending on the deposit conditions. However, the growth process has been difficult due to different factors, like the chemical stability. Cu(In,Ga)$_{1-x}$Se$_x$ (CIGS) received more attention due the adequate stability and the possibility to increase its band gap according to the amount of Ga in the film. There are different techniques to prepare CIGS thin films, such as the sputtering reactive (Yamamoto 1996), the co-evaporation (Hwang et al. 1980), the combination of the evaporation and sulfurizacion (Sheer et al. 1997, 2001; Gossla et al. 2001; Kanzari et al. 1997), spray pyrolisis (Krunks et al. 1999, 2002; Ortega and Morales 1998; Olvera et al. 1999), and chemical deposition (Bini et al. 2000).

This type of cells were developed during 1980's, when Boeing Company in Seattle WA, was developing the first cell based mainly on CuInSe$_2$ thin films. After some time certain work was made with the incorporation of Ga and S into the film in order to increase the absorption coefficient. Also, Bell Laboratories were developing this work and they formed a large working party. At the same time in Europe, another group was formed, known as "EuroCIS" giving origin to a consortium that develop more efficient solar cells, as it is shown in the Fig. 4. This figure shows that the efficiency has been increased drastically from 6 %, during eighties up to near 21 % in 2005.

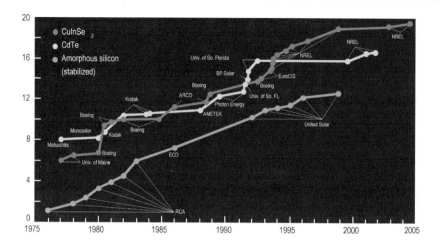

Fig. 5. Cell Efficiency vs. time curves for thin films solar cells (NREL, courtesy of M Contreras)

There are two methods to obtain high efficiency solar cells using CGIS thin films, one is to combine Cu, In, Ga Se and S in adequate quantities to form an stequiometric film, and the second is to prepare tandem layers of binaries compounds, such Cu_xSe, In_2Se_3, Ga_2Se_3. These methods need to be optimized in order to improve the crystalline structure, stability and optoelectronic properties. The co-evaporation is one of the most used techniques to prepare these compounds. This technique allows incorporating elements and binary compounds into the film and the possibility to adjust the band gap. Figure 6 shows different processes used to prepare CGIS absorbers and their corresponding solar cell efficiency. It can be seen that a Mo/glass substrate was used in all cases. However in some cases, it is possible to replace Ga for Al still obtaining an adequate efficiency for the corresponding solar cell, as it shown in Table 2.

Table 2. Efficiency values for different configuration of solar cells

Configuration	Area [cm^2]	V_{OC} [V]	J_{SC} [mA/cm^2]	FF [%]	Efficiency [%]
CIGSe/CdS	0.41	0.97	35.1	79.52	19.5
CIGSe/ZnS(O,OH)	0.402	0.67	35.1	78.78	18.5
Cu(In,Ga)S$_2$/CdS	0.409	0.83	20.9	69.13	12.0
Cu(In,Al)Se$_2$/CdS	-	0.621	36	75.5	16.9

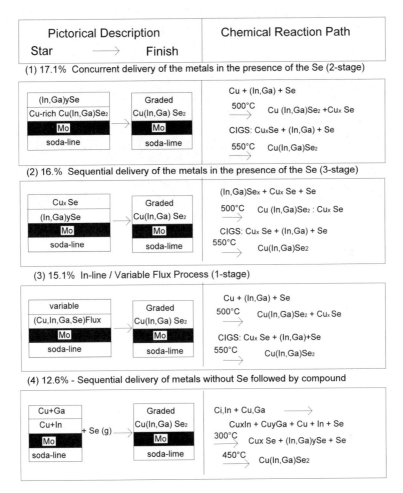

Fig. 6. Co-evaporation processes used to grow CIGS thin film absorber for different efficiency solar cells

Since 1996, the Centro de Investigación en Energía-UNAM-Mexico started working in the development of the electrodeposition technique, to growth CGIS thin films. This method consists in growing a 1–2 μm thickness layer from a chemical solution prepared from metallic salts, using a three electrodes system. This layer receives a thermal annealing process to form an adequate crystalline structure. After that, it is possible to build the photovoltaic structure shown in Fig. 7.

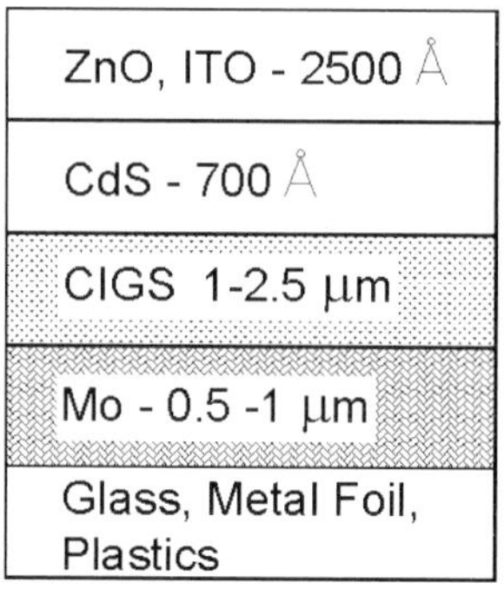

| ZnO, ITO - 2500 $\overset{\circ}{A}$ |
| CdS - 700 $\overset{\circ}{A}$ |
| CIGS 1-2.5 μm |
| Mo - 0.5 -1 μm |
| Glass, Metal Foil, Plastics |

Fig. 7. CIGS solar cell structure

CGIS film is grown on a Mo/glass substrate making with the latter an ohmic contact. On the top of this absorber layer, a 700 $\overset{\circ}{A}$ thin film of CdS is placed by chemical deposition. This film is the *n*-type semiconductor material. With the intention of reducing the "dangerous" risk of employing CdS, there are some research groups working on the substitution of this layer for another material just as the In(OH)$_3$ case. On top of the cell there is a ZnO layer acting as window layer allowing the incidence of solar radiation with wavelengths longer than 0.36 microns. The electric contact is formed with a deposit of Ag.

Table 3. State of art of the photovoltaic technology at the market

STATE OF ART OF PHOVOLTAIC TECHNOLOGY		
Technology type		Situation at the market
Homounion design	Monocrystaline Silicion (thick)	Available
	Polycrystalline Silicion (thick)	Available
	Amorphous Silicion (thin films)	Available
	Monocrystalline thin films	
	Galium Asenic (GaAs)	Space application
Heterounion design	Polycrystalline Thin films	Available
	Copper-Indium diselenide	
	Cadmiun teluride	
Multiple junction design	a-SiC/a-Si	Available
	a-Si/a-Si	Available
	a-Si/a-SiGe	Available
	a-Si/poli-Si	Available
	a-Si/CuInSe$_2$	Low development
	GaAs/GaSb	Low development

A CGIS film was grown on a Mo/glass substrate with which an ohmic contact was made. On the top of the absorber layer (p-type CGIS) was deposited of CdS thin film (700 Å of thickness) by chemical deposition as use as semiconductor n-type material (Martinez et al. 2004).

Table 3 summarizes the development of the photovoltaic technology based on each one of the different solar cells above explained and also their present state in the market.

5 Solar photovoltaic powered water pumping systems

One of the main problems in countryside communities which restrict their life quality is the availability of water. It is well known that there is underground water and also that there are several ways of getting it to the surface. In order to have availability of water, people must build wells (drilled or hand made), or another kind of reservoirs. Water can be extracted either with the use of a simple machine "manual extraction with a bucket", or hand made pumps, or the assistance of animals, or electromechanical power sources. Using these kinds of "technologies", people have gotten their daily water requirements. In all cases, it is necessary to expend energy to extract water to ground level. For example, if a hydraulic potential is needed for irrigation purposes, water must be withdrawn to ground level. All this process is known as water pumping.

Water pumping systems are very important for the development of countryside areas because water is one of the essential resources to support life as well as for agriculture production, but in order to execute the water supplying process it is necessary to waste energy. The quantity of energy required for a water pumping system depends on the water volume per day (m^3) and the hydraulic head (m) across water must be pumped. The product of the volume and the hydraulic head is called de hydraulic energy or hydraulic cycle, for which the unit m^4 ($m^3 \times m$) is often used.

Solar photovoltaic powered pumping systems have demonstrated to be the best choice, from an economical point of view, to pump water for applications in which the hydraulic energy is less than 1,500 m^4. They have many advantages over more traditional technologies, among them; the available solar electricity produced by PV systems is well matched to the water demand, its reliability is high because the maintenance is minimum due to there are only a few moving parts, the system does not require operator neither oil or diesel consumption (Barlow et al. 1993).

PV water pumping systems are similar to those represented by the conventional pumping systems, exempt for the power source. Conventional systems commonly available in the market have been developed for working under a constant electrical voltage. On the other hand, the power that produces the PV modules is directly proportional to the availability of the solar radiation: as the sun changes its position during the day, the irradiation of the PV modules change, and the generated power by the modules varies; consequently, it changes the power given to the pump. Therefore, special water pumping systems have been developed to work with variable voltage and current. The requirement is that the motor-pump system must be able to work fairly efficiently over a range of voltage and current levels.

While PV water pumping system use the solar power, conventional pumping systems use oil or wind to produce mechanical power. A typical PV water pumping system is shown in Fig. 8 Its principal components are the PV array formed by several modules connected in series and parallel, the pump controller, DC-AC inverter to convert DC in AC current, the tracking system to mount and orient the modules to the sun, water level sensors, float switches, water conduction lines, and a storage tank.

A PV array must provide daily enough electricity to the pumping system in order to pump the volume of water required per day. A tracking system is necessary to follow the apparent movement of the sun and to increase almost 30 % the energy production, which allows getting a higher volume of pumping water. The pumping system is composed by an electrical motor and a pump. While alternating (AC) or direct current (DC) motors can be used to supply the mechanical power to the pump, water pumps can be of centrifugal or of positive displacement.

The centrifugal pumps have an impeller that drags water by its axis due to its high speed expelling it radially. These pumps can be submersible or not submersible (pumps for surface). They are capable to pump water 60 meters away or more, depending on the number and type of impellers. The superficial centrifugal pumps are installed at ground level and have the advantage that they can be inspected and repaired easily. However, if the suction takes place deeper than 9 meters, the pumps don not work. This kind of problem is not present in submergible pumps. Some of the submergible pumps have a built-in engine directly to the impellers and they can be completely submerged. Others have the motor on the surface, whereas the impellers are completely submerged and united by a shaft.

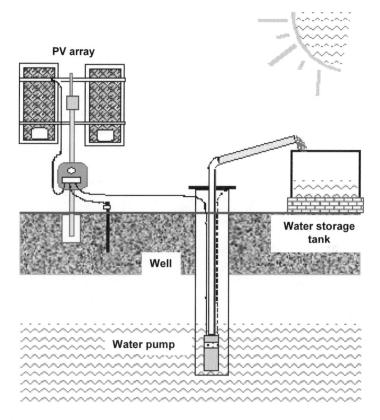

PV array

Water storage tank

Well

Water pump

Fig. 8. Typical diagram of a PV water pumping system that has been used for most of the water pumping projects in countryside areas

Since PV systems do not generate electricity at night, it is necessary to storage the electrical energy in an electrochemical battery. In a hydraulic system, like a water pumping system, the energy produced by the PV system is represented by the volume of water pumped daily. Since there is not pumping process at night, and in order to have the availability of water at any time, the storage of water is needed. Generally, the water is stored in a container or a tank. Water storage in a tank is cheaper than energy storage in an electrochemical battery. The storage container provides water requirements at night or when there is not enough solar energy for pumping. The container can also provide an enough hydraulic potential for irrigation purposes. Its volume size must be greater than the daily water consumption. The storage of electrical energy in electrochemical batteries is normally justified only when the peak efficiency of the well during the hours of sun is insufficient to satisfy the daily water needs and when it is required to pump water during the night.

At the moment, there are thousands of photovoltaic pumping systems in operation in farms and ranches around the world. PV systems can satisfy a wide range of necessities that go from small cattle ranches (less than 20 heads of cattle) to moderate requirements of irrigation. The solar pumping systems are simple, reliable and require of a little maintenance. Fuel is not required either. These advantages must be considered carefully when the initial costs of a conventional and a solar photovoltaic powered pumping system are compared.

Fig. 9. PV water pumping system of 1 kW peak installed in a rural community of the State of Guerrero, Mexico

Due to its advantages, the PV water pumping system has become the best choice to pump water in countryside communities of Mexico. With more than 1600 systems installed under two National Programs, it has been at present time, the technology that has more reliability in the field. Figure 9 shows the setting up of a 1 kW PV module for water pumping, supplying daily 10 m^3 of water, in the countryside of the State of Guerrero, Mexico.

6 Costs and challenges in the photovoltaic technology

The major obstacle for the use at a great scale of the photovoltaic technology based on the silicon technology is the high cost for the initial investment. Nowadays, the generation costs of electricity vary between US$ 0.50 and US$ 1.0 per kW-p depending on the kind of systems configuration and locations (country, city, etc.). As a result of the high investment price, the use of the photovoltaic generators is this limited mainly to applications where the connection cost to the grid and consumption of the electricity is high, or, where it is necessary a generation of clean, silent and reliable electricity. This is the case of remote areas (rural electrification, telecommunications, pumping of water, signaling, etc.) or when it is necessary to reduce the high peaks consumption in factories or buildings. Without considering the initial cost of investment, the popularity of the photovoltaic technology is based on its easy use, dependability, relative low maintenance and an immense readiness of the solar resource.

It is expected that the market of the photovoltaic generators can expand quickly when the PV generated electricity can be connected to the grid, at comparable prices to the costs of electricity generation by means of fossil fuels (between $0.50 and $1.25 Mexican pesos per kW-p). This goal can be reached with the actual crystalline silicon technology only if the production costs of photovoltaic systems decrease significantly, and if the social costs of conventionally generated electrical energy with fossil fuels are considered.

Another alternative is the use of the thin film technology of CGIS solar Cells, which allow us to obtain economical saving by expending less material. For example, if we want to build PV modules with an electric power of 150 W/m^2, whose material cost is US$ 50/m^2 and with efficiency close to 15 %, then one would have a cost near to US$ 0.30/W-p, which means US$ 50/m^2/150 W/m^2, 10 times less than current photovoltaic systems. Using these values, it is possible to predict that the cost of the photovoltaic systems based on thin film technologies will be in order of US$ 1–1.5/W-p that compared with actual costs of silicon PV systems (US$ 6–10 /W-p), it represents a considerable saving. With this prognosis, it is possible to expect a reduction cost in the next generation of the thin film solar cells technology. Logically, it depends on the opening of the market for CGIS solar cells as those existing at the moment for the silicon PV modules. The main markets that could allow reaching this reduction correspond to the grid electric connection.

A substantial growth in the photovoltaic technology must be accompanied by 1) an increase in the conversion efficiency of solar cells and as well as of photovoltaic modules and by 2) a reduction of the production costs, which can be achieved by application of new materials based on thin films. However, the photovoltaic devices have also demonstrated that they are ideal to generate electricity: they don't require maintenance, they don't produce waste and the alone electricity generated depends only on the sun. The PV technology is one of the cleanest productions of electrical energy.

Acknowledgement

This work was partially financed by projects DGAPA-UNAM: IN112206-2, CONACYT: 49895-Y and OAS: AE/141/01.

References

Barlow R, McNelis B, Derrick A (1993) Solar pumping: an introduction and update on the technology, performance, costs, and economics. World Bank Technical Paper No 168, UK

Bini S, Bindu K, Lakshmi M, Sudha Kartha C, Vijayakumar KP, Kashiwaba Y, Abe T (2000) Renewable Energy 20:405–413

De Buen O (2006) Estudio de Mercado, 2a fase. Proyecto de energías renovables. FIRCO-SAGARPA, Mexico

First Solar (2006) http://www.firstsolar.com/index.html

Gossla M, Metzner H, Mahnke HE (2001) Thin Solid Films 387:77–79

Green MA (2001) Cristalline Silicon Solar Cells. In: Archer MD, Hill R (eds) Clean electricity from photovoltaic. Imperial College Press, UK, pp 149–198

Green MA, Emery K, King DL, Igari S, Warta W (2005) Solar Cell Efficiency Tables (Version 25). Progress in Photovoltaic Research Applications 13:49–54

Hwang HL, Tu CC, Maa JS, Sun CY (1980) Solar Energy Materials 2:433–446

Kanzari M, Abaab M, Rezig B, Brunel M (1997) Materials research Bulletin 32:1009–1015

Krunks M, Bijakina O, Varema T, Mikli V, Mellikov E (1999). Thin Solid Films 338:125–130

Krunks M, Kijatkina O, Rebane H, Oja I, Mikli V, Mere A (2002) Thin Solid Films 403–404:71–75

Lorenzo E (1994) Solar Electricity: Engineering of Photovoltaic Systems. PROGENSA, Spain

Martinez AM, Arriaga LG, Fernandez AM, Cano U (2004) Materials Chemistry and Physics 88:417–420

Matulionis I, Sijin H, Drayton JA, Price KJ, Compaan AD (2001). Cadmium telluride solar cells on molybdenum substrates. In: Proceedings of the Materials

Research Society symposium on II-VI compound semiconductor photovoltaic materials, Warrendale, PA, USA, Vol 668 H8.23:1–6

McCandless BE, Youm I, Birkmire RW (1999) Optimization of vapor post-deposition processing for evaporated CdS/CdTe solar cells. Prog Photovolt Res Appl, 7 pp 21–30

Meese JM, Manthuruthil JC, Locker DR, Bulletin (1975) American Physical Society 20:696

NREL (2001) http://www.nrel.gov/news/press/2001/1501_record.html

Olvera M, Maldonado A, Asomoza R (1999) Superficies y Vacío 8:109–113

Ortega LM, Morales AA (1998)Thin Solid Film 330:96–101

Panicker MPR, Knaster M, Kroger FA (1978) Cathodic deposition of CdTe from aqueous electrolytes (large area solar cells). J Electrochem Soc 125:556–572

Rajaman P, Sharma AK, Raza A, Agnithori OP (1983) Thin Solid Films 100:111

Romeo A, Arnold M, Bätzner DL, Zogg H, Tiwari A (2002) Development of High Efficiency Flexible CdTe Solar Cells. In: Proceedings of PV in Europe from PV Technology to Energy Solutions Conference and Exhibition, Rome, Italy, pp 377–381

Romeo A, Khrypunov G, Kurdesau F, Bätzner DL, Zogg H, Tiwari AN (2004) High Efficiency Flexible CdTe Solar Cells on Polymer Substrates, In: Technical Digest of the International PVSEC-14, Bangkok, Thailand, pp 715–716

SENER (2005) Secretaría de Energía. Balance Nacional de Energía 2004. México

Sheer R, Alt M, Luck I, Lewerenz HJ (1997) Solar Energy Materials and Solar Cells 49:423–430

Sheer R, Luck I, Kanis M, Matsui M,. Watanabe T, Yamamoto T (2001) Thin Solid Films 392:1–10

Wronski CR, Carlson DL (2001) Amorphous Silicon Solar Cells. In: Archer MD, Hill R (eds) Clean Electricity from Photovoltaic. Imperial College Press, UK, pp 99–244

Yamamoto Y, Yamaguchi T, Demizu Y, Tanaka T, Yoshida A (1996). Thin Solid Films 281:372

Some Recent Research on Solar Energy Technology

Camilo A Arancibia-Bulnes, Antonio E Jiménez, Oscar A Jaramillo, Claudio A Estrada

Departamento de Sistemas Energéticos y Departamento de Materiales Solares, Centro de Investigación en Energía, Universidad Nacional Autónoma de México, Privada Xochicalco s/n, Colonia Centro, A. P. 34, Temixco, Morelos 62580, México. E-mail: ajg@cie.unam.mx

1 Introduction

Mexico is located in the Earth's sunbelt, where solar energy is plentiful for potential applications of solar energy conversion systems. According to several estimations (Renné et al. 2000), the average insolation over the country's surface amounts to 5 kWh/day, which puts Mexico in a privileged situation for the deployment of solar energy technologies. Other renewable energy sources such as: wind, biomass, geothermal, large and small hydrological, the ocean, play a role in the national energy supply system or have the potential for being an important factor, but none of them has the potential of solar energy, which could easily satisfy all of the country's energy requirements. In spite of this, the development of solar energy in México is still marginal. There are several small industries that have been producing and installing flat plate solar collectors for household hot water production and swimming pools for several decades, and the installed capacity has been growing steadily for some years now (68,725 m^2 of collectors were installed during 2004). Also, PV panels are imported and installed in a regular basis but not in very large quantities (9,923 kW were installed during 2004) (SENER 2005).

Solar energy is still is not very noticeable in the national energy balance (SENER 2005). However, interest in the subject is growing and the Mexican authorities are starting to allocate research funds specifically targeted to solar and other renewable energy sources. A 25 MW parabolic trough solar power plant is planed for deployment at Agua Prieta, in Sonora, under the auspices of GEF and the Comisión Federal de Electricidad, the state owned Mexican utility. Moreover, a bill for a law supporting renewable energy is being discussed in the Mexican Congress.

Research on solar energy has been carried out in Mexico for more than four decades. The National Autonomous University of Mexico (Universidad Nacional Autónoma de México, UNAM) has played an important role in the development and research in this area, and also in training experts in the field through its graduate programs in engineering. In this document some recent research carried out in the field of solar concentrating systems by the authors and other collaborators at UNAM's Centro de Investigación en Energía (Energy Research Center; CIE-UNAM) is reviewed.

2 Calorimetric evaluation of concentrated radiative fluxes

Despite of the many achievements in solar concentrating systems, which have taken place in the last decades, the seemingly simple task of measuring highly concentrated radiative fluxes remains a problem (Ballestrín et al. 2003; Ballestrín et al. 2004; Ballestrín et al. 2006; Kaluza and Neumann 2001; Ulmer et al. 2004). The methods that have been employed for this measurement include both radiometric and calorimetric techniques. There are three kinds of devices that have been used to measure the concentrated solar flux, namely:

- *Calorimeters* (Groer and Neumann 1999; Ferrier and Rivoire 2000). The operation principle of any instrument of this type consists on determining the heat absorbed by a cooling fluid circulating through the device, which is usually water. By carrying out a balance between the energy absorbed by the fluid and the incident solar energy on the concentrating system, it is possible to estimate the concentrated radiative flux that reaches the calorimeter. This technique requires reducing at a minimum the uncertainties in the mass flow rate measurement of the cooling fluid, and in the variation between its entrance and the exit temperature. It is very important to know the thermal properties of the material of the receiver, the properties of the cooling fluid, and to carry out a good estimation of the heat losses to the environment.
- *Radiometers* (Gardon 1953; Ferrier and Rivoire 2002). In this case the concentrated radiative flux is estimated through a variation of temperature measured by a transducer at a given location on the sensor.

Thermocouples or thermopiles are commonly used to carry out this task. Photo-sensors are also used in some devices to measure the radiative flux in a direct manner. All radiometers require calibration.

- *Calorimeter-Radiometers* (Ballestrín et al. 2004; Ballestrín et al. 2006; Pérez-Rábago et al. 2006). They are sensors that allow measuring the concentrated flux by simultaneous calorimetric and radiometric methods. This type of device operates normally as a radiometer; nevertheless, the calorimetric component of the system is included to facilitate a reliable calibration of the radiometer in the same environment where the system is to be operated.

In 1995, a solar concentrating system called DEFRAC (Spanish acronym for *Device for the Study of Highly Concentrated Radiative Fluxes*) (Estrada et al. 1998) was developed at CIE-UNAM in order to support basic and applied research on solar concentrating technologies. DEFRAC (Fig. 1) is a point focus solar concentrator with an equatorial solar tracking system. It consists of two frames: One is used as the main structural support and the other one, of hexagonal shape, holds 18 first surface parabolic mirrors. Each mirror measures 30 cm in diameter and is made of aluminized glass with a reflectivity of 0.95. They are arranged in three sets of six mirrors each and this sets have three different focal distances with values close to 2 m. The total area of the surface reflectors is 1.27 m^2. The axis of the hexagonal frame is supported by two lateral bearings, attached to the main frame. The main frame has an electric motor, sensors and a control mechanism to follow the sun with a high degree of accuracy. DEFRAC concentrates a mean flux close to 3,000 suns with a peak above 4,750 suns (Estrada et al. 1998).

Fig. 1. Photo of the solar concentrator DEFRAC

Cold Water Calorimetry (CWC) is one of the techniques that have been explored with the purpose of evaluating the concentrated solar flux that arrives at the receptor of DEFRAC (Estrada et al. 2006). Usually calorimetry is carried out by measuring the inlet and outlet temperatures of water, and performing a thermal balance. It is important to indicate that CWC technique requires a very good heat transfer from the solar flux to the fluid flow in order to have the receiving surface of the calorimeter at a temperature close to ambient. This can be achieved varying the fluid flow rate or increasing the conductive heat transfer rate by using a material with high thermal conductivity. The aim is to reduce the temperature of the receiving plate in order to reduce radiative and convective losses, hopefully eliminating the need for a precise estimation of these losses. In principle this should simplify very much the task of carrying out the heat balance.

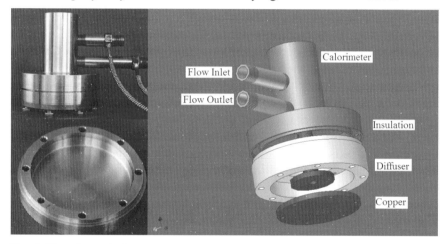

Fig. 2. Flat plate calorimeter

A photograph and a drawing of the FPC that was developed for the determination of the concentrated radiative power produced by DEFRAC are depicted in Fig. 2. The body of the calorimeter is made of AISI 316 stainless steel (Fe/Cr18/Ni10/Mo3) and consists essentially of two concentric cylindrical conduits. A circular steel plate, which acts as a flow diffuser, is located at the lower end of these conduits. The concentrated solar radiation arrives onto a 5 mm thick circular receiving plate, which is made of copper. Water enters the calorimeter from one side, and washes the inner surface of the receiving plate, removing heat. At this point the flow moves outwards between the receiving and diffuser plates. Finally, water exits at one side of the device. Note that there are several heat fluxes on the system: on the inner face of the receiving plate, convective heat transfer occurs by the radial flow of water, while in the outer face heat is lost by convection and radiation to the ambient. The wall of the calorimeter is insulated to prevent heat transfer to the surroundings.

A theoretical and experimental study of the thermal behavior of the FPC under concentrated solar flux was developed in order to estimate the temperature of the receiving plate. The mathematical model allows the calculation of the temperature field in the plate and the results are validated against the experimental results. Based on the thermal study and analyzing the heat balance in steady state it is possible to verify the supposition of the CWC technique for the FPC since the heat losses to the ambient from the front face of the receiving plate can be neglected (Jaramillo et al. 2006).

In order to improve the calorimetric measurements of DEFRAC's concentrated solar power, a cavity calorimeter named CAVICAL was built and tested. This device allows a better control over thermal emission and makes measurements less dependent on the absorptivity of the surface that collects the concentrated solar flux. Figure 3 shows a schematic view of CAVICAL. It is a calorimeter which acts as a conic heat exchanger with water as a cooling fluid. It consists of two concentric cones. The inner one, which is made of copper, receives the concentrated solar radiation coming from the mirrors and has a vertex angle of 15°, a height of 16 cm, a base diameter of 8.57 cm, a base aperture diameter of 3.24 cm and a wall thickness of 0.3 cm. The outer cone is made of stainless steel and its wall is 0.8 cm thick. There is a separation gap of 1.0 cm between the cones. Water enters the calorimeter at the vertex, flows between the cones and exits the device at a base aperture. Flow direction can also be reversed. All calorimeter dimensions were obtained optimizing the apparent absorptance and acceptance efficiency of the cavity (Pérez-Rábago 2003).

In CAVICAL the concentrated radiative energy that comes from the mirrors and passes through the calorimeter's aperture is absorbed by the surface of the inner cone and is then transmitted to the interior of the wall by conduction. The water circulating between the cones absorbs that energy by forced convection. Also, part of the irradiance entering through the calorimeter aperture is lost by reflection and thermal emission back through the aperture, by heat conduction from the calorimeter to the surroundings, through the insulation, and by natural convection into the air through the aperture. It is important to point out that the cavity calorimeter design has as a primary objective to diminish these losses.

In order to analyze the thermal behavior of CAVICAL a theoretical and experimental study was carried out. The theoretical heat transfer study was focused on the determination of the temperature distributions on the different components of the device as well as heat losses from the calorimeter to the surroundings by using a Computational Fluid Dynamic (CFD) commercial code. The simulation was validated through a comparison between experimental and theoretical wall temperature profiles for the CAVICAL, having as a maximum difference less that 2K or less than 6 %.

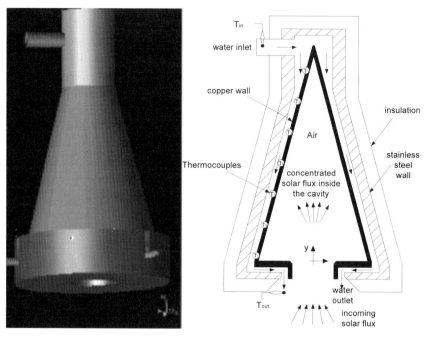

Fig. 3. Schematic view of the CAVICAL

3 Direct solar steam generation for soil disinfection

Other project that is being developed is the study of a sustainable soil pasteurization system. In the intensive agriculture, some plagues and diseases deteriorate the quality of the crops. Methyl bromide (MeBr) is a broad spectrum pesticide that is used to control nematodes, fungi, other pathogens, insects and weeds. Its primary agricultural use is fumigation of soils prior to planting. It is safer than other chemicals in some respects because it leaves no toxic residue in the soil. However, this toxic gas rises into the atmosphere and depletes stratospheric ozone. On the other hand, non-chemical alternatives are steam soil disinfection, non-soil cultivation, solarization, and bio-fumigation. In particular, steam disinfection is highly efficient, requires a short processing time, does not leave toxic residues, and does not present risks to human health.

In steam disinfection, steam is injected, either from above or below, in soil beds of 65 cm in depth, which are covered with an impermeable material. Earth is heated to 85°C during 30 minutes, in order to inactivate nematodes, and 95 % of the spores of the pathogenic fungi. However, steam is generated for this process by using conventional boilers, which burn gas, diesel, or other fuels. Therefore, high maintenance and combustible costs, and generation of polluting gases (CO, CO_2, NO_x, and S_2) that contribute to the atmospheric greenhouse effect are the drawbacks of this method.

The use of steam generated direct from solar energy by means of parabolic trough collectors (PTC) is being studied. Firstly, a single low-cost and easy-to-build parabolic trough module was constructed, and characterized. In Fig. 4 the PTC prototype is depicted. Secondly, a parabolic trough solar field to achieve a steam pasteurization process was modelled. This theoretical study was based on experimental results obtained from the prototype. Currently, the investment cost of the soil pasteurization system, the savings in maintenance costs, and the potential for reduction of polluting gases emissions with this technology are being estimated.

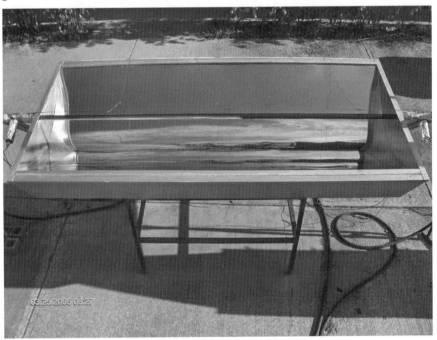

Fig. 4. Parabolic trough collector prototype

The PTC was built in aluminum, since this material offers low cost and a high resistance to adverse environmental conditions. A commercial aluminum sheet with dimensions 1.22×2.44 m was used to construct the reflective surface of the PTC. This surface presents a reflectance close to 0.92 %. The collector was designed with a rim angle of 90°, in order to place a glass cover on the top for reducing the convective heat losses to the ambient. The focal distance was fixed to 0.25 m and the aperture of the collector is around 1 m in width by 2.44 in length. The size of the cylindrical receiver was determined by a ray tracing method. The prototype can heat water in the temperature range from 90 to 100°C, with a pressure of 2 kg/m^2 of and a mass flux of 0.5 L/min.

4 Advanced oxidation processes for water detoxification

One of the principal problems in the world for the next millennium will be the scarcity of drinking water. Dramatic growth in the contamination levels of rivers, lakes, lagoons are observed. Discharges from agriculture, farms, housings, extractive industries, sanitary wells, garbage dumps, and deposits of dangerous residues, have been identified as major sources of water contamination. These contamination sources originate pollution in the form of nutrients (nitrates and phosphates), sediments, bacteria, viruses, organic enrichments, toxic organic chemicals, and heavy metals (Cheremisinoff and Graffia 1996; Callahan et al. 1979). In particular, the industrial sector generates a high variety of contaminants that have their origin in refineries, chemical and textile industries, soap and paper factories, etc. In addition, the use of pesticides, fertilizers, detergents and aerosols have drastically contributed to the deterioration of soils and water (Pelizetti et al. 1994).

The standard methods for residual waters treatment include sedimentation, activated muds, decantation, ozonisation, and chlorination. However, these processes are not effective for the remotion of recalcitrant toxic contaminants, even worse; some of them can lead to degradation intermediates which are even more toxic than the original pollutants (Stachka 1984). Even the traditional treatments of biological type are considerably limited when they are applied for the depuration of persistent contaminants.

During the last three decades, researchers in several areas such as environmental engineering, chemistry, photochemistry and physics have contributed to develop Advanced Oxidation Processes (AOP). In general, AOP's include all possible oxidation reactions that tend to transform toxic organic and inorganic compounds in more simple compounds (not toxic) like CO_2 and mineral acids.

Some of the AOP's involve electromagnetic radiation as excitation source. In particular, two of these processes that have raised considerable interest are heterogeneous photocatalysis, and photo Fenton. They are particularly attractive from the environmental point of view, because they are able to work with a clean radiation source like the sun.

In heterogeneous photocatalysis (Blesa and Sánchez 2004; de Lasa 2005), a semiconductor solid in contact with the water to be treated is illuminated. The material is able to absorb photons with energies above the band gap, and electron-hole pairs are generated during the irradiation. Electron and holes are capable to induce reduction and oxidation reactions, respectively. As a consequence of these oxidation and reduction reactions, one of the generated species on the surface of the photocatalyst is the hydroxyl radical ˙OH (Alfano et al. 1997). Since OH radicals have a very high oxidation power (they are the second most oxidizing chemical species known), they can oxidize the majority of the organic compounds. Because of this, photocatalysis is able to treat complex mixtures of pollutants. Very often, it is possible to use hydrogen peroxide (H_2O_2) together with the photocatalyst in order to favor the formation of extra ˙OH radicals to achieve total degradation of contaminants. The capacity of the TiO_2 semiconductor utilizing UV radiation to degrade a vast variety of organic compound has been already reported (Blesa and Sánchez 2004).

The second very important AOP activated by radiation is photo-Fenton, which is based in the process discovered by Fenton (1884). Fenton demonstrated that aqueous solutions of hydrogen peroxide and iron salts, are capable to generate ˙OH radicals via reactions of the iron ion, $Fe(II/III)/H_2O_2$, at a pH value of 3, and to oxidize several acids and other organic compounds (Nesherwat and Wanson 2000; Huang et al. 1993; Walling 1975). In general, Fenton and photo-Fenton reactions have been worked in the homogeneous phase, nevertheless, lately it has been possible to develop immobilized iron catalysts and to utilize them in processes of heterogeneous photocatalysis.

The implementation of the AOP for the treatment of residual waters involves research in three areas: photocatalytic materials, photoreactors, and applications. Recent work on these three areas will be discussed in the following sections.

5 Photocatalytic materials

A great variety of semiconductor materials with photocatalytic properties exists nowadays that can be used for the oxidation of organic compounds.

There are in general two conditions for the election of such materials: i) the material must be photosensitive against an electromagnetic excitation with adequate wavelength. Not all materials are photosensitive. During the absorption of radiation a quantum coupling must exist for the electron transition from the valence band to the conduction band. ii) The potential redox ε of a redox couple (A/A$^-$) in aqueous solution must remain within the band gap of the photocatlyst, and near to the valence band maximum, if the specie represented by the redox couple is to be oxidized or near to the conduction band minimum, if it is to be reduced. Several metallic oxides and chalcogenides satisfy these conditions.

Since the beginning of the research on these water treatment methods, people have utilized catalysts in homogeneous as well as heterogeneous phases. The use of TiO_2 as catalyst in heterogeneous photocatalysis is widespread. TiO_2 has a great stability in acidic, alcoholic and aqueous environments. In order to improve its catalytic activity, it is necessary to manipulate adequately the chemical and physical properties of TiO_2. In addition to TiO_2, the use of ZnO and CdS for photocatalytic degradation of organic compounds has been reported.

In powder form, it is possible to commercially obtain TiO_2 catalyst from Degussa, J. T. Backer, Adrich, Merck, etc. This material is suspended in the polluted water and exposed to radiation to carry out the degradation process. At the end of the process it is necessary to separete the catalyst from the treated water. Another possibility is the immobilization of TiO_2 thin film in glass and metallic substrates, with different geometries. The immobilized catalyst offers the advantage of easy recovery and reuse in later processes of degradation. This is more difficult with the catalysts in suspension.

Although diverse techniques exist for the preparation of the TiO_2 catalyst, the sol-gel technique is very adequate for this task and is well controlled. Sol-gel allows preparing TiO_2 in powder form or as films immobilized on glass, metals and ceramics (Gelover et al. 2004). Fig. 5 for instance, shows a picture of Pyrex glass tubes covered with a sol-gel deposited TiO_2 film. The thickness of the films is 800 nm. After the sol-gel deposition, the films were treated in air at 350°C during one hour.

Fig. 5. Immobilization of the TiO$_2$ photocatalyst on Corning glass tubes to be used in CPC solar concentrators for photocatalytic degradation processes

6 Photocatalytic reactors

A necessary condition for the excitation of a photocatalyst is that the energy of the incident photons must be higher or equal to the band gap of the material. Thus, the use of ultraviolet lamps or solar radiation is very common for this type of processes. For each kind of irradiation source, a specific photocatalytic reactor design is required. During this work, we will restrict ourselves to the use of solar radiation as excitation source for photocatalytic degradation process.

When solar radiation is used for the degradation processes, it is possible to use both non concentrated as well as concentrated irradiation. In the first case, several photocatalytic reactors have been designed, such as the fixed bed reactors, horizontal or inclined, where solar radiation illuminates directly a catalyst covered by a water film.

When the concentrated radiation is used, it is necessary to use solar concentrators to focus the energy in the reactors. The reactors are generally glass tubes located along the focal line of a trough concentrator, and where water is circulated. Two kinds of solar concentrators have been developed for photodegradation processes (Fig. 6): parabolic trough concentrators (PTC) and compound parabolic concentrators (CPC). The PTC's require a sun tracking system to follow the apparent movement of the sun in such form that the aperture plane of the collector is always perpendicular to the solar beams (Jiménez et al. 2000). This type of concentrators is adequate to achieve concentration ratios in the range from 15 to 50 suns.

Fig. 6. Pictures of PTC (CIE-UNAM) and CPC (developed by AO SOL, Portugal) solar collectors used as reactors for photo degradation processes of organic compounds

CPC's offer with the best optics for systems of low concentration (Blanco 2002). Due to its low concentration, they do not require sun tracking. The reflective surface in these collectors plays mainly the role of helping to distribute radiation better around the reactor perimeter, rather than concentrating it by an important factor. The precise shape of the concentrator depends on the concentrating factor of the system, with sections that follow an involute curve around the cylindrical reactor and others that follow parabolic profiles. CPC's without concentration (C=1) have been studied in the past (Blanco 2002; Giménez et al. 1999; Malato et al. 1997).These low concentration systems have the advantage of making a more efficient use of the collected solar energy because, as will be discussed below, in many cases photocatalytic reaction rates increase not proportionally to radiation intensity, but as the square root. Therefore, higher concentration systems are less energy efficient even though they can be much faster to carry out pollutant degradation. In addition to this a CPC solar collector has the ability to use diffuse as well as beam solar radiation, while a PTC uses only the latter. Some studies have compared the performance of concentrating and non-concentrating collectors (Malato et al. 1997; Bandala et al. 2004).

Modeling is a useful tool to achieve better design of reactors and scaling-up of processes. In particular, photocatalysis is based on the absorption of UV light. Therefore, to model photocatalytic reactors, radiation transfer effects are as important as chemical kinetic or mass transfer effects. A careful design of the geometry and catalyst concentration, position of the light sources, etc., is important to achieve the best performance of a reactor. To this end, detailed optical modeling of the system is necessary (Cassano and Alfano 2000). In contrast, heat transfer effects are not so important; because temperature is not a relevant parameter in the range were photocatalytic detoxification processes usually work.

In particular, the reaction rate depends on the number of photons absorbed by the catalyst. There are two situations observed in practice: linear dependence when intensity of radiation is small and square root dependence when intensity is high. The first case is generally observed for non concentrating solar photoreactors, while the second occurs for concentrating photoreactors. Physically, the latter case corresponds to a situation where a large number of photons are available, and their efficient use is limited by the recombination of electron hole pairs in the semiconductor catalyst. Because of the generally nonlinear dependence of the local reaction rates with radiation intensity, the global reaction rate in a reactor does not depend only on the amount of radiative energy absorbed, but rather on the way this absorbed energy is distributed in the reaction space.

To achieve better reactor efficiencies it is generally necessary to have the semiconductor distributed in the reaction volume. This can be achieved in two ways: by depositing the catalyst as thin films supported on a large number of surfaces (Gelover et al. 2004), or suspending it in water as micro particles (Arancibia-Bulnes et al. 2002). Radiative transfer modeling is different depending in the way catalyst is used in the reactor; in the first case, essentially ray tracing of light between multiple surfaces is required, while the second situation is more involved. Because the particles scatter and absorb radiation, the optical calculations require the use of models which take scattering into account (Cassano and Alfano 2000).

To consider scattering effects, a radiative transfer model known as the P1-approximation has been employed. This approximation greatly simplifies the problem of modeling suspended catalyst reactors (Villafán-Vidales et al. 2006; Arancibia-Bulnes and Cuevas 2004; Arancibia-Bulnes et al. 2002).

The P1 approximation assumes that the angular distribution of radiation intensity is almost uniform within the medium, so for each point, radiative energy comes almost equally from all directions. To establish such a regime requires a fair amount of scattering, in order to erase from the propagating radiation the directional behavior originated from its sources.

Fig. 7, for instance shows the volumetric distribution of absorbed radiation in a cross section of the cylindrical reaction space located in the focus of a parabolic trough photoreactor (Arancibia-Bulnes and Cuevas 2004).

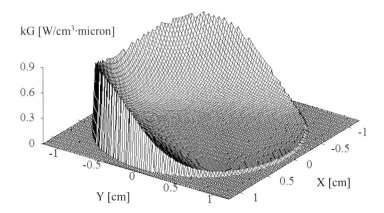

Fig. 7. Distribution of absorbed radiation in a cross section of the reactor tube in a parabolic trough photocatalytic reactor, at a wavelength of 325 nm and for catalyst concentration of 0.15 g/L

Figure 7 corresponds to a relatively high catalyst concentration, from the optical point of view. This causes that most radiation absorption is localized in volumes close to the tube wall from which the radiation is entered; a shading of the inner part of the reactor is caused by the high concentration of particles. A situation like this is not desirable, because a large portion of the reactor volume becomes ineffective for lack of illumination.

Generally, increasing catalyst concentration leads to improvement of the reaction rate until a maximum is achieved. In many cases a decrease of the reaction rates occurs if catalyst concentration is further increased (Jiménez et al. 2000). In some cases this decrease can be explained simply in terms of the optics of the reactor: an increased catalyst load can produce shadowing of the inner portion of the reactor leading to a reduction on the reaction rate. This is true for the case where the reaction kinetics behave nonlinearly (Arancibia-Bulnes et al. 2002; Arancibia-Bulnes and Cuevas 2004).

When the photocatalytic degradation of dyes is carried out (Villafán-Vidales et al. 2007) a competition for the absorption of radiation takes place between the catalyst and the dye. Although the absorption peaks of dyes are located in the visible region of the solar spectrum, their absorption on the UV region, where the TiO_2 catalyst is active, is not negligible. This may lead to a dramatically reduced efficiency of the degradation process when water is too heavily colored.

In Fig. 8 the reaction rates as function of catalyst concentration are shown for the degradation of an Azo dye in a non concentrating CPC photoreactor.

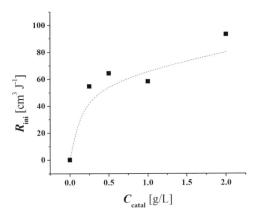

Fig. 8. Initial degradation rate of the Acid Orange 24 Azo dye, as a function of catalyst concentration, in a CPC photocatalytic reactor. Theoretical model (line), and experimental results (points). Initial dye concentration 12ppm, and catalyst concentration 0.1 g/L

The experimental results were fit to a theoretical model including the effects discussed above, and with radiation modeling based on the P1 approximation.

7 Application of advanced oxidation processes

Using the several types of photoreactors, it has been possible to degrade different families of toxic organic compounds by means of heterogeneous photocatalysis. Some of them are alkenes, aliphatic alcohols, carboxylic acids, aromatics compounds, phenols, detergents, VOC's, etc., which are already very well documented in the report by Blake (1999).

Figure 9 shows the photocatalytic degradation curve of the plaguicide carbaryl using a PCT solar concentrator and films of TiO_2 immobilized on Corning glass tubes as catalyst (Gelover et al. 2004). The contaminated aqueous solution was prepared taking carbaryl commercial grade at an initial concentration of 20 mg/l and to which hydrogen peroxide (3 g/l), as oxidant agent, was added. When the immobilized catalyst of TiO_2 was utilized, the degradation behavior was very similar to that case when the catalyst in powder form Degussa P-25 was taken. In accordance to the experimental results, both forms of the catalyst, in powder or immobilized, are equally efficient. This result is important because the operational advantages of the immobilized catalyst are of great value, without any diminishing in its catalytic efficiency in comparison to powder catalysts (Gelover et al. 2004). The addition of hydrogen peroxide improves the efficiency of the films TiO_2, reaching degradation levels of 93 %, as it can be seen in Fig. 9.

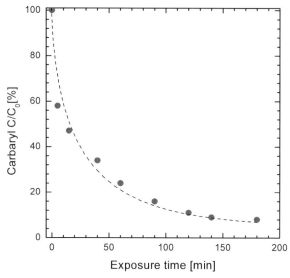

Fig. 9. Photocatalytic degradation curve of the plaguicide carbaryl using a CP solar concentrator of 41 suns and TiO_2 immobilized on glass tubes as catalyst

Figure 10 shows to the degradation curve of the industrial azo yellow lanasol dye using Fe_2O_3 thin films immobilized on corning glass substrates. The films were prepared by the sol-gel technique and treated thermally at 100°C in air. For the degradation process, an aqueous solution of the azo yellow lanasol dye was prepared at an initial concentration of 50 mg/l, and to which the pH was adjusted to a value of 3 and H_2O_2 was added as oxidant agent. As it can be seen in Fig. 10, the degradation of the dye was carried out in a very short period of irradiation; during the first hour a degradation of 90 % is achieved and after 3 hours, the dye is completely degraded.

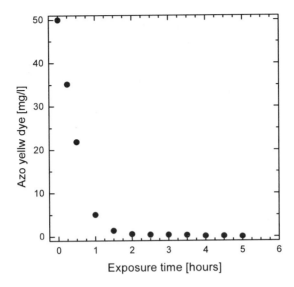

Fig. 10. Photocatalytic degradation of the Azo yellow dye using thin films of Fe_2O_3 immobilized on Corning glass as catalyst

Acknowledgements

We acknowledge technical support from Ing. José de Jesús Quiñones Aguilar and Ing. Rogelio Morán Elvira. The research reported has been partially supported by DGAPA-UNAM, through grants IN104205, IN113805-2, and IN112206-2; by CONACYT, through grants 37636-U, J36640-E, and 49895-Y; and OAS trough grant OAS AE/141/01.

References

Alfano OM, Cabrera MI, Cassano AE (1997) Photocatalytic reactions involving hydroxil radical attack. I. Reaction kinetics formulation with explicit photon absorption effects. J Catal 172:370–379

Arancibia-Bulnes CA, Cuevas SA (2004) Modeling of the radiation field in a parabolic trough solar photocatalytic reactor. Solar Energy 76:615–622

Arancibia-Bulnes CA, Bandala ER, Estrada CA (2002) Radiation absorption and rate constants for carbaryl photocatalytic degradation in a solar collector. Catal Today 76:149–159

Ballestrín J, Estrada CA, Rodríguez-Alonso M, Pérez-Rábago CA, Langley LW, Barnes A (2004) High-heat-flux sensor calibration using calorimetry. Metrologia 41:314–318

Ballestrín J, Estrada CA, Rodríguez-Alonso M, Pérez-Rábago CA, Langley LW, Barnes A (2006) Heat flux sensors: Calorimeters or radiometers?. Solar Energy 80:1314–1320

Ballestrín J, Ulmer S, Morales A, Barnes A, Langley LW, Rodriguez M (2003) Systematic error in the measurement of very high solar irradiance. Solar Energy Mater Solar Cells 80:375–381

Bandala ER, Arancibia-Bulnes CA, Orozco SL, Estrada CA (2004) Solar Photoreactors Comparison Based on Oxalic Acid Photocatalytic Degradation. Solar Energy 77:503–512

Blake DM (1999) Bibliography of work on the heterogeneous photocatalytic removal of hazardous compounds from water to air. Technical report NREL/TP-570-26797. NREL, USA

Blanco J (2002) Desarrollo de colectores solares sin concentración para aplicaciones fotoquímicas de degradación de contaminantes persistentes en agua. Ph.D. Tesis, Universidad de Almería, Spain

Blesa MA, Sánchez B (eds) (2004) Eliminación de contaminantes por fotocataálisis Heterogénea. CIEMAT, Madrid, Spain

Cassano AE, Alfano OM (2000) Reaction engineering of suspended solid heterogeneous photocatalytic reactors. Catal Today 58:167–197

Callahan MA, Slimak M, Gbel N, May I, Flower C, Freed R, Jennings P, DuPree R, Whitmore F, Maestri B, Holt B, Gould C (1979) Water-related environmental fate of 129 priority polutants. EPA-44014-79-029a,b, NTIS

Cheremisinoff NP, Graffia M (1996) Handbook of Pollution and Hazardous Materials Compliance. Marcel Dekker, Chapter 4

De Lasa H, Serrano B, Salaices M (2005) Photocatalytic Reaction Engineering. Springer, New York

Estrada CA, Cervantes JG, Oskam A, Cruz F, Quiñones JJ (1998) Thermal and optical characterization of a solar concentrator for high radiative flux studies. In: Campbell-Howe R, Cortez T, Wilkins-Crowder B (Eds) Proceedings of the 1998 Annual Conference of the American Solar Energy Society, vol 1 ASES, USA, pp 259–266

Estrada CA, Jaramillo OA, Acosta R, Arancibia-Bulnes CA (2006) Heat Transfer Analysis in a Calorimeter for Concentrated Solar Radiation Measurements. Solar Energy. In press

Fenton HJJ (1894) J Chem Soc 65:899–910

Ferrier A, Rivoire B (2000) Measurement of Concentrated Solar Radiation: The Camorimeter ASTERIX. In: 10th International Symposium on Concentrated Solar Power and Chemical Energy Technologies, Australia

Ferrier A, Rivoire B (2002) An Instrument for Measuring Concentrated Solar-Radiation: a Photo-Sensor Interfaced With An Integrating Sphere. Solar Energy 72:187–193

Gardon R (1953) An instrument for the direct measurement of intense thermal radiation. Rev Sci Instruments 24:366–370

Gelover S, Mondragón P, Jiménez A (2004) Titanium dioxide sol-gel deposited over glass and its application as a photocatalyst for water decontamination. J Photochem Photobiol A: Chem 165:241–246

Giménez J, Curco D, Queral MA (1999) Photocatalytic treatment of phenol and 2,4–dichlorophenol in a solar plant in the way to scaling-up. Catal Today 54:229

Groer U, Neumann A (1999) Development and test of a high flux calorimeter at DLR Cologne. J Phys IV 9:643–648

Huang CP, Dong Ch, Tang Z (1993) Waste Manag 13:361–377

Jaramillo OA, Estrada CA, Arancibia-Bulnes CA, Pérez-Rábago CA (2006) A Calorimetric Evaluation of a Solar Concentrator System. J Solar Energy Eng, Submmited

Jiménez AE, Estrada CA, Cota AD, Román A (2000) Photocatalytic degradation of DBSNa using solar energy. Solar Energy Mater Solar Cells 60:85–95

Kaluza J, Neumann A (2001) Comparative measurements of different solar flux gauge types. J Solar Energy Eng 123:251–255

Malato S, Giménez J, Richter C, Curco S, Blanco J (1997) Low concentrating CPC collectors for photocatalytic water detoxificattion with a medium concentrating solar collector. Water Sci Technol 35:157

Nesherwat K, Wanson AG (2000) Clean Contaminated sites using Fenton's reagent. Chem Eng Prog

Pelizetti E, Minero C, Vincenti M (1994) Technologies for Environmental Clean up: Toxic and Hazardous Waste Management; pp 101–138; ECSC, EEC, EAEC, Brussels

Pérez-Rábago CA, Marcos MJ, Romero M, Estrada CA (2006) Heat transfer in a conical cavity calorimeter for measuring thermal power of a point focus concentrator. Solar Energy 80:1434–1442

Pérez-Rábago, CA (2003) Master's Degree Thesis. Universidad Nacional Autónoma de México

Renné D, George R, Brady L, Marion B, Estrada-Cajigal V (2000) Estimating Solar Resources in México Using Cloud Cover Data. In: Estrada CA (ed) Proceedings of the ISES Millennium Solar Forum 2000. Asociación Nacional de Energía Solar, México, pp 627–632

SENER (2005) Secretaría de Energía. Balance Nacional de Energía 2004. México

Stachka JHY, Pontinuos FW (1984) J Am Water Works Assoc 76:73

Ulmer S, Lupfert E, Pfander M, Buck R (2004) Calibration corrections of solar tower flux density measurements. Energy 29:925–933

Villafán-Vidales HI, Cuevas SA, Arancibia-Bulnes CA (2007) Modeling the Solar Photocatalytic Degradation of Dyes. J Solar Energy Eng. In press

Walling Ch (1975) Acc Chem Res 8:125–131

Wind Energy: an opportunity for diversifying electricity generation in Mexico

Marco Antonio Borja

Marco A. Borja, Instituto de Investigaciones Eléctricas, Calle Reforma No. 113, Col. Palmira, Cuernavaca, Morelos, C.P. 62490, México.E-mail: maborja@iie.org.mx

1 Introduction

At present, wind energy is one of the most promising alternatives for diversifying electricity generation towards the sustainable development worldwide. Wind power technology is one of the most mature renewable energy technologies. In fact, over the last 20 years, almost all of the industrialised countries progressed towards the implementation of wind energy, looking to attain significant levels of contribution for satisfying the electrical demand at the national level. An increasing number of developing countries are already following the example.

Mexico possess abundant wind energy resource and has a number of niches of opportunity for taking advantage of all and each one of its potential benefits, including not only those related with the diversification of energy sources, but also most important challenges at the national level: employment creation and regional development.

This introduction gives some indicators of the deployment of wind energy worldwide as well as the most relevant for specific countries.

- *The World*. By July 2006, the wind power capacity installed worldwide was more than 63,000 MW. Sixty-one countries around the world have already installed certain wind power capacity, ranging from more than 19,000 MW in Germany to a few MW in a number of developing coun-

tries. Over the past six years, the wind power capacity installed worldwide has an average growth rate slightly lower than 30 %. During 2004
alone, the global wind energy industry has installed more than 8,000
MW of new wind power capacity (GWEC 2005), representing an investment of over USD 8,000 millions and an average installation speed
of more than 22 MW per day.

- *Europe*. The implementation of wind power in Europe is progressing
  quite quickly. In 1997, the European Commission established a strategic
  goal of 40 GW of installed capacity by 2010 (EC 1997); the goal was
  fulfilled in 2005. At present, around 75 % of the worldwide generating
  capacity and 90 % of the generating equipment are European. By 2005,
  wind power supplies more than 2 % of gross electricity consumption in
  Europe (GWEC 2005).

- *Germany*. Wind is the leading renewable energy in Germany and a
  growing factor in the national electrical energy market. In autumn 2005,
  the German government acknowledged the importance of renewable
  energies, and among them wind energy. By the end of 2005, the wind
  power installed capacity was 18,428 MW. During 2005, 1,808 MW of
  wind power were installed, which means that around 5 MW of wind
  power were installed every day. In 2005, wind power supplied 4.3 % of
  the total national electric demand. At this stage, wind energy is potentially able to produce 6.7 % of the German annual net electrical energy
  market. At present, there is a shortage of good wind areas onshore;
  therefore, the national wind-offshore strategy focuses on installing 2,000
  MW by 2010. At the end of 2005, the federal authorities approved 11
  pilot wind farms with 777 wind turbines and 3,331 MW within the
  German Exclusive Economic Zone (EEZ). Most of them still require
  permission for the transmission cable (Kutscher and Christmann 2006).

- *Denmark*. During 2005 electricity generation from wind energy was
  6,614 TWh covering 18.5 % of the total electricity consumption. Technical availability of new wind turbines on land is usually in the range of
  98 % to 100 % (Lemming et al 2006).

- *Spain*. By the end of 2005, the installed wind power capacity in Spain
  was 10,028 MW. During 2005, wind energy has supplied 20,236 TWh
  corresponding to 7.78 % of the national electric demand. The new installed wind power capacity in 2005 was 1,630 MW, which means that
  around 4.5 MW of wind power were installed every day. The new Spanish Renewable Energy Plan 2005–2010 has the goal of meeting 30 % of
  total electricity demand from Renewable Energy Sources at the end of
  2010. The specific target for wind energy was 13,000 MW, but now is
  increasing to 20,000 MW. A Royal Decree consolidated the support

scheme for renewable energy. The price paid for electricity generated by wind farms is guaranteed during the life of the installation and is strongly related to a very stable economic indicator, the Average Electricity Tariff (Soria 2006).

- *United States*. The year 2005 was a record-breaking regarding the installation of new wind power capacity in the U.S. New wind energy capacity for 2,431 MW was installed (an average of 6.75 MW every day); this brought the total national wind energy capacity to 9,149 MW. The wind energy industry expects this unprecedented growth to continue in 2006 with installations topping 3,000 MW. However, during 2005, electricity from wind contributed with less than 1 % of the national electrical demand, since the national electrical system is huge (O'Dell 2006).

- *India*. By the end of 2005, the wind power capacity installed in India was 5,200 MW. The Government of India, thought the Centre for Wind Energy Technology, supports a Wind Resources Assessment Programme, which up to now has estimated a wind power technical potential of 12,875 MW, based in the measurements of 59 anemometric stations (INWEA 2006).

- *China*. By the end of 2004, the wind power capacity installed in China was 760 MW, based in 43 wind power stations. The strategic goal for wind energy deployment is 20 GW by 2020 (NREL 2006).

Table 1. Indicators of the wind power capacity installed by the end of 2006, in some countries

Country	Wind-energy installed capacity by the end of 2006 [MW]	Country	Wind-energy installed capacity by the end of 2006 [MW]
Germany	19,500	Canada	950
Spain	11,200	France	950
United States	10,000	Australia	800
India	5,500	China	800
Denmark	3,200	Greece	600
Italy	2,000	Brazil	110
United Kingdom	1,800	Mexico	85
The Netherlands	1,400	Costa Rica	80
Portugal	1,400	Caribbean	60
Japan	1,150	Colombia	20
Austria	1,000		
		Total expected	68,000

Source: Wind RD+D Agreement of the International Energy Agency

Table 1 shows indicators of the wind power capacity installed by the end of 2006 in some countries.

2 Mexico's wind energy resource

At present, there is a constructive environment for launching the implementation of wind power in Mexico. More than a few of official documents recognize that Mexico's most important wind resource would be sufficient for the installation of at least 5,000 MW of wind power. In a topical workshop carried out in May 2005, the most important promoters of wind power agreed, by common consent, a shared vision of 6 % of wind power penetration at the national level for the year 2030. The workshop included representatives of the secretariats of energy, environment, and economy, as well as representatives of the Federal Electricity Commission, the Regulatory Commission for Energy, the National Chamber of Electrical Manufacturers, the Mexican Wind Energy Association, and the major research and academic institutions. Usually, it is acknowledged that the wind energy resource in Mexico is abundant and that it would be sufficient for the potential installation of more than twenty thousands of MW wind power capacity. However, all the participants of the workshop agreed that a strategic goal of 6 % for 2030 would be congruent and attainable, based on the development rate of the implementation of wind energy in industrialised countries.

Mexico's largest wind energy resource is found in the Isthmus of Tehuantepec in the State of Oaxaca; average annual wind speed in this region ranges from 7 to 10 m/s, measured at 30 meters above the ground. Given the favourable characteristics of this region, particularly its topography, it is estimated that more than 2,000 MW of wind power could be commercially tapped there. In fact, a 1.6 MW pilot plant located in one of the best sites in the region (La Venta), has operated for slightly more than 10 years at annual average capacity factors ranging from 30 to 40 %, which compares favourably to wind power plants located in the best inland sites in the world.

Detailed wind resource assessment at the national level is still insufficient. The Isthmus of Tehuantepec in Oaxaca is the region where more than 25 anemometric stations have been installed, and wind mapping has been carried out. In other regions of the country, an unknown number of anemometric stations have been installed in promising areas. From this, data acquired by companies of the private sector, as well as the one acquired by the Federal Electricity Commission (CFE), is not publicly available. By the end of 2004, the Electrical Research Institute (IIE), with the economic support of the Global Environment Facility (GEF) through the United Nations Development Programme (UNDP), started the installation of 20 anemometric stations in promising areas; this information is openly available and nowadays more than 80 users are using it.

In addition to the Isthmus of Tehuantepec in Oaxaca, some of the promising areas are located in the states of: Tamaulipas, Veracruz, Zacatecas, Hidalgo, Baja California, Baja California Sur, Quintana Roo, Yucatán, Chiapas, Chihuahua, Sinaloa, and Puebla. Based on anemometric measurements, the IIE has estimated that capacity factors of inland wind power stations in some of these areas could range from 18 to 30 % (considering the installation of the hub of the wind turbines at 80 meters above ground). An especial case is the Isthmus of Tehuantepec in Oaxaca where wind power stations installed in the best wind-lands could reach capacity factors up to 50 %.

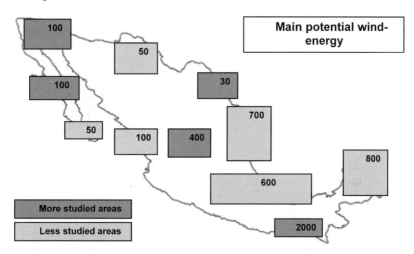

Fig. 1. Potential areas for the installation of up to 5,000 MW of wind power

Figure 1 shows a draft illustrating the potential areas for the installation of up to 5,000 MW of wind power capacity, notwithstanding neither the availability of electrical transmission lines nor environmental limitations. Figure 2 shows the wind area for the Isthmus of Tehuantepec in Oaxaca.

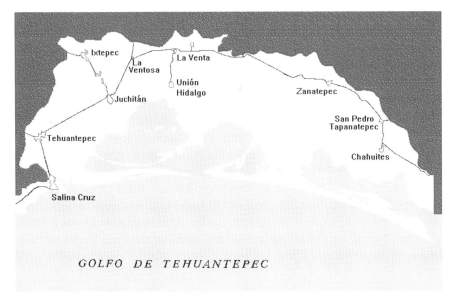

Fig. 2. The wind area at the Isthmus of Tehuantepec, Oaxaca

3 National Policy

At present, it is clear that both the energy and the environmental policies in Mexico consider renewable energy as a fitting way for diversifying energy supply within a sustainable development framework. The current National Development Plan as well as the Energy Sector Programme takes into account the promotion of renewable energy, giving an important role to wind power. On the other hand, energy supply in Mexico focuses to secure projected economic development. Therefore, the Government of Mexico is formulating the official instruments that will lead to a suitable balance for the mix of conventional and renewable power, considering the short, medium, and long terms.

National consumption of electricity is expected to increase at an average annual rate of 5.2 % for the period 2005 to 2014. This growth results in a projected requirement of 305 TWh of electricity generation for 2014, representing an increase of 122 TWh and equivalent to 22.1 GW of additional new generating capacity. Of this, 6.2 GW is already under construction or planned, the majority using combined-cycle gas-turbine technology, in addition to several new hydro and geothermal plants. The remaining 15.9 GW will come from new projects (SENER 2005). Therefore, an opportunity niche exists for supplying a reasonable portion of the non-committed 15.9 GW of new capacity using Mexico's wind energy resource.

At present, by constitutional mandate, the generation of electricity for the public service corresponds exclusively to the State (the federation). With the purpose to allow the private sector for contributing to satisfy electricity generation requirements, since 1992, the law of electricity for the public service (Ley del Servicio Público de Energía Eléctrica – LSPEE), does not consider some modalities as public service. Therefore, under these modalities, the law allows to private investors to construct and operate any kind of power plants (excepting nuclear power plants, which are restricted to the CFE).

Consequently, at present, the construction and the operation of wind power plants can be carried out for public service purposes (owned and operated by the CFE), as well as for non-public service purposes (owned and operated by private investors). However, LSPEE does not allow private investors to sell electricity to anybody different to the CFE. The basic rules for the non-public service modalities are as follows:

- Independent Power Producer (IPP)
  - Applies for capacity plants ≥ 30 MW.
  - The owner of the plant is obligated to sale the whole plant's electricity production to CFE under a long-term contract.
  - For issuing a public tender in this modality, it is a requisite that the IPP project is previously included in the CFE's official expansion plan.
  - By law, CFE is obligated to buy the lowest price electricity.
- Small Power Producer (SPP)
  - Applies for capacity plants under 30 MW.
  - The owner of the plant is obligated to sale the whole plant's electricity production to the CFE.
  - The buy-back price for electricity production is 95% of the electricity marginal cost of the region if the production is properly scheduled, and 85 % if the production is not scheduled.
  - By November 2006, the model of power purchase agreement has not been officially issued.
- Self-Supply (SS)
  - There is no limitation for the capacity of the power plant, but the regulatory framework does not allow selling electricity to CFE coming from capacity in excess of 20 MW of the one corresponding to the accumulated capacity of the self-supply loads. However, after a constitutional controversy related with the corresponding regulatory instrument, the CFE adopted an internal criterion limiting the capacity

of the self-supply projects to that equivalent to the capacity of the electrical loads.
- The law allows the integration of partnerships for electricity self-supply.
- CFE apply economic charges for electricity transmission, back-up capacity, and ancillary services.
• Exportation
- There is no limitation of the capacity of the plants.
- CFE apply economic charges for electricity transmission, and ancillary services.

One may argue that since 1992, there is a general legal and regulatory framework under which wind power projects could be carried out. However, the reality is that this general legal and regulatory framework was designed for conventional power plants, and do not take into consideration the particularities of renewable energy.

4 Progress towards the improvement of the legal, regulatory, and institutional frameworks affecting wind power development in Mexico.

Numerous stakeholders from both the public and private sectors are promoting the use of renewable energy in Mexico, considering an important role for wind energy. However, within both the political and business scenarios, some pending issues have been delaying the progress on this direction. Frequently, promoters of renewable energy state that the main constraint is the lack of a fitting legal and regulatory framework, which materialise in sound programmes with strategic goals and impelling measures. Over the last years, the political situation for the energy sector was very controversial; therefore, the legislative power did not passed initiatives of substantial reforms. However, promoters of renewable energy achieved some amendments to existing official instruments, as follows.

• In September 2001, the federal government through the Regulatory Commission for Energy (CRE) issued the first incentive for renewable energy. Embedded in the existent legal and regulatory framework, this incentive consisted in a model of agreement for the interconnection of renewable energy power plants to the national electrical grid. It allowed self-supply generators to interchange electricity between different billing periods (e.g., base to peak). In this fashion, self-suppliers do not necessarily had to sell surplus electricity to the CFE, because generation delivered to the grid during certain period could be credited to compensate for the electricity extracted from the grid during a different period. The interchange is allowed based on the ratio of the marginal costs between different billing periods; therefore, it is required to generate more than one kWh during a base period to match one kWh required in a peak period. This administrative incentive intended to improve the economic feasibility of some self-supply wind power projects, especially those for municipal public lighting, where the plants generate a considerable quantity of electricity during the daylight period when no electricity is required. Furthermore, before this incentive, electricity transmission charges for a renewable energy self-supply project were computed based on its rated capacity; today, these charges are reduced to the power plant capacity factor level. However, this incentive was not fully effective since capacity charges were computed based on five minutes period. Therefore, if a specific wind power plant for self-supply purposes does not generate any power during just five minutes over one month, then the regulatory instrument allowed to CFE to use the full contracted back-up capacity to compute capacity billing charges. During 2005, the Secretariat of Energy (SENER), the CRE, and the Mexican Wind Energy Association (AMDEE), with the support of technical studies developed by the IIE, carried out an intensive negotiation with CFE in order to achieve the recognition of certain capacity credit for wind power. By the end of 2005, they agree to amendment of the regulatory instrument by recognising capacity credit of renewable energy technologies, based on its average capacity factor computed during the whole electrical-system's peak hour. The CRE issued the new regulatory instrument in early 2006.

- The federal law for income tax (Ley Federal del Impuesto sobre la Renta) allowed deducting only 5 % in one year (20 years depreciation) for investments in any kind of equipment for electricity generation. By December 2004, the Law was amended allowing accelerated depreciation of investments in renewable technologies (wind energy is specifically included). Today, the Law allows investors in renewable energy to deduct up to 100 % of the investment in one year (1-year depreciation). In order to take advantage of this incentive, the equipment shall operate at least for the next five years following the tax deduction declaration; otherwise, complementary declarations are obligatory.

The two improvements are worthy but still incipient for establishing the fitting conditions for the substantial development of renewable energy. At present, there are other important initiatives as follows:

- The main official instrument aimed at including renewable energy in Mexico's electricity generation mix, is an initiative of *Law for the Use of Renewable Energy*. In December 2005, after introducing some amendments related with social development, the Federal Congress approved it. However, by November 2006, the initiative is still pending within the Senate. Upon positive decision, the Law would come into force by early 2007. The initiative of Law includes the creation of a green fund to improve the economic feasibility of wind power projects under the current constitutional and legislative mandates. The green fund would grant a price premium for electricity generation (kWh) based on renewable energy. The initiative includes the obligation of the national electrical system to take the electricity from renewable energy at any time of generation. A transitory article introduces a strategic goal of 8 % of penetration from renewable energy for the year 2012, not including large hydropower plants. It also instructs official institutions to formulate and issue, regulations, programmes, methods, and any other necessary instruments.
- The Secretariat of Environment and Natural Resources (SEMARNAT), through its Direction of Energy, prepared a proposal for a National Standard for the Protection of the Environment during the Construction, Operation, and Decommissioning of Wind Power installations; in agricultural, cattle rising, and uncultivated lands. This standard is now under consideration of the Federal Commission for Regulatory Improvements (COFEMER). If the standard moves forward, Mexico will become the first country having an official instrument of this kind. In addition, Semarnat, through its Direction of Environmental Impact, is progressing on the systematic and comprehensive evaluation of wind power projects.

Other relevant efforts are under progress, as follows:

- The National Commission for Energy Conservation (CONAE) issued a *Guide on Official Steps for the Construction and Operation of Renewable Energy Projects*, which has a specific section for wind energy. The guide will become an important tool for assisting to wind project developers in fulfilling, by the appropriated channel; all and each one of a number of official requirements for constructing a wind power project in Mexico. In addition, CONAE is also an important player regarding the promotion of wind energy in Mexico.
- The Government of the State of Oaxaca is considering the implementation of wind power in the Isthmus of Tehuantepec as a strategic project to improve regional development. To this end, since the year 2000, it has organised year by year *Colloquiums on Opportunities for the Devel-*

opment of Wind Power in the Isthmus of Tehuantepec. Up to now, this Forum has been the most important one on the wind energy subject at the national level. Besides, the Government of Oaxaca has facilitated to land owners the regularisation of land tenure in order they are legally able to sing land leasing contracts with wind-energy project developers. Governments of other states are also starting to promote wind energy.

- CFE included four 101 MW wind power stations in its programme for works and investments 2004–2013 (CFE 2004). Within CFE, the Unit of New Sources of Energy has been the major promoter of wind power projects. In fact, this Unit formulated La Venta I, as well as La Venta II, wind power projects. At present, this Unit is formulating La Venta III wind power project, as well as carrying out feasibility studies for evaluating potential projects in other regions of the country.

- By the end of 2003, the Electrical Research Institute (IIE), together with the United Nations Development Programme (UNDP), achieved that the Global Environmental Facility (GEF) approved the sponsorship of the project *Action Plan for Removing Barriers to the Full-Scale Implementation of Wind power in Mexico*. The first phase of this project (2004–2007) is addressing capacity building, wide promotion of wind energy at the national and regional level, human resource development, strategic studies and actions, and assessment of wind energy resource in promising areas. It also includes actions for contributing to the analysis and formulation of proposals for improving the legal, regulatory, and institutional framework for the implementation of wind power. The construction of a Regional Wind Technology Centre is also one of the main goals of this project. Furthermore, the IIE will carry out preparatory activities geared towards the formulation of business-demonstration wind power plants. Phase 2 of the wind-power action plan (2008–2010), will launch a competitive bidding process for three prototype projects. GEF will consider supporting these projects by emulating temporary production incentives. Next, the IIE will monitor and document the technical and economical performance of the commercial wind power plants, and will conduct a national campaign – based on lessons learned and best practices – for promoting the progressive replication of successful projects. This project has organised workshops on relevant topics, including: a) Workshop on technological routing of wind power in Mexico; b) Workshop on the analysis of the legal, regulatory, and institutional framework affecting the development of wind energy in Mexico; c) Workshop on birds life and wind power in the Isthmus of Tehuantepec. The implementing agency for this project is the UNDP while the prime mover and execution agency is the IIE. In addition, at present, the IIE is carrying out a feasibility study of a 30

MW wind power plant for CFE, as well as a viability study for a wind energy based self-supply project for a private company. The IIE is the major organisation in Mexico that has a permanent communication with R&D institutions worldwide; in fact, the IIE is representing Mexico in the Agreement for Co-operation in the Research, Development and Deployment of Wind Turbine Systems, which is carried out under the auspices of the International Energy Agency.

- The wind-power action plan is paving the way to a more extensive project originated by the SENER: the *Large Scale Renewable Energy Development Project*. The Global Environmental Facility will also be the sponsor of this project; the World Bank will be the implementing agency, and SENER will be the execution agency. The *Large Scale Renewable Energy Development Project* started formally in 2006. This project focuses at launching an IPP renewable energy market by means of the creation of a green fund, targeted to complement the buy-back price that CFE would pay according with on force regulations. This project will address the amendment of specific subjects on the regulatory topic, including the methodology for estimating the cost of electricity from different energy sources.

- A number of prestigious private companies formally integrated a Mexican Association of Wind Power (AMDEE). During 2005, this association became an important stakeholder in the negotiation and lobbying of legislative and regulatory instruments. All the members of this association are promoting their own wind power projects. In addition, the National Solar Energy Association (ANES) continued a more than fifteen years labour on the promotion of renewable energy.

5 Progress on the implementation of wind power

La Venta I was the first wind power station installed in Mexico. It was carried out within the context of the public service; therefore, CFE owns and operate it. CFE considered *La Venta I* as a pilot or demonstrative plant, which was commissioned in mid 1994. It has a total capacity of 1.575 MW, integrated by seven wind turbines of 225 kW each, manufactured by the Danish company Vestas (model V-29, 225 kW).

CFE reported that during the first six months of operation *La Venta I* reached an average capacity factor close to 50 % (the period included the months with more intense and persistent winds). During one of these months, the capacity factor of *La Venta I* surpassed 80 %. In that time, the wind turbines of *La Venta I* were informally competing for the world record regarding capacity factor, against a wind turbine of the same model installed in Wellington, New Zealand.

For the next six years, CFE reported an average capacity factor of 40 %. Later on, the capacity factor was decreased to 31 % during 2005. The capacity factor of this plant decreased with respect to previous years because of the *premature degradation or recurrent failures* of some components of the wind turbines. During 2005, the wind turbine with maximum capacity factor reached 36 % while the one with minimum capacity factor only reached 24 %. The seven wind turbines are exposed to the same wind regime (plain terrain, unidirectional wind, 3 diameters of separation between wind turbines). Therefore, a possible reason for such operational differences, in only a little more than 50 % of its commercial useful life, is that the wind turbines are not fully appropriated for the strong and persistent winds in La Venta. Nevertheless, *La Venta I* demonstrated to the general public, and to some skeptic people, the technical viability of wind power generation in the Isthmus of Tehuantepec. Furthermore, *La Venta I* allowed to a few CFE's engineers to gain understanding on the general topics of the technology and how to solve specific problems. Figure 3 shows a view of *La Venta I* facility in operation.

Fig. 3. *La Venta I* wind power station

In 1995, a private company of the cement industry installed a 550 kW wind turbine in the state of Coahuila (from the American manufacturer Zond Systems Inc.). According to the owner of the wind turbine, the technical performance of this machine was satisfactory, but it operated at low capacity factors, due to the relatively low wind regime in the site. Unfortunately, a lighting stroke hit the wind turbine burning the entire nacelle and, consequently, the owner decided to let it out of service.

Next, in November of 1998, CFE issued a contract for the supply and installation of a 600 kW wind turbine (from the Spanish manufacturer Gamesa Eolica). The machine was interconnected to the electrical grid of a diesel power plant in Guerrero Negro, in the state of Baja California Sur. CFE reported 25 % of capacity factor for the first 18 month of operation. However, CFE has not released detailed information about the operational performance of this machine. It is known that some technical constraints had to be solved.

In spite that there were a number of public and private initiatives for the construction of wind power projects, during eight years (from 1998 to 2006) nobody was able to materialise any project. Therefore, regrettably, the wind power capacity installed in Mexico remained below 3 MW during almost 12 years. Taking into account the size of the national electrical system it is evident that the contribution from wind energy to the national electricity demand was negligible.

By 2000, promoters of wind energy within both the public and private sectors intensified diverse actions for promotion, lobbying, and negotiations. The main attention was focused to the Isthmus of Tehuantepec in Oaxaca, were several private companies formulated wind power projects for more than 500 MW; a number of these companies secured the necessary lands and obtained permissions from the CRE.

In 2005, CFE decided to build a 101 MW wind power station in La Venta, within the modality of Financed Public Works (OPF). The bidding process was declared deserted, because all economic proposals surpassed the available budget. Therefore, CFE issued the public tender again reducing the capacity of the plant to around 80 MW. In this occasion, there was a bidding winner: the Spanish consortium Iberdrola-Gamesa, who offered to finance and construct an 83.3 MW wind power station based on 98 wind turbines rated at 850 kW (Model G-52). The indicator of the investment cost for this power plant is USD 1,370 per kW installed, not including the cost for the electrical transmission line.

Commercial implementation of wind power plants in La Venta and contiguous areas requires a carefully selection of wind turbines. The suitable installation of wind turbines over 40 meters height above ground could require special class wind turbines, in accordance with IEC standards. Otherwise, a 20 years useful life would be unlikely to comply. However, within the Isthmus of Tehuantepec there is plenty of land where Class I, as well as Class II wind turbines could be properly installed.

Therefore, *La Venta II* will be the first significant wind power project in Mexico. At present it is under construction and will be commissioned by the end of 2006. La *Venta II* will be the largest wind energy project in Latin America and it is expected to become one of the most productive among the inland wind energy projects worldwide. Figure 4 shows a view of *La Venta II* wind power plant in construction.

It is expected that CFE issue the public tender for *La Venta III* by the end of 2006 or by early 2007. In principle, this new project could be carried out under the IPP modality, applying the production premium to be granted by the *Large Scale Renewable Energy Development Project* (SENER-GEF-WB).

Fig. 4. *La Venta II* wind power station

Meanwhile, according to Table 2, the CRE has issued a permit for the construction of ten wind-energy private owned projects; but, by November 2006, none of them has started construction yet. The accumulated capacity of these projects is 1,258 MW, from which 898 MW are planned for the Isthmus of Tehuantepec in Oaxaca. Other 300 MW are planned for the state of Baja California. All of the projects for the state of Oaxaca are focused on the self-supply modality, while 300 MW for the state of Baja California are focused on exportation to the United States. In 2006, the

CRE granted permits for two projects (BiiNee Stipa Energía Eólica, and Eurus). In 2005, the CRE granted permits for other two projects (Vientos del Istmo, and Eoliatec del Istmo). Permits for the other six projects were granted from 1998 to 2002.

Table 2. Wind energy projects with permits already granted by the CRE

Company	Capacity [MW]	Date	State	Modality
BiiNee Stipa Energía Eólica	26	September, 2006	Oaxaca	Self-supply
Eurus	250	July, 2006	Oaxaca	Self-supply
Vientos del Istmo	120	December, 2005	Oaxaca	Self-supply
Eoliatec del Istmo	163	March, 2005	Oaxaca	Self-supply
Parques ecológicos de México	102	September, 2002	Oaxaca	Self-supply
Eléctrica del Valle de México	180	September, 2001	Oaxaca	Self-supply
Electricidad del Sureste	27	March, 1996	Oaxaca	Self-supply
Fuerza Eólica del Istmo	30	January, 1998	Oaxaca	Self-supply
Sub-total Oaxaca	898			
Fuerza Eólica de Baja California	300	July, 2002	B.C.	Exportation
Baja California 2000	60	January, 1998	B.C.	Self-supply
Sub-total Baja California	360			
Total National	1,258			

Source: Regulatory Commission for Energy (CRE)

One of the main constraints for the projects in the Isthmus of Tehuantepec, in Oaxaca, is the one related with the grid interconnection. At present, the electrical transmission line for evacuating wind power generation in the Tehuantepec Isthmus is about 500 MW. Therefore, CFE's plans virtually obstructed the interconnection of the first potential private projects. However, there have been positive negotiations among CFE, SENER, CRE, and AMDEE, in order to allow the interconnection to the existing electrical grid of up to 300 MW of private owned wind-energy projects. In addition, negotiations and specific commitments are progressing well enough in order to construct a new electrical transmission line for evacuating up to 2,000 MW of wind energy. The construction of most of the projects for the Isthmus of Tehuantepec is planned in different stages; therefore, certain part of at least four of them could be interconnected to the existing transmission line. Likely, the first private owned wind-energy project in Mexico could start construction by early 2007.

6 Industry

A Mexican company has manufactured a number of 750 kW electric generators for an international wind turbine manufacturer. According to the status of Mexican industry, a number of wind turbine components – including towers, generators, gears, conductors, and transformers – could all be manufactured in Mexico using existing infrastructure. More than 200 Mexican companies have been identified as having the capacity for manufacturing parts required for wind turbines and for wind power plants. The country also has excellent technical expertise in civil, mechanical and electrical engineering, which could be tapped for plant design and construction. In fact, the 98 towers for the wind turbines installed in *La Venta II* wind power station, were already manufactured in Mexico. In addition, a Japan manufacturer of wind turbine installed a wind-turbine blade factory at the North of Mexico.

7 Remarks

1. At the international level, Mexico is considered as one of the countries having great opportunities for the implementation of wind energy.
1. Mexico posses abundant wind energy resources, which surpass the one that have several of the industrialised countries.
2. Wind energy is an important niche of opportunity for the creation of employment and the regional development.
3. Important efforts for improving the legal, regulatory, and institutional frameworks affecting the development of wind power in Mexico, have been carried out by several promoters of wind energy, but still are incipient for establishing the conditions for achieving a significant contribution of wind energy to the energy mix.
4. The implementation of wind energy in Mexico has already started with the construction of *La Venta II* wind power station. However, for achieving a progressing and significant deployment it is necessary to implement fitting legal, regulatory and institutional frameworks; otherwise, taking into account the expected growth of the national electrical system, the contribution of wind energy would remain negligible.
5. A shared vision of 6 % of wind energy contribution to the national electrical demand for 2030 is congruent with the deployment of wind energy in industrialised countries over the last 15 years.

6. It is fundamental that within the legislative power (both the Congress and the Senate) arise awareness regarding the potential benefits of wind energy, not only regarding energy diversification for sustainable development, but also on its potential for jobs creation and regional development.

References

CFE (Federal Electricity Comisión), (2004), Programa de Obras e Inversiones del sector Eléctrico 2004–2013

EC (1997) European Commission, Energy for the Future: Renewable Sources of Energy. White Paper for a Community Strategy and Action Plan COM(97) final (26/11/1997), Brussels, Belgium

GWEC Global Wind Energy Council (2005), Wind Force 12: a blueprint to achieve 12 % of the world's electricity by 2020. http://www.ewea.com

INWEA (2006) Indian Wind Energy Association, http://www.inwea.org/windenergy.html

Kutscher J, Christmann R (2006) Chapter 15: Germany, IEA RD&D Wind 2005, International Energy Agency, June 2006. http://www.ieawind.org/ pp 133–138

Lemming J, Oster F, Nielsen R. (2006) Chapter 12: Denmark. IEA RD&D Wind 2005. International Energy Agency. June 2006. http://www.ieawind.org/ pp 101–114

NREL (National Renewable Energy Laboratory) (2006) Renewable Energy in China: Grid connected Wind Power. www.nrel.gov/docs/fy04osti/35789.pdf

O´Dell K (2006) Chapter 29: United States, IEA RD&D Wind 2005. International Energy Agency. June 2006. http://www.ieawind.org/ pp 261–278

SENER (Secretaría de Energía) (2005) Documento de Prospectiva del Sector Eléctrico 2005–2014, Dirección General de Planeación Energética, http://www.energia.gob

Soria E (2006) Chapter 25: Spain, IEA RD&D Wind 2005. International Energy Agency. June 2006. http://www.ieawind.org/ pp 211–224

Development of geothermal energy in México and its energetic potential for the future

Peter Birkle

Instituto de Investigaciones Eléctricas, Gerencia de Geotermia, Calle Reforma 113, Col. Palmira, Cuernavaca, Morelos, 62490 México. E-mail: birkle@iie.org.mx

Abstract

The current contribution of 3.1 % of geothermal energy to México's electricity supply is relatively elevated in comparison to a global offer of primary geothermal energy of 0.442 % (ISES 2002), or 0.8 % by renewable technologies, including geothermal, solar, wind and tide/wave/ocean technology (IEA 2006), but potential favorable national resources are still underexploited. Feasibility studies proved a potential for national reserves of 3,650 MW, whose generation (20,460 GWh) could provide more than 12 % (20,460 GWh) of the total electricity generation. The exploitation of middle- and high-enthalpy geothermal reservoirs could represent a major contribution towards a more environmental friendly electricity production in México, as the combustion of more than 40 million barrels of fossil fuels per year and the emission of approximately 16.5 million tons of CO_2 into the atmosphere could be avoided by geothermal expansion strategies. On an economic point of view, the average generation costs of 3.98 USD cents per kWh for geothermal electricity are currently lower than for any other renewable energy type, and competitive with most conventional energy types. Low- to middle temperature sites ($< 170°C$) are still undeveloped in México, although an estimated heat potential of several thousands of megawatt could be used for private and industrial consumption, such as district and greenhouse heating, spas, and aquaculture. The implantation of decentralized small-scale plants (< 5 MW) with binary-cycle or heat pump technology could be fundamental for the energetic development of remote rural areas. As a basic condition to expand the national geothermal energy market, the legal and regulatory frame has to appropriate in order to fortify the renewable energy sector, as well as market incentives and innovative financing schemes have to be established to promote private inversions.

1 Introduction to geothermal technology

Geothermal exploitation is defined as "the energy derived from heat, being extracted from fluids, emerged by natural or artificial accumulation and heat processes from the subsoil" (CEPAL 1999). From a thermo-dynamical point of view, the use of geothermal energy is based on temperature differences between the rock mass and groundwater and the mass of water or air at the earth's surface. The temperature gradient between both environments allows the direct use of geothermal energy (<150°C) or its conversion into mechanical or electrical energy. Geothermal energy is extracted from the subterranean reserves by means of production wells, drilled up to a depth of 3,000 m with a bottom temperature of up to 310°C. Vapour is separated from the liquid phase to transform kinetic to electrical energy through turbine generators. The extracted geothermal fluid is subsequently reinjected into peripheral parts of the reservoir to maintain pressure conditions of a closed and renewable production cycle.

Three different types of power plant technologies are being used to convert reservoir fluids to electricity:

- Dry steam power plants: Steam from the reservoir goes directly to a turbine.
- Flash power plants: Used for hydrothermal fluids above 175°C. High pressured reservoir fluids are vaporized rapidly ("flash" to steam) by being sprayed into a low pressure tank.
- Binary-cycle power plants (< 175°C): The geothermal fluid yields heat to a secondary fluid (with low boiling point and high vapour pressure at low temperatures) through heat exchangers, in which this fluid is heated and vaporises; the vapour produced drives a normal axial flow turbine, is then cooled and condensed, and the cycle begins again.

Hot water as output from the mentioned electricity generating processes as well as geothermal resources with medium to lower enthalpy characteristics (Temperature < 150°C) can be used for direct use purposes, especially for heating buildings (either individually or whole towns), raising plants in greenhouses, drying crops, heating water at fish farms, and several industrial processes, such as pasteurizing milk, paper and textile manufacturing, and production of sulfuric acid. The classical Lindal diagram (Lindal 1973) in Fig. 1 show possible applications of geothermal fluids at different temperatures.

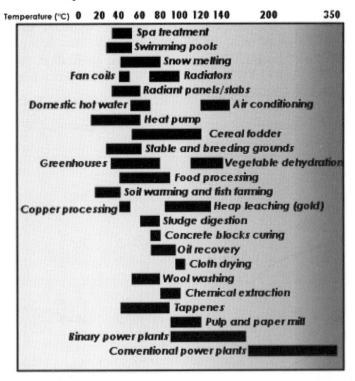

Fig. 1. Diagram showing the utilization of geothermal fluids (derived from Lindal 1973)

The main aim of this paper is to describe geothermal energy generation in México and its global context, as well as to point out future technologies and strategies in the geothermal industry that could make México's energy demand less dependent from conventional energy types.

2 Global use of geothermal energy

2.1 Electricity generation

For their energy content, geothermal fluids have been used for industrial purposes since the 19th century. In 1827, Francesco Lardarel developed a system for using the heat of boric fluids in the evaporation process of hot fluids in the zone known today as Lardarello in Italy. At the same site, the first successful commercial project for generating electricity from geothermal steam was initiated in 1904, followed by a 250 kW geothermal power plant in 1913. The first commercial geothermal power plant using a liquid-dominated, hot-water reservoir, initiated operation in 1958 in Wairakei, New Zealand (IEA 2003). At present, US leads the world's geothermal electricity production with an installed generation capacity of 2,543 MW, followed by Philippines (1,930 MW) and México is positioned at third place with 953 MW (see Table 1).

Table 1. Installed capacity for geothermal electricity generation in December 2004 (in MWe) (modified from IGA 2006)

	1982	1990	1995	2000	2004
AFRICA:					
Ethiopia	0.0	0.0	0.0	8.5	7.3
Kenya	15.0	45.0	45.0	45.0	128.8
NORTH AMERICA:					
México	205.0	700.0	753.0	755.0	953.0
USA	932.0	2,774.6	2,816.7	2,228.0	2,543.0
CENTRAL AMERICA + CARIBBEAN:					
Costa Rica	0.0	0.0	55.0	142.5	162.5
El Salvador	95.0	95.0	105.0	161.0	116.0
Guadeloupe	0.0	4.2	4.2	4.2	15.0
Guatemala	0.0	0.0	33.4	33.4	32.8
Nicaragua	30.0	35.0	70.0	70.0	77.0
SOUTH AMERICA:					
Argentina		0.7	0.7	0.0	0.0
ASIA:					
China	2.0	19.2	28.8	29.2	27.6
Philippines	501.0	891.0	1,227.0	1,909.0	1,930.0
Indonesia	32.0	144.8	309.8	589.5	797.0
Japan	220.0	214.6	413.7	546.9	534.3
Russia	11.0	11.0	11.0	23.0	79.0
Thailand	0.0	0.3	0.3	0.3	0.3
Turkey	0.5	20.6	20.4	20.4	20.0
EUROPE:					
Iceland	41.0	44.6	50.0	170.0	202.1
Italy	446.0	545.0	631.7	785.0	791.0
Portugal (Azores)	0.0	3.0	5.0	16.0	16.0
OCEANIA:					
Australia	0.0	0.0	0.17	0.17	0.15
New Zealand	202.0	283.2	286.0	437.0	434.5
TOTAL:	2,732.5	5,831.7	6,833.38	7,974.06	8,867.35

Geothermal development in US and México experienced its major expansion during the 80's (from 932 to 2,775 MW, and from 205 to 700 MW, respectively), whereas Philippines had its fastest growth during the 90's (from 891 to 1,909 MW between 1990 and 2000). In contrast to stagnant generation capacities for most geothermal producing countries at present, Indonesia is still increasing national geothermal electricity production from 589.5 to 797 MW between 2000 and 2004.

Other countries with significant geothermal electricity generation are Italy (791 MW), Japan (534 MW) and New Zealand (435 MW). Worldwide, geothermal power plants are producing over 8,800 MW of electricity in 21 countries supplying about 60 million people (GEO 2004).

Major efforts to make independent the national energy market from the importation of fossil fuels were especially reached by the Philippines, El Salvador and Costa Rica with a geothermal contribution of 21.5, 22.2 and 14.1 % (CIA 2001), respectively, of the total electricity generation.

2.2 Direct use

In general, the global potential for non–electrical applications is estimated to be various orders higher than high temperature applications. Bathing, space and district heating, agricultural applications, aquaculture and some industrial uses are the best known forms of utilization for the direct use of geothermal energy, whereby heat pumps are the most widespread (12.5 % of the total energy use in 2000). Worldwide, 58 countries reported the direct use of geothermal energy in 2000 (IGA 2006). Table 2 shows a list of 33 countries with an installed capacity of more than 100 MW for non-electrical applications of geothermal energy. The most common non-electric use world-wide (in terms of installed capacity) is heat pumps (34.80 %), followed by bathing (26.20 %), space-heating (21.62 %), greenhouses (8.22 %), aquaculture (3.93 %), and industrial processes (3.13 %) (Lund and Freeston 2001). An exemplary case for the application of space and district heating is given by Iceland, where the total capacity of the operating geothermal district heating system of 1,350 MW_t is used by 87 % of the houses in the country (Ragnarsson 2005). China, the world leader in direct use technology, shows a wide distribution of applications, including district heating (550 MWt and 6,391 TJ/yr); greenhouse heating (103 MWt and 1,176 TJ/yr); fish farming (174 MWt and 1,921 TJ/yr); agricultural drying (80 MWt and 1,007 TJ/yr); industrial process heat (139 MWt and 2,603 TJ/yr); bathing and swimming (1,991 MWt and 25,095 TJ/yr); other uses (monitoring) (19 MWt and 611 TJ/yr); and heat pumps (631 MWt and 6,569 TJ/yr) (Lund et al. 2005).

Table 2. Countries with an installed geothermal capacity and direct use generation of more than 100 MWt in December 2004 (modified from IGA 2006).

Country	Installed capacity [MWt]	Direct use [GWh/year]
Algeria	152.3	671.4
Australia	109.5	824.5
Austria	352.0	2,229.9
Brazil	360.1	1,839.7
Bulgaria	109.6	464.3
Canada	461.0	707.3
China	3,687.0	12,604.6
Croatia	114.0	189.4
Czech Republic	204.5	338.9
Denmark	821.2	1,211.1
Finland	260.0	541.7
France	308.0	1,443.4
Georgia	250.0	1,752.1
Germany	504.6	808.3
Guadeloupe	308.0	1,443.4
Hungary	694.2	2,205.7
India	203.0	446.2
Iceland	1,791.0	6,615.3
Italy	606.6	2,098.5
Japan	413.4	1,433.8
Jordan	153.3	427.8
México	164.7	536.7
Norway	450.0	642.8
New Zealand	308.1	1,968.5
Netherlands	253.5	190.3
Poland	170.9	232.9
Romania	145.1	789.2
Russia	308.2	1,706.7
Slovak Republic	187.7	842.8
Sweden	3,840.0	10,000.0
Switzerland	581.6	1,174.9
Turkey	1,177.0	5,451.3
USA	7,817.4	8,678.2

3 Geothermal potential in México

3.1 Electricity production

México represents one of the global pioneers in the exploration and exploitation of geothermal reservoirs. The first geothermal plant on the American

continent was installed in 1956 in Pathé, State of Hidalgo. Four high enthalpy sites are currently producing commercial electrical power in México
(2005) (Fig. 2): Cerro Prieto in Baja California Norte with 720 MWe
(starting in 1973), Los Azufres in Michoacán state with 188 MWe (starting
in 1982) (Fig. 3), Los Humeros in Puebla state with 35 MWe (1990) and
the recent project of Las Tres Vírgenes in Baja California Sur with 10
MWe (2001) (Gutiérrez-Negrín and Quijano-León 2005). More technical
details of the development of the individual fields are given in Birkle and
Verma (2003). The activities at the geothermal field in La Primavera, Jalisco, were suspended by request of the local government, in order to carry
out a complete environmental restoration program of the area. Thirty six
power plants of several types (condensing, back pressure and binary cycle)
between 1.5 MWe and 110 MWe are fed by 197 wells with a total output
of 7,700 metric tons of steam per hour (t/h) and a total generation of 6,282
GWh in 2003. In June 2006, a total of 3.02 % of the national electricity
generation of 109,047 GWh were contributed by a total installed geothermal capacity of 959.5 MW.

Fig. 2. Location of potential hydrothermal sites and geothermal fields on Mexican
territory (modified after Martínez-Estrella et al. 2005)

Fig. 3. Geothermal production unit at the Los Azufres geothermal field, Michoacán. The clouds consist of almost pure water steam from the separation process

Cerro Prieto, the second largest global geothermal field, produces 94.64 % of the electricity distributed within the network of the isolated Baja California state in NW-México (CFE 2006).

For the time period from 2005 to 2014, the strategic plan of the Secretary of Energy for the geothermal sector considers the installation of an additional 2×50 MW (813.2 GWh/year) and 25 MW (207.1 GWh/year) for the Cerro Prieto V and Los Humeros II development, respectively (SENER 2006). Additional long-term technical options are considered for Cerrito Colorados (1. Phase: 26.9 MW, 2. Phase: 26.9 MW), and Los Humeros III (55.0 MW), resulting in a total increase of 220.0 MW (1,656.3 GWh/year) of installed capacity between 2005 and 2014. Quijano (2005) reported the possible installation of a 21 MWe binary cycle in Los Humeros in 2007 and 25 MWe at the Cerritos Colorados site (Cerritos Colorados 1 in 2008), the amplification of the Los Azufres field by 50 MW (Los Azufres 3 in 2010), 100 MW in Cerro Prieto VI (substitutes 75 MW in 2011) and 50 MW in Cerritos Colorados 2 (2011).

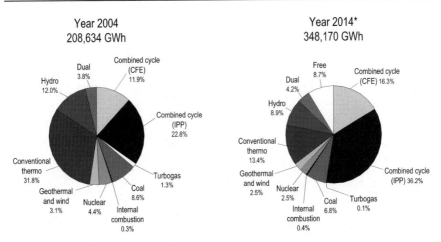

Fig. 4. Primary electricity generation for México in 2004 and prognostics for 2014 (SENER 2006)

On the other hand, the expansion of geothermal capacities is not proportional to the fast rising national energy demand, as a decrease of the geothermal contribution from 3.1 % in 2004 to 2.5 % is expected for the year 2014, see Fig. 4 (SENER 2006).

3.2 Non-electrical use

Favorable geological-tectonic natural conditions explain the widespread distribution and abundance of hydrothermal sites in México. Dominant vertical pathways of fracture and fault systems allow the convective circulation of geothermal fluids within the reservoir, and shallow located magma chambers beneath the volcanic chains provide the required conductive heat for the exchange process. Hydrothermal outcrops at the surface represent witnesses of the partial rise of thermal fluid from underlying potential reservoirs. More than 2,300 geothermal localities with low- to medium-temperature conditions (28–200°C) were identified in 27 of 32 Mexican States (Torres et al. 1993; Torres et al. 2005; Martínez et al. 2005; see also Fig. 2). From 1,380 studied manifestations as part of the *Carta de manifestaciones termales de la República Mexicana*, 808 correspond to warm-hot thermal springs, 526 to hot wells, 25 fumaroles, 6 mud volcanoes, 11 bubbling springs and 3 hot soils. 68 of them represent high enthalpy sites with temperatures above 150°C (Herrera and Rocha 1988).

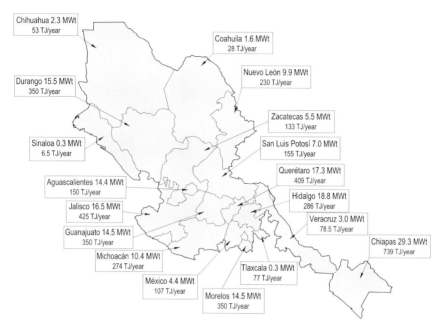

Fig. 5. Currently used thermal bathing sites in México and their estimated heat generation (modified from Hiriart Le Bert 2004)

On the other hand, geothermal energy is used mostly to generate electric energy, with some isolated direct-uses restricted to small pilot projects in the Los Azufres (fruit drying, timber drying, greenhouse heating and space heating of a conference center and small cabins) and Los Humeros geothermal fields. Few of them are used for balneology purposes, most of them are without coordinated efforts to promote it. Geothermal heat pumps are undeveloped in México, except for some private and isolated cases. Of the direct-use developments, 99.6 % are represented by 160 bathing and swimming sites (164.0 MWt and 1,913.4 TJ/yr), followed by space heating (0.5 MWt and 13.2 TJ/yr), greenhouse heating (0.004 MWt and 0.1 TJ/yr), agricultural drying (0.007 MWt and 0.2 TJ/yr), and mushroom growing (0.2 MWt and 4.9 TJ/yr) for a total of 164.7 MWt and 1,931.8 TJ/yr (Gutiérrez-Negrin and Quijano-León 2005; Lund et al. 2005). Major spa sites and their energetic potential are shown in Fig. 5.

In general, the energetic potential of low- to middle temperature resources in Mexican territory is several orders higher than the potential of high enthalpy resources, but there is no national strategic plan for their respective development.

4 Comparison with conventional energy types

The use of geothermal energy technology is favored by following technical, environmental and economic aspects in comparison to conventional
energy types and other renewable resources:

- *Autochthonous energy:* With an average capacity factor of 89 to 97 %
  (measure of the amount of real time during which a facility is used) is
  much higher than for other renewable technologies, such as wind energy
  (26–40 %) or solar (22.5–32.2 %) (DOE 1997). Therefore, geothermal
  energy has the potential to provide reliable contributions for the electricity market comparable to baseload carbon or nuclear power sources
  (Kagel et al. 2005). Geothermal energy can partially liberate the national
  energy sector from the dependency on fossil fuels and their related unstable economic situation by fluctuating fuel prices. Regarding the limited petroleum reserves on Mexican territory, an economic house hold-
  ing of the limited natural resources is recommended.
- *Renewable energy:* Geothermal systems can be considered renewable
  "on timescales on technological/societal systems and do not need geological times for regeneration as fossil fuel reserves do" (Rybach et al.
  1999). The underground water or steam used to convert heat energy into
  power will never diminish if managed properly by maintaining a close
  loop between extraction and reinjection of fluids. The world's first geothermal site in Lardarello, Italy, is currently celebrating its 100th anniversary of commercial electricity production.
- *Environmental friendly:* Geothermal power plants do not burn fuel like
  fossil fuel power plants, thus they release virtually no air emissions.
  Modern coal plants updated with scrubber and other emissions control
  technologies emit 24 times more CO_2, 10,837 more SO_2, and 3,865
  times more NO_x per megawatt hour than a geothermal steam plant
  (Kagel et al. 2005). In 2000, México (Quintanilla 2004) emitted more
  than 363 million tons (Mt) of CO_2 with mayor contribution of the transport (116.2 Mt) and electrical sector (111.7 Mt). Concerning the land
  occupation for the installation of the power sites, geothermal plants require an average between 0.1 and 0.3 hectar/MW (404 m^2/GWh), in
  comparison to 0.3 to 0.8 ha/MW for gas combustion, 0.8 to 8.0 h/MW
  (3,642 m^2/GWh) for coal combustion, 2.4 to 1,000 ha/MW for hydroelectric plants (GEA 2006). Other renewable energies, such as wind
  (1,335 m^2/GWh), solar thermal (3,561 m^2/GWh), and photovoltaics
  (3,237 m^2/GWh), require more land space than geothermal installations
  (Kagel et al. 2005).

- *Economically competitive:* The generation costs for geothermal electricity in México are very competitive in comparison with conventional energy types. The average investment costs for geothermal plants in México are approximately 1,400 USD/kW, with 1,181 and 1,219 USD/kW for Cerro Prieto and Los Azufres, respectively. These costs are only superior to industrial turbo gas (342 USD/kW), combined cycle (493 USD/kW) and conventional vapor (973 USD/kW), but more economic as nuclear (2,670 USD/kW), mini hydroelectrics (1,600 USD/kW), and coal plants (1,560 USD/kW). Due to the high capacity factor (> 90 %), the average generation cost of geothermal electricity (3.986 cUSD/kWh) is lower than for any other conventional or renewable energy type, except combined cycles (3.59 cUSD/kWh) (SENER 2004). Figure 6 shows the average inversion per installed kilowatt (USD per kW) and the cost for each produced kilowatt hour (USD cents per kWh), respectively, for the generation of electricity in México. Prognostics of the International Energy Agency (IEA) indicate a further global decrease of the inversion costs for geothermo-electrical plants from a range between 1,200 and 5,000 USD/kW (in 2002) to prices between 1,000 and 3,500 USD/kW for the year 2010 (IEA 2003). Potential environmental impacts, such as the contamination of surface waters and groundwater by elevated concentrations of B, F, NH_3, Si, H_2S, As and some metals in geothermal fluids, as well as noise disturbance at the production zone can be minimized by maintaining a closed extraction-reinjection cycle and noise silencers. Cases of land subsidence and the drawdown of piezometric levels in groundwater systems have been observed, especially in New Zealand, by the excessive extraction of geothermal fluids (Brown 1995).

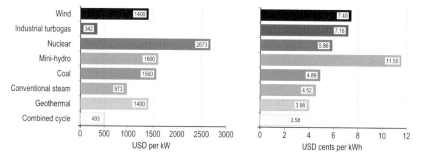

Fig. 6. Comparison of the (a) average inversion requirements (USD per kW) and b) electricity generation costs for different technologies in México (USD cents per kWh)

5 Strategies and Future options

5.1 Feasibility and exploitation of additional high-enthalpy geothermal fields

In 1992, CFE selected 41 priority zones with potential for electricity generation to perform pre-feasibility and feasibility studies of a total of 545 identified potential hydrothermal zones (Pantoja and Gómez 2000). As of December 2003, 35 wells with a combined depth of 46,337 m were drilled in several parts of the country, such as in Laguna Salada (Baja California), Acoculco and Las Derrumbadas (Puebla), and Los Negritos (Michoacán) (Gutiérrez-Negrin and Quijano-León 2005). As a result of the feasibility studies, 21 sites were defined as potential zones for the extraction of high-enthalpy vapour to generate electricity, and 20 zones were classified as low-enthalpy sites (CFE and UNAM 1993). The additional energy potential of the currently drilled sites is estimated to comprise 600 MWe (Fig. 7). Estimations for the probable reserves for the known Mexican geothermal reservoirs vary between 2,340 MW (CEPAL 1999) and up to 8,000 MW for feasible generation (GEA 2006). A realistic total capacity of 4,600 MW, based on information from CFE, can be subdivided in proved 1,350 MW for the geothermal fields of Cerro Prieto, Los Azufres, Los Humeros and Primavera, and probable reserves of 3,250 MW for the sites of Las Tres Vírgenes, Ceboruco, Araró, Ixtlán de los Hervores and Los Negritos (SENER 2004). Tables 3 and 4 give details of the proved and probable reserves of high enthalpy sites.

The geothermal potential of high enthalpy sites could allow a minimum generation of an additional 2,400 MW (SENER 2006). Considering a present national electricity capacity of 46,552 MW and a total generation of 217,793 GWh in 2004 (SENER 2006), an expansion from current 960 MW towards a total of 3,650 MW for the installed geothermal capacity (SENER 2004) could contribute 12.51 % (20,460 GWh) to the total electricity generation. Other organizations, such as the Geothermal Energy Association considers even a potential contribution of 20 % of geothermal generation for the total electricity demand of México (Gawell et al. 1999). To reach this target, a) the production in existing fields, such as Cerro Prieto (Baja California Norte), must be expanded, b) defined potential sites, especially in Nayarit (Ceboruco) and Michoacán, have to be developed, and c) further feasibility studies should be performed in defined potential sites.

Fig. 7. Potential geothermal sites with additional geothermal capacity (modified from Hiriart Le Bert 2004)

Table 3. Proved reserves of high-enthalpy geothermal reservoirs in México

Geothermal field	Location	Proved reserves [MW]
Cerro Prieto	Baja California Norte	840
Los Azufres	Michoacán	300
Los Humeros	Puebla	110
La Primavera	Jalisco	100
Total		1,350

Source: SENER 2004

Table 4. Probable reserves of high-enthalpy geothermal sites in México

Geothermal field	Location	Probable reserves [MW]
Las Tres Vírgenes	Baja California Norte	10
Volcán Ceboruco	Nayarit	802
Araró	Michoacán	816
Ixtlán de los Hervores	Michoacán	813
Los Negritos	Michoacán	809
Total		3,250

Source: SENER 2004

The generation of 20,460 GWh of geothermal electricity could avoid the emission of approximately 16.5 million tons of CO_2 (SENER 2004), and could therefore be a principal component to accomplish with the reduction of greenhouse gas emission into the atmosphere. The total national emission of generated CO_2 by burning fuels of about 350 million tons in 1998 (SEMARNAT 2005) could be lowered significantly by about 5 %.

5.2 Use of medium to low-enthalpy reservoirs

Electricity generation: High enthalpy reservoirs are less abundant as lower temperature systems, and most known high-temperature fields in México have undergone pre-feasibility and feasibility studies. In contrast, geothermal fields with medium enthalpy conditions are underestimated and not yet considered for national electricity generation, although feasible by the direct use of vapour or by heat transfer towards a secondary working fluid. Mercado et al. (1985) estimated a total energy potential for low- to medium-enthalpy sites of 31,498 MWe for the Transmexican Volcanic Belt in the central part of México and 14,317 MWe in the NW of México.

Non-electrical applications: No significantly improvements were obtained in the exploitation of low to middle enthalpy sites (< 150–200°C) during the last few years as a stagnant installed capacity of 164 MWt reflect the lack of expansion of these abundant energy resources. In contrast, Iglesias and Torres (2003) calculated thermal energy reserves of 2.14×10^{10} to 2.39×10^{10} MWth for 276 assessed geothermal sites in Mexican territory, whereby an average reserve of 8.15×10^{16} kJ is equivalent to 2.14×10^{15} m^3 of natural gas or 1.9×10^{9} barrel of Arabian light crude oil (Iglesias et al. 2005). Most likely temperature for the geothermal localities ranges from about 60 to 180°C. Some researchers have roughly estimated that hydrothermal reserves of low enthalpy (temperatures below 180°C) to be of at least 20,000 MWt. Concrete application could be the use of heat pumps for heating or cooling of districts and greenhouses, production of hot water for domestic use, spas with therapeutic or recreational purposes, aquaculture and agricultural facilities, such as fish farming, fruit drying, crop production, and as industrial process heat. Especially remote rural areas could benefit from the enhanced use of geothermal energy.

5.3 Proposed technology changes

Mini plants – Small size facilities: The national consumption patron in our époque is sustained by large-scale and centralized plants for energy generation. Geothermal power plants tend to be in the 20 to 60 MW range, and a standard 50 MW plant supplies between 350,000 and 500,000 persons with electricity. The large-scale dimension of the Mexican territory with a surface of 1,964,375 km² favours the utilization of renewable technologies with a decentralized distribution of small-scale plants, approaching energy production sites towards remote consumers. Binary-cycles and heat pumps with a capacity range from several hundreds kilowatts to some megawatts (mostly up to 5 MW) are most prospective to provide a small-scale energy supply in local sites.

Binary plant technology: The connection of rural zones to the national electric grid is an extremely expensive and labour-extensive task: In 2002, the electrification of 786 rural populations and 342 popular districts with 160,000 inhabitants caused an estimated inversion of 311 million pesos (about USD 30 million). Therefore, binary plant technology is a very cost-effective and reliable means of converting into electricity the energy available from water-dominated geothermal fields. Small mobile plants can be used to increase standard of living in isolated local communities with elevated costs to connect to the national electric grid. A pioneer project represents the geothermal micro-plant Piedras de Lumbre, which provides 600 kW for 800 inhabitants from the rural community of Maguarichic, State of Chihuahua, liberating the community from diesel combustion. A total of 50 sites were identified on Mexican territory where electricity could be produced via small binary cycle systems (Hiriart Le Bert 2004).

By comparison, the capital cost of a binary unit is in the order of 1,500–2,500 USD per installed kW (excluding drilling costs), whereas the local distribution of electricity, at 11 kV using wooden poles, costs a minimum of USD 20,000 per kilometre. A 1,000 kWe binary plant provides electricity for about 1,000 to 5,000 people (Entingh et al. 1994).

Coupling of conventional units with binary-cycles: Generating electricity from low-to-medium temperature geothermal fluids and from waste hot waters coming from the separators in water-dominated geothermal fields has made considerable progress since improvements were made in binary fluid technology. At present, only two 1.5 MWe binary units are installed in Mexican geothermal fields (Los Azufres). The coupling of binary systems with existing flash and back-pressure units in producing geothermal fields represents a feasible technical option to take advantage of the remnant heat of geothermal fluids within a temperature range of 85–170 °C. The flash/binary system have been successfully operating in Hawaii since 1991 (Puna Geo Venture facility), as well as in three power projects in New Zealand, and in the Philippines since 1995 (Upper Mahiao geothermal facility).

5.4 Legal aspects – Private contribution

Constitutional and regulatory limitations hinder the private participation in the energy sector. For the specific case of electricity generation, the Public Service Act for Electrical Energy (*Ley del Servicio Público de Energía*

Eléctrica) defines an exclusive privilege for the Nation to produce electricity, which only allows private generation for specific cases, and mandates public electricity enterprises to acquire energy at the lowest available short term costs (SENER and GTZ 2006). As a logical consequence, all of the current geothermal power plants are operated by the public utility, the *Comisión Federal de Electricidad* (CFE). The following key points are proposed in order to enhance the contribution of geothermal resources:

- A general reformation of the electrical sector in order to avoid discrimination of renewable energies.
- To make appropriate the legal and regulatory frame, as well as market incentives and innovative financing schemes to fortify the renewable energy sector and to promote private inversions. In this sense, the Law for the Use of Renewable Sources of Energy (*Ley para el Aprovechamiento de las Fuentes Renovable de Energía, LAFRE*) would contribute to the development of renewable energies, that if approved by the Senate (SENER and GTZ 2006).
- Legal frame to regulate the quotation mechanism of the certificates, assignations, taxation, property regime of the certificates and transfer procedures (IILSEN and CIE-UNAM 2004).
- Fiscal and contractual incentives for producers of renewable energies will assure the profitability of the renewable energy project. The LAFRE Initiative plans the creation of a trust fund that would grant temporal incentives to projects that generate, through renewable energy sources, electricity for public service (SENER and GTZ 2006).
- Include economic effects of environmental impacts into the cost-benefit balance for different energy types.

5.5 Future consciousness and long-term investments

As for most renewable energy systems, the costs of a geothermal plant are heavily weighted towards up front investments. Costs may vary from USD 50 to 150 million for a 50 MW power plant, including exploration and leasing (USD 3 to USD 6 million), project development and feasibility studies (USD 0.25 to USD 2.3 million), well-field development (up to USD 32 million) and project finance, plant construction, and start-up (IEA 2003). Between 11 and 14 production and injection wells are required for the generation of 50 MW, with average drilling costs of USD 2 million per well. In México, an average investment cost of USD 84 million for a standard 50 MW plant is subdivided in USD 32 million for field development and USD 52 million for generation and transmission (Hiriart Le Bert 2004).

Most benefit-costs calculations for the development of energy infrastructure ignore the long-term availability of renewable energy sources. Geothermal reservoirs can be exploited for several decades, with a guarantee for the reimbursement of elevated initial investment costs with time. Non-reversal environmental damages, such as climatic changes by the combustion of fossil fuels, as well as the natural limitation of the availability of hydrocarbon resources – at a current production rate of 4 million barrels, the proven 1P fossil fuel reserves of 16,500 million barrels (PEMEX 2006) will be exploited within 11.3 years – requires a pronounced mentality change of decision makers and within the population in order to transform the present short-term energy management towards a sustainable development. An exemplary case for the environmental consciousness and future vision of the population is given for the city of Unterhaching in Bavaria, Germany, where approximately USD 64 million are going to be invested into a high-technology geothermal drilling program to make independent 20,000 people from fossil fuel supply. A total of 5 MW will be produced for the local district heating and electricity demand, and national subvention programs guarantee an income of 0.19 USD for each kWh delivered to the regional public network. An estimated time period of 15 to 20 years has been calculated as required time period to reimburse the inversion costs (SPIEGEL 2006).

6 Conclusions

Geothermal exploitation and other renewable energy types represent an insufficient assumed option for global energy production (CEPAL 1999). In the case of México, after a period of stagnation of geothermal production during the 90's, increased investments from 2000 to 2004 (USD 415.0 million in comparison to USD 245.6 million from 1990–1994) allowed the expansion of geothermal electricity production from 755 to 959.5 MWe between 2000 and 2006. On the other hand, the expansion of geothermal capacities is not equal to the fast rising national energy demand, as a decrease of the geothermal contribution from 3.1 % in 2004 to 2.5 % is expected for the year 2014 (SENER 2006).

Favorable geological-tectonic conditions in several regions of the Mexican territory could allow the expansion of the actual geothermal electricity production towards additional potential hydrothermal sites. The present installed geothermal capacity could be increased from a current generation of 959.5 MW to 3,650 MW, which would rise the geothermal contribution for the national electricity market from a present 3.1 % to over 12 %, depending on the implementation of enhanced technology and enforced exploration research. In contrast to technical-feasible geothermal expansion strategies, the future energy plan from the national provider CFE for 2010 considers a low-budget geothermal capacity of 1,078 MW for the year 2010 (Gutiérrez-Negrín and Quijano-León 2004).

In contrast to countries with low geothermal temperature gradients, the Mexican energy market with abundant high- and medium- enthalpy reservoirs do not yet require the application of specific high-technology methods, such as hot dry rock technology (HDR), or the development of geopressurized or marine reservoir systems. The amplification of geothermal electrification could save considerable amounts of fossil fuels, e.g. the installation of the 100 MW geothermal facility in Los Azufres (2005–2006) will save 1.2 million barrels of combustible per year (NOTIMEX 2006).

Major attention should be given to direct use options, where an enormous energy potential of several thousands of megawatts in the underground could be used for district and greenhouse heating, spas, aquaculture, and other low-temperature applications.

References

Birkle P, Verma M (2003) Geothermal Development in México. In: Chandrasekharam D, Bundschuh J (eds) Geothermal Energy Resources for Developing Countries. A.A. Balkema Publishers, Lisse, The Netherlands, pp 385–404

Brown KL (1995) Environmental aspects of geothermal development. World Geothermal Congress 1995, Pre-Congress Courses, Pisa-Italy, 18–20 May 1995, International Geothermal Association, 145 pp

Central Intelligence Agency (CIA) (2001) World Fact Book 2001. Central Intelligence Agency, Washington D.C.

Comisión Federal de Electricidad (CFE) (2006) Generación de Electricidad. Geotermoeléctrica. 30.06.2006. http://www.cfe.gob.mx/es/LaEmpresa

Comisión Federal de Electricidad (CFE), Universidad Autónoma Nacional de México (UNAM) (1993) Documento de análisis y prospectiva del Programa Universitario de Energía. Geotermia en México. Facultad de Ingeniería de la UNAM

Comisión Económica para América Latina y el Caribe (CEPAL) (1999) Uso de la geotermia como fuente de energía debería aumentar en América Latina. 24.03.1999, http://www.eclac.cl

Department of Energy (DOE U.S.) (1997) Renewable energy technology characterizations. EPRI Topical Report No. TR-109496, December 1997, Energy

Efficiency   and   Renewable   Energy   (EERE),
    http://www.eere.energy.gov/consumerinfo/tech_reports.html
Entingh DJ, Easwaran E, McLarty L (1994) Small geothermal electric systems for
    remote powering. U.S. DoE, Geothermal Division, Washington, D.C., 12 pp
Gawell K, Reed M, Wright PM (1999) Preliminary report: Geothermal energy, the
    potential for clean power from the earth. Geothermal Energy Association
    (GEA), Washington D.C., 7 pp
Geothermal Education Office (GEO) (2004) Geothermal energy facts. 13.01.2004,
    http://geothermal.marin.org/geoenergy.html
Geothermal Energy Association (GEA) (2006) La geotermia: Energía confiable y
    limpia para Las Américas, http://www.geo-energy.org/publications
Gutiérrez-Negrin LCA, Quijano-León JL (2005) Update of geothermics in
    México. Proc. of the World Geothermal Congress 2005, Antalya, Turkey, 24–
    29 April 2005, 10 pp
Gutiérrez-Negrín LCA, Quijano-León JL (2004) Update of geothermics in
    México. Geotermia 17:21–30
Herrera J, Rocha V (1988) Carta de manifestaciones termales de la República
    Mexicana. Informe 20/88, CFE, Gerencia de Proyecto Geotermoeléctricos
Hiriart Le Bert G (2004) Generación de energía eléctrica con centrales geotérmi-
    cas. UNAM Ciclo de Conferencias sobre el Sector Eléctrico, 24.03.2004
Iglesias ER, Arellano AG, Torres RJ (2005) Estimación del recurso y prospectiva
    tecnológica de la geotermia en México. Instituto de Investigaciones Eléctri-
    cas, Informe IIE/11/3753/I 01/P, 63 pp
Iglesias ER, Torres RJ (2003) Low- to medium-temperature geothermal reserves
    in México: a first assessment. Geothermics 32:711–719
Instituto de Investigaciones Legislativas del Senado de la República (IILSEN),
    Centro de Energia (CIE-UNAM) (2004) Nuevas energías renovables: Una al-
    ternativa energética sustentable para México, Aug. 2004, 183 pp
International Energy Agency (IEA) (2006) Renewables in global energy supply.
    An IEA Fact Sheet, Sept. 2006, 13pp
International Energy Agency (IEA) (2003) Renewables for power generation.
    Status & Prospects, 189 pp
International Geothermal Association (IGA) (2006) Geothermal in the World.
    http://iga.igg.cnr.it
International Solar Energy Society (ISES) (2002) Transitioning to a renewable en-
    ergy future, Germany
Kagel A, Bates D, Gawell K (2005) A guide to geothermal energy and the envi-
    ronment. Geothermal Energy Association (GEA), Washington D.C., 87 pp
Lindal B (1973) Industrial and other applications of geothermal energy. In: Arm-
    stead HCH (ed) Geothermal Energy. UNESCO, Paris, pp 135–148
Lund JW, Freeston DH, Boyd TL (2005) World-Wide Direct Uses of Geothermal
    Energy 2005. Proc. of the World Geothermal Congress 2005, Antalya, Tur-
    key, 24-29 April 2005, 20 pp
Lund JW, Freeston D (2001) World-wide direct uses of geothermal energy 2000.
    Geothermics 30:29–68

Martínez-Estrella JI, Torres RJ, Iglesias ER (2005) A GIS-based information system for moderate to low-temperature Mexican geothermal resources. Proc. of the World Geothermal Congress 2005, Antalya, Turkey, 24–29 April 2005, 8 pp

Mercado S, Sequeiros J, Fernández H (1985) Low enthalpy geothermal reservoirs in México and field experimentation on binary-cycle systems. Geothermal Resources Council Transactions 9:523–526

NOTIMEX (2006) Agencia de Noticias del Estado Mexicano. 21.09.2006. http://www.notimex.com.mx

Pantoja JA, Gómez AC (2000) Géiseres y manantiales termales en México. Ciencias 59:23–25

PEMEX (2006) Reservas de hidrocarburos al 31 de diciembre del 2005, 20 pp., www.pemex.com, March 16, 2006

Quijano L (2005) Situación actual y perspectivas de la energía geotérmica en México. VI Foro Regional. Impacto Estratégico de la Energía Geotérmica y Otros Renovables en Centreo América. Managua, Nicaragua, October 2005

Quintanilla JM (2004) Escenarios de emisiones futuras en el sistema energético mexicano. In: Martínez J, Fernández A (eds) Cambio climático. Una visión desde México. Instituto Nacional de Ecología (INE), 525 pp

Ragnarsson A (2005) Geothermal development in Iceland 2000–2004. Proc. of the World Geothermal Congress 2005, Antalya, Turkey, 24–29 April 2005, 11 pp

Rybach L, Mégel T, Eugster WJ (1999) How renewable are geothermal resources? Geothermal Resources Council Transactions 23:563–567

Secretaría de Medio Ambiente y Recursos Naturales (SEMARNAT) (2005) El mundo ambiente en México. Un resumen

Secretaría de Energía (SENER) (2006) Prospectiva del sector eléctrico 2005–2014. Dirección General de Planeación Energética, 134 pp

Secretaría de Energía (SENER) (2004) Balance Nacional de Energía 2003: Potencial de capacidad y generación de energía eólica, geotérmica y minihidráulica en México, pp 127–144

Secretaría de Energía (SENER), Gesellschaft für Geotechnische Zusammenarbeit (GTZ) (2006) Energías renovables para el desarrollo sustentable en México. Renewable energies for sustainable development in México. Secretaría de Energía (México) and Gesellschaft für Geotechnische Zusammenarbeit (Germany)

SPIEGEL (2006) Unterhaching will sich von Öl und Gas befreien. Spiegel Online, www.spiegel.de, 10.07.2006

Torres RJ, Martínez-Estrella JI, Iglesias ER (2005) Database of Mexican medium-to low-temperature geothermal resources. Proc. of the World Geothermal Congress 2005, Antalya, Turkey, 24–29 April 2005

Torres VR, Venegas SS, Herrera JF, González EP (1993) Manifestaciones termales de la República Mexicana. In: Torres VR (ed) Geotermia en México. Universidad Autónoma de México

Energy and Activated Carbon Production from Crop Biomass Byproducts

Stanley E. Manahan[1], Manuel Enríquez-Poy[2], Luisa Tan Molina[3], Carmen Durán-de-Bazúa[4]

[1]ChemChar Research, Inc. 123 Chem. Bldg. Univ. of Missouri-Columbia, Columbia, Missouri 65211, USA. [2]Cámara Nacional de las Industrias Azucarera y Alcoholera, Calle Río Niágara 11, 06500 México D.F. México. [3]Molina Center for Energy and the Environment, 3262 Holiday Court, Suite 201, La Jolla, California, 92037, USA. [4]UNAM, PECEC, Paseo de la Investigación Científica s/n, 04510 México D.F. México. E-mail: mcduran@servidor.unam.mx

Abstract

The proposed research is designed to upgrade sugarcane biomass byproducts (bagasse) to high-grade fuel (hydrogen), activated carbon, and chemical feedstocks (synthesized gas). It is shown how gasification can significantly enhance the value of sugarcane through the production of energy, synthesis gas, and carbon, which is required in the sugar purification and decoloring process. Furthermore, it can act as a valuable associate to sugar fermentation processes used to produce ethanol employed as a substitute for hydrocarbon fuels and raw materials. Through this project one can make the large sugarcane-growing regions of Mexico and other Latin American, African, and Asian countries more economically diverse and self-sufficient, increasing local employment and enabling support of larger rural populations. A key part of the proposed research is the training of highly skilled professionals to guide the Mexican sugarcane agro-industry toward production of high added value products in addition to sugar.

1 Introduction

As worldwide reserves of petroleum dwindle and become more expensive, biomass will have to take a role in providing fuel and other products now made from petroleum and natural gas. Although biomass can be grown from crops, such as trees, high-yielding grass (switchgrass), and algae, resources from which to produce such biomass are limited and they are required for food production. A source of biomass largely overlooked in the past is crop byproduct biomass, such as rice or wheat straw, cornstalks, and bagasse from sugarcane production. Although a fraction of these materials should be returned to soil to maintain soil quality and fertility, an excess is generally produced and, for some crops is simply burned in the fields. An advantage of using crop byproduct biomass to make other products is that the crops have to be processed by a machine to harvest grain or, in the case of sugarcane, sap from which the sugar is made. Therefore, it adds little incremental cost to collect the excess biomass for further use. Furthermore, crops can be bred to produce additional plant biomass, if there is a market for it.

Utilization of crop byproduct biomass can be done in several ways. One of the simplest ways is to burn the material in boilers to produce steam and generate electricity. Although already practiced, for example, in sugarcane industry, burning biomass is not a very effective means to use the energy in it. Another approach is to employ enzymes to hydrolyze sugars from it that can be fermented to make alcohols. However, crop byproduct biomass is a complex lignocellulose material that yields a mixture of sugars requiring different yeasts for their fermentation plus large quantities of lignin that is almost completely resistant to biological action.

Gasification provides a means to use crop byproduct biomass to produce high-value products utilizing all of the biomass material. Gasification consists of the partial combustion of combustible material at high temperature in an oxygen-deficient environment to produce carbon dioxide and a combustible mixture of carbon monoxide, elemental hydrogen, and varying amounts of methane. For the generation of fuel and synthesis gas (carbon monoxide together with elemental hydrogen), gasification is carried out under conditions such that essentially all the carbon is consumed. The ChemChar process described in this work differs from other common gasification processes in that it can be operated under conditions that maximize production of elemental carbon residue as well as of synthesis gas.

This paper describes the application of ChemChar gasification to crop byproduct biomass, specifically bagasse residue remaining from the processing of sugarcane, to produce a maximum amount of solid carbon residue and of synthesis gas. The solid material can be further processed to yield powdered activated carbon, a high-value product with a growing worldwide market for industrial applications (clarifier of sugarcane syrups to produce refined sugar), air pollutant removal, and water purification. Activated carbon from this source presents a great advantage over that from petroleum coke and especially coal because it generally contains only innocuous mineral constituents, such as calcium, potassium, sodium, phosphorus, and silicon and does not contain arsenic and toxic heavy metals that are present in carbon from coal and in coal coke, especially that from heavy crude oils that are increasingly being used.

2 Proposal

Excess crop residues, designated here as crop biomass byproduct, are huge potential sources of fuel and can serve as sources of carbon for the generation of powdered activated carbon as well. By tapping this material for fuel and carbon, significant economic advantages could be gained by the agricultural sector of the economy. The means of using crop byproduct biomass for fuel and carbon remain underdeveloped. The obvious use of these materials as a fuel burned in boilers to generate electricity by steam-driven generators is handicapped by their generally high water content, which consumes large amounts of energy. Furthermore, the high moisture content of crop biomass byproduct adds to the expense of transport and can make the material subject to decay in storage.

Problems with the utilization of crop biomass byproduct can be overcome by partial gasification (Robertson et al. 2002) in which the organic material is subjected to a gasification process that conserves the carbon content of the fuel while driving off water and volatile matter, leaving a dry, carbon-rich, high calorific content material that can be economically transported and safely stored. Furthermore, partial gasification generates a synthesis gas byproduct rich in combustible elemental hydrogen, methane, and carbon monoxide that can be used for electricity generation and co-generation at the gasifier site. The research outlined in this document proposes an investigation of partial gasification for the treatment of crop biomass byproduct to give valuable fuel and activated carbon.

3 Background and rationale

Historically, biomass from wood has been a significant source of energy for heating and electricity generation and remains so today. With uncertain energy reserves, serious consideration has been given to growing crops that produce high amounts of biomass that can be utilized for energy production. Energy crops considered for biofuel production include hybrid willow, switchgrass, and hybrid poplar. However, to make a significant contribution to worldwide necessities of energy, such an approach would require large amounts of farmland that can more profitably be used for growing commodity crops, such as corn and other edible crops. Aside from high-value fuels produced from grain (ethanol from corn fermentation and biodiesel fuel from soybeans), the price that can be paid for biomass fuel from "energy plantations" is generally not high enough at present to make such enterprises economically attractive.

An alternative source of biomass energy is that of crop production byproducts such as cornstalks (corn stover). These kinds of materials fall into the classification of lignocellulosic biomass (Wyman 2003). The US Energy Information Administration estimates that as much as 600 million wet tons of biomass per year is grown in the United States. Although some of the crop biomass byproduct must be returned to soil each season to maintain healthy soil quality, it is generated in significant excess over that requirement. Indeed, with modern conservation tillage practice, some problems, such as excess growth of mold, can be encountered leaving all of the crop byproduct biomass in the field and removal of the excess biomass would be beneficial. Considering that crops such as corn must be processed in the field to harvest grain, the biomass byproduct in the form of straw, stalks, leaves, husks, and corn cobs is available for minimal cost above the cost of transporting it from the field. In addition to its low cost and abundance, biomass provides an energy source that is generally low in pollution potential. Furthermore, since all of the carbon from this source is fixed from atmospheric carbon dioxide, biomass fuel does not add any net amount of greenhouse gas carbon dioxide to the atmosphere. As a carbonaceous fuel, biomass can readily substitute for existing fossil fuel sources in the power generating infrastructure and can be cofired with fossil fuels, such as coal.

Direct combustion of biomass for power production is not necessarily the best means of using the biofuel resource in place of fossil fuels. The US Department of Energy's (DOE) Office of Energy Efficiency and Renewable Energy (EERE) recognizes as three leading approaches to the utilization of biomass: (1) fermentation of sugars, widely practiced on grain in the production of fuel ethanol, (2) production of biodiesel fuel from plant oils, and (2) thermochemical gasification.

Gasification offers opportunities for higher overall efficiencies, better control over emissions, and the production of a wide range of products including synthetic fuels, methanol, acetic acid, ethers, alcohols, and activated carbon (van der Grift et al. 2001).

Furthermore, hot carbon from biomass may be reacted with steam to produce elemental hydrogen, the "fuel of the future" that is finding increased use in fuel cells. At present, the most efficient approach to use biomass for power production appears to be through biomass integrated gasification combined-cycle (BIGCC) plants.

In addition to its direct use as a fuel, biomass can serve as a source of carbon for powdered activated carbon. This material has proven to be extremely useful, particularly for the treatment of water, including water processed by the agricultural sector. In most cases, spent activated carbon can be used as a fuel, adding to its value.

One of the biggest barriers to the use of biofuel is its low density, which means that a container such as a truck or rail car can carry only a relatively small mass of fuel per unit volume. The density can be approximately doubled by baling the material and doubled again by chopping the biofuel and compressing it into pellets. Both of these operations add expense and time to the collection and processing of biofuels.

Another major detriment to the use of biofuel is its high moisture content (Turnbull et al. 1996). Safe storage of biomass generally requires a moisture content of less than 15 %. High moisture levels in biomass make its combustion or gasification much less efficient. For combustion or gasification processes, the energy required for evaporation is substantial. For combustion, fuel moistures of approximately 50 % (wet basis) require nearly 10 % of the energy content of the dry material simply to evaporate excess water, and moistures of approximately 60 % require nearly 20 % of the energy content of the dry material. Relatively small decreases in moisture levels can lead to dramatic energy gains for a given combustion or gasification process.

According to Prasertsan and Krukanont (2003), a decrease in moisture level from 60 % to 40 % produces a power increase by a factor of approximately 4, while increasing the boiler thermal load by a factor of approximately 1.7. Important benefits of biomass drying for combustion boilers are more flexible and stable boiler operation, using less support fuel, increased steam generation in existing boiler, smaller size of a new boiler due to decreased flue gas flow, reduced fan power, improved combustion efficiency, lower emissions (usually), and reduced need for support fuel.

The process of drying biomass has proven to be problematic. The use of heated pipes in a bed of biofuel is inefficient because of the limited rate of heat transfer from hot gases or steam inside the pipes through pipe walls to the biofuel in which the pipe is imbedded. Hot gas forced through a bed of biofuel in a fluidized bed or entrained flow configuration is an efficient means of heat transfer and drying.

However, the exhaust gas from such an operation tends to contain an aerosol of so-called "blue haze" composed of organic particles that can cause air pollution problems (Brammer and Bridgewater 1999). Furthermore, if hot air is used, accidental combustion of the biofuel is always a problem. Technology and economics dictate that drying biomass be considered along with the production of power, synthetic fuel, and other products of biomass (Raiko et al. 2003).

4 Research and development

4.1 Technical, economic, social, and other benefits to the communities

The present research is designed to provide the technical basis for the application of partial gasification to the economic use of crop biomass byproduct for fuel and activated carbon production. It will positively impact the local economy and particularly the agricultural sector by adding a domestic stream of energy to existing supplies. Partial gasification units sited in numerous locations in agricultural regions would add income, employment, diversity, and stability to these regions, thereby promoting their economic and social well-being. Each gasifier-based biofuel processing facility could be designed to generate electricity in a number of decentralized power generating stations tied to the power grid and making it less vulnerable to disruption, such as ice storms in rural areas. The byproduct heat generated could be used to dry grain grown in surrounding agricultural areas and could be used for district heating in small towns, thereby reducing the demand on expensive propane now widely used for heating and grain drying in many rural communities. Besides, if the local communities receive carbon dioxide bonuses because of crops cultivation, this would promote human settlements in the rural areas with adequate living conditions, thus alleviating the need for migration. By using renewable biomass, the proposed process is greenhouse-gas-neutral and hence desirable from a social point of view. As a member of the international community and major industrial country, Mexico and other countries must also give consideration to greenhouse gas emissions.

The process described in this work is designed to be integrated with existing technologies including small scale combustion boilers, combined heat and power (CHP) generators, integrated biomass gasification-fuel cells (IBGFC), synthetic fuel generation, chemical synthesis, or carbon production. It is this integration of biomass drying, power production, and carbon production that allows for the maximum economic advantages of this process over existing technologies alone.

4.2 The estimated total cost of the proposed approach relative to benefits

Although exact figures cannot be given at this stage of the research, it may be assumed that the total costs of bringing this technology to economic fruition would be in a very few millions of US dollars whereas the economic benefits of successful use of the technology could be estimated to amount to US$100s millions per year.

4.3 Any specific policy issues or decision which might be affected by the results

Effects on policy should be minimal. No major changes in policy should be required by major use of the proposed technology.

5 Technical objectives

The proposed research addresses the problems inherent to biomass utilization discussed above by application of partial gasification of crop biomass byproduct to dry and enrich the material to a high-energy fuel or high-value powdered activated carbon. The partial gasification process to be used is an outgrowth of the long-established ChemChar gasification system (Manahan 1990). It differs from most other biofuel gasification processes by operating in a manner that minimizes consumption of the biofuel and maximizes production of dry carbonized biofuel that can be transported to power plants economically. In addition to being burned directly as a fuel, the carbon product can serve as a feedstock for additional gasification in the presence of steam to produce synthesis gas or elemental hydrogen. During gasification, only the minimum amounts of biofuel will be gasified to produce the energy required for drying additional biofuel. The ChemChar process has proven useful in the past to reactivating spent granular activated carbon (Manahan 1992), and the partial gasification process is proposed to make powdered activated carbon as well.

The following major technical objectives of the proposed study are considered in more detail in the phase I of the working plan:

- Design, construction, and testing of a pilot and prototype scale continuous-feed cocurrent flow gasifier optimized to run on biofuel with corn stover (mixed stalks, leaves, husks, and cobs from corn production) as the major test material for the pilot equipment, and with sugarcane bagasse for the prototype unit.
- Design, construction, and testing of a pilot and prototype scale fluidized bed drying unit using hot synthesis gas from the gasifier to dry ground biofuel.
- Characterization of the gasifier/drying unit with respect to mass and energy balance.
- Characterization of solid and gaseous products including activated carbon.
- Economic analysis of the system for commercial viability.
- Analysis of the system with respect to environmental, sustainability, and social considerations.

6 Working plan

All of the construction work for the prototype system is proposed to be performed by personnel of the cooperating sugarcane mills with the support of the shops available there.

The starting-up, follow-up, and maintenance tasks of the prototype system is proposed to be performed by personnel of ChemChar Research, Inc., UNAM, MCE2, and the collaborating sugarcane mill.

6.1 Task 1

Design, construction, and testing of pilot and prototype scales continuous feed cocurrent flow gasifiers optimized to run on biofuel with corn stover and sugarcane bagasse as the major test materials.

An advanced-design ChemChar gasification system is proposed to be constructed and tested. This system is an outgrowth of previous ChemChar gasification systems based upon very recent laboratory studies of the gasification of chopped cornstalk biomass (Haney 2003). Past research on these systems has concentrated on coal-based granular carbon feedstocks mixed with various kinds of wastes including hazardous waste sludges.

The design of the gasification unit is shown in Fig. 1 and referred to subsequently as "the gasifier." Basically, the gasification reactions occur inside the gasifier chamber in an incandescent thermal zone (ITZ) that forms the surface of a short column of finely divided carbon produced by gasification of the biofuel. Biofuel is fed continuously from the hopper on top of the gasifier chamber through a central pipe and spread with a spreader uniformly on top of the ITZ. The carbon is continually removed from the gasification chamber by a feed mechanism and is collected in a container. Hot synthesis gas product is withdrawn from the bottom of the gasification chamber.

A slight positive flow of carbon dioxide (oxygen-free combustion gas in a commercial unit) is introduced into the biofuel hopper to prevent any oxygen from getting into this container, which might pose a fire hazard. Once the ITZ is established by an ignition process, biofuel and oxygen inlet flows and char product outlet flows are continuously adjusted to maintain an ITZ that is not quite completely covered by a continuously re-plenished inflow of biofuel fed through the central pipe from the feed hop-per and spread by rotating spreader rods. These feed rates can be adjusted within a range from a relatively high oxygen to fuel ratio that converts es-sentially all the biofuel to ash and gases to a much lower ratio that maxi-mizes production of carbon.

In order to understand the operation of the gasifier, it is necessary to consider the chemical processes that occur in the ITZ. These are summa-rized in the following reactions in which $\{CH_2O\}$ represents the empirical formula of the high-oxygen-content biomass:

$$C + O_2 \rightarrow CO_2 + heat \quad \Delta H = -94.05 \text{ Kcal/mol}$$

$$\{CH_2O\} + O_2 \rightarrow CO_2 + H_2O + heat \text{ (major heat-yielding reaction)}$$

$$C + 1/2O_2 \rightarrow CO + heat \text{ (yields combustible CO)} \quad \Delta H = -26.42 \text{ kcal/mol}$$

$$C \text{ (hot)} + CO_2 + heat \rightarrow 2CO \text{ (yields combustible CO)} \quad \Delta H = 41.2 \text{ kcal/mol}$$

$$H_2O + C + heat \rightarrow H_2 + CO \text{ (yields combustible } H_2 \text{ and CO)} \quad \Delta H = 31.4 \text{ kcal/mol}$$

$$C + 2H_2 \rightarrow CH_4 \text{ (yields small quantities of combustible } CH_4)$$

The water-gas shift reaction can be used to convert most of the synthesis gas product mixture of CO and H_2 to elemental hydrogen, which is valuable for applications in fuel cells and in chemical synthesis:

$$CO + H_2O \rightarrow H_2 + CO_2$$

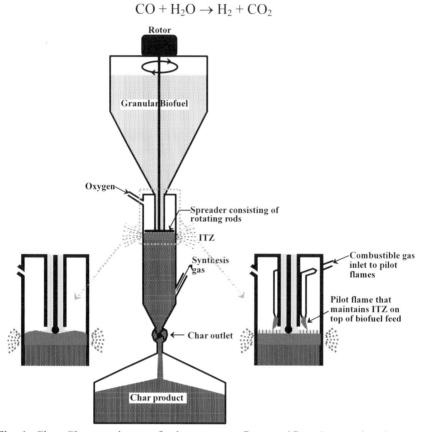

Fig. 1. ChemChar continuous feed co-current flow gasifier. Conventional operation in which the incandescent thermal zone (ITZ) is maintained only partially covered by the biofuel on top of a column of char product. A spreader is used to maintain a uniform ITZ surface and operation entails maintaining appropriate flow rates of biofuel, oxygen, and char outlet from the gasifier chamber. Pilot flames fueled by combustible gas maintain the surface of the biofuel feed so that fuel starts to gasify as soon as it is introduced into the gasification chamber and extinguished when it becomes covered with a fresh layer of biofuel.

It should be noted that the cost of oxygen will be a significant factor in the economics of the system. Commercial tank oxygen will be used for the relatively small gasifiers in the Phase I studies.

For Phase II and commercial units, vacuum swing sources of 90 % oxygen should prove to be cost effective. Fortunately, the oxygen does not have to be pure; 90 % oxygen would be totally adequate and is appreciably less expensive than more pure grades.

Although this was a partial gasification mode, the rate of movement of the ITZ relative to the column of product and feedstock was controlled by the rate of ignition of the feedstock. This resulted in a high degree of conversion of carbon in the feedstock to elemental carbon, and the rate of ITZ movement and the degree of carbonization were self controlled and could not be altered significantly by operational changes. It has since been demonstrated on a batch basis that maintaining the ITZ on the surface of a column of char product enables significant control of the degree of carbonization and gives a gas product that is richer in combustible H_2, CO and CH_4 relative to CO (Haney 2003). Figure 1 shows the configuration for the proposed continuous feed gasifier in which a configuration, a slow rate of biofuel addition relative to oxygen flow results in a relatively high degree of biomass burnout and low production of product, although the product is highly carbonized. Faster feeding of fuel relative to the oxygen flow results in gasification of a smaller fraction of the biomass and higher production of dried carbonized fuel.

There are limitations to the operation of the gasifier as described above. If the biofuel is fed too fast in an effort to gasify even less of the biofuel, a point is reached in which the ITZ becomes buried and control of the rate of ITZ movement and hence the degree of gasification is determined by the rate at which the ITZ moves spontaneously upward through the overlying layer of biomass. In order to circumvent this problem, the system shown in Fig. 1 will be employed. It is proposed to adapt this system to continuous feed.

The second mode of gasifier operation is one in which pilot flames are maintained inside the gasifier burning just at the biofuel outlet from the center feed pipe from the biofuel hopper. We propose to operate two of these flames in the gasifier used in the Phase I research and, for convenience, to power them by natural gas; in a commercial unit, such as the one that would be constructed for a Phase II project, the pilot flames would be fueled by a side stream of the synthesis gas.

The function of the pilot flames is to continuously ignite the incoming charge of biofuel. At a relatively higher biofuel flow rate, the biofuel will have a short residence time in contact with oxygen before being covered with fresh biofuel charge.

The net result of this is that partially gasified and partially charred biofuel will be buried continuously so that a relatively small fraction of the biofuel will be gasified. The larger fraction that does not react will be dried by the hot synthesis gas produced.

The product will be a significantly dried biofuel material mixed with some carbon that can be used as a fuel or as a charge to a gasifier designed to produce maximum amounts of synthesis gas.

The gasifier will be machined from steel. The main gasifier chamber will be 8 cm inside diameter and 40 cm long. These dimensions are sufficient to provide a meaningful scale that can lead to scaleup for later phases of the research, but small enough to enable convenient testing on a laboratory scale with reasonable quantities of material. Control will be provided for the following gasifier operational parameters:

- Oxygen inflow.
- Biofuel inflow to the top of the gasifier.
- Char outflow from the bottom
- Carbon dioxide flow into the top hopper.
- Carbon dioxide flow into the bottom char collection hopper.

We propose to monitor the gasification process by thermocouples embedded in the gasifier wall in the vicinity of the ITZ. Continuous adjustment of the biofuel inflow, char outflow, and oxygen inflow will enable maintainenance of the ITZ at the appropriate level.

Carbon dioxide flows into the top hopper and into the bottom char collection hopper will be maintained at 5 % of the oxygen inflow as a safety measure to prevent oxygen from getting into these chambers and posing a fire or explosion hazard.

As an additional safety measure, oxygen sensors will be installed and monitored in the biofuel feed hopper and char collection hopper.

In the biofuel drying unit shown in Fig. 2, the key parameter is inflow of moist biofuel as a function of the flow of hot exhaust gas through the dryer. Various relative flow rates will be tried and the products tested to determine the optimum drying rate.

We proposed to analyze the major products of the system. These are the following:

- Analyze char product of the gasifier for carbon content, volatile matter content, surface area (for activated carbon), and ash content.
- Analyze gas product of the gasifier before and after passing through the biofuel drier to be analyzed for CO, H_2, CO_2, H_2O, N_2, H_2S, NH_3, CH_4, and condensed liquids.
- Analyze biofuel before and after passing through the dryer to be analyzed for moisture

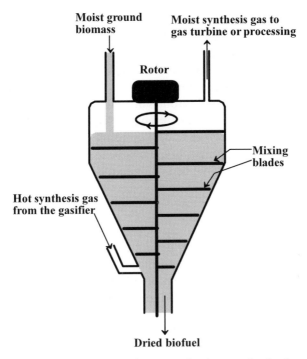

Fig. 2. Agitated bed drier for ground biomass using hot synthesized gas as a drying agent

6.2 Task 2

Design, construction, and testing of pilot and prototype scale fluidized bed drying units using hot synthesis gas from the gasifier to dry ground biofuel.

We propose to investigate two approaches to drying biofuel. The first is the partial charring of biofuel by operating the gasifier with pilot flames as discussed above. This approach is essentially a matter of operating the gasifier in a partial gasification mode at appropriate flow rates. We propose to determine the efficacy of drying by simply measuring the moisture content of the charge before and after partial gasification. The formation and removal of aerosol produced during gasification is likely to be a consideration in a prototype and later on in a commercial scale unit, and will be investigated. The product gas will be filtered through a bed of carbon produced by gasifying biofuel to partially remove the aerosol. The gas will also be pressurized and collected in a container and allowed to stand, which will allow aerosol to coagulate and settle. It is anticipated that the aerosol will not cause any problems in those cases in which the synthesis gas product is burned in a power gas turbine.

We propose to use a mechanically stirred drying chamber, as shown in Fig. 2, for drying biofuel with synthesis gas from the gasifier. The hot synthesis gas will be pumped directly from the gasifier into the drying chamber equipped to feed biofuel through at a constant rate that can be varied to test the drying efficiency at different contact times. The biofuel will be analyzed for moisture before and after going through the dryer. The temperature and moisture content of the synthesis gas will be measured before and after passing through the bed of biofuel in the dryer.

6.3 Task 3

Characterization of the gasifier/drying units with respect to mass and energy balance. One has to measure masses and compositions of solids and gases at all stages of the gasification and drying process. Data from these measurements have been analyzed to determine a mass and energy balance for the system as a whole.

6.4 Task 4

Characterization of solid and gaseous products including activated carbon. We propose to characterize the solid products for their potential commercial application. These materials will consist of 3 major kinds of products. The first is relatively highly carbonized biofuel made by operating the gasifier as shown in Fig. 1. The second is the partially gasified biofuel made by operating the gasifier. In both cases, the pertinent parameters are moisture content, volatile matter other than moisture, ash content, and calorific value or heat content. These parameters are primarily important for determining the fuel value of the solids. The third kind of solid that we propose to characterize is activated carbon. Research to date indicates that a good grade of powdered activated carbon product can be made by a second gasification of carbonized biofuel to which 20 % water has been added, without the harsh chemical additives commonly used in making activated carbon (Haney 2003).

Activated carbon samples prepared under different conditions of added water and degree of gasification (loss of carbon solid converted to gas) will be tested for surface area by a commercial laboratory and for efficacy in absorbing phenol from water in the ChemChar laboratory pilot unit, and for sugar syrup clarification in the sugarcane mill prototype unit.

Gas products have to be analyzed by a gas chromatograph to determine the products of hydrogen, carbon monoxide, methane, and carbon dioxide. These include the gases produced in the initial gasification of biofuel as well as the gases produced in the activation of carbon product by gasification.

6.5 Task 5

An economic analysis of the system for commercial viability. Based upon the data obtained for mass and energy balance, costs of materials, costs of transportation, and markets for dried biofuel, carbon, synthesis gas, and electricity, we propose to perform an economic analysis of the systems (pilot and prototype scales) to determine its commercial viability.

One has to extrapolate future costs and benefits of the system with respect to commercial viability. For example, it is necessary to consider the potential effects of a significant rise in fossil energy prices as a consequence of decreased supply and increased demand.

6.6 Task 6

An analysis of the system with respect to environmental, sustainability, and social considerations. We propose to conduct an environmental/sustainability assessment of the process to determine if its widespread use would be environmentally acceptable and sustainable as well as societal acceptable.

The Phase I performance schedule is summarized in Fig. 3.

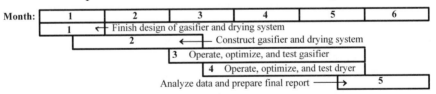

Fig. 3. Calendar of activities

7 Related research and development

The use of waste and byproduct biomass for fuel and for chemical synthesis is the topic of intense current interest as summarized in a number of publications, such as the papers by Chum and Overend (2001) and by Haq (2003). At present, biomass provides more renewable energy in the U.S. than hydropower. In 2001, US biomass energy consumption was 3.055×10^{18} J or 3055 PJ (2.9 quadrillion Btu, quads) compared to 2.53×10^{18} J or 2530 PJ (2.4 quads) for hydropower (Anonymous 2003). However, for environmental and other reasons, hydropower may be near its maximum utilization, whereas the potential for biomass utilization of energy is very high. Utilization of biomass for energy is by no means a new phenomenon, and it anticipates use of fossil fuel by thousands of years. In the US, there are more than a thousand electrical generating facilities and cogeneration plants that utilize biofuel. Because of subsidies and political mandates, more than 10 % of the corn grain produced in the US each year is used to produce ethanol by fermentation processes. The price of producing fermentation ethanol is estimated to become competitive with gasoline from petroleum through the use of hydrolyzed lignocellulose biomass residues, which would otherwise be waste material (National Research Council 1999).

The four major sources of biofuels are the following: (1) Energy crops, including hybrid poplar, hybrid willow, and switchgrass may be grown solely or primarily for biofuel. This source is limited by availability of land, but may be economic when grown on idled land to prevent food crop surpluses. (2) Forestry residues composed of logging residues, rough rotten salvageable dead wood, and excess small pole trees that are byproducts of all logging operations and are a steady source of biofuel. (3) Residues and wastes from manufacturing operations that employ wood, old wood pallets, waste wood from lumber yards, construction waste, and demolition debris that would be landfilled or simply burned can be used as biofuel. (4) Agricultural residues generated after each harvesting cycle of commodity crops represent the largest unused source of biofuels in the US. The majority of these crop residues available for use consist of wheat straw and corn stover. Agricultural land should not be completely stripped of agricultural residues, roughly half of which should remain on the land to maintain fertility and soil quality. For efficiency in harvesting, crop residues can be harvested for energy in alternate years. The cost of acquiring and delivering crop residues to the processing plant include (1) cost of collecting in the field, (2) cost of transportation from the field to the processing plant, and (3) a premium paid to farmers to encourage participation. Even in the US where fossil fuels have long been favored, there are examples of biofuel installations. Biomass gasification is practiced at the 50 MW McNeil Generating Station demonstration project in Burlington, Vermont, which supplies electricity to the City of Burlington. This plant became fully operational in 2000. In the Burlington plant a low-pressure wood gasifier with a capacity of 200 tons per day of wood chips was added to an existing wood combustion facility fueled by waste wood and forest thinning from nearby forestry operations and discarded wood pallets. Operating at capacity, fuel from the gasifier generates slightly more than 20 % of the output of the electrical plant. To this existing wood combustion facility a low-pressure wood gasifier has been added that is capable of converting 200 tons per day of wood chips into fuel gas. The fuel gas, fed directly into the existing boiler augments the McNeil Station's capacity by an additional 12 MW. Other recent projects funded by the US Department of Energy that involve biofuels include an integrated gasification and fuel cell power plant that uses segregated municipal solid waste, animal waste, and agricultural residues operated by Emery Recycling in Salt Lake City, Utah; an atmospheric gasifier with gas turbine at a malting facility using barley residues and corn stover operated by Sebesta Blomberg in Roseville, Minnesota; a combined-cycle system using a fluidized-bed pyrolyzer and corn stover feedstock operated by Alliant Energy in Lansing, Iowa; a biomass gasifier coupled with an aero-derivative turbine with fuel cell and steam turbine options using clean wood residues and natural gas as feedstocks operated by United Technologies Research Center in East Hartford, Connecticut; and a wood-residue-fueled gasification system producing a reburning fuel stream for utility boilers operated by Carolina Power and Light in Raleigh, North Carolina. A biofuel gasification and electrical generating system based upon sugar cane bagasse is illustrated in Fig. 4.

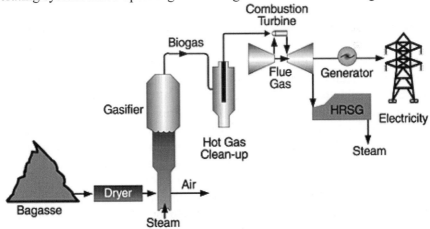

Fig. 4. A biofuel gasification and electrical generating system based upon sugar cane bagasse. Source: "Gasification Technology for Clean, Cost-Effective Biomass Electricity Generation," US Dept. of Energy, 2000, http://www.eren.doe.gov/biopower/bplib/ library/li_gasification.htm

For tropical countries (those located between the Cancer and Capricorn Tropics), where sugarcane is widely developed since it grows in very poor lands where no other crops can be cultivated, and has the highest agronomical yield of any land plant (80 to 150 tons per hectare), the use of bagasse as fuel is very important, both for the sugarcane processing plants and for electricity cogeneration. For Mexico, for example, energy comes primarily from hydrocarbons (more than 50 % of the total, that is 2843 PJ from hydrocarbons of a total energy productions of 5647 PJ, for year 2002), in spite of the fact that most of Mexico's territory is between the Equator and the Tropic of Cancer, where photosynthetic productivity is very high, with sugarcane bagasse providing only 88 PJ and wood 255 PJ of energy (SENER 2005).

The ChemChar gasification process has been the subject of numerous investigations over the last 20 years. Much of this work has been concentrated upon gasification of various kinds of hazardous wastes, but it has included some studies of gasification of biological materials as well. One study involves production of carbon from the gasification of sewage sludge and subsequent use of the carbon for drying and conditioning moist sewage sludge (McAuley et al. 2001). ChemChar gasification has been used to prepare an activated carbon product from milo grain (Bapat et al. 1999). An investigation of the fate of organic nitrogen during ChemChar gasification has shown that the nitrogen is evolved as elemental N_2 and no nitrogen oxides are produced (Medcalf et al. 1997). Gasification of organosulfur compounds has been shown not to generate sulfur oxides (Medcalf et al. 1998). The ability of the ChemChar process to produce activated carbon and to reactivate spent activated carbon has been investigated (Kinner et al. 1991).

The significance and methodology of drying biofuels are discussed elsewhere in this proposal. Biofuel that typically contains 30–60 % moisture must be dried to 15–20 % moisture before it can be burned or gasified economically. Proper feedstock drying is crucial for biomass gasification (Brammer and Bridgwater 2002). Drying requires (1) a heat source, (2) means of removing water evaporated from biofuel, and (3) a means of agitating the material being dried. The most costly of these is often the heat source. Fortunately, facilities that process biofuel for energy also produce large amounts of heat that can be used to dry biofuel (Brammer and Bridgewater 1999). A biomass gasification plant possesses the following sources of heat that can be used for drying:

- Hot synthesis gas product to be cooled.
- Hot engine or gas turbine exhaust.

- Dedicated combustion of additional or byproduct biomass, or diverted product gas.
- Hot air from the air-cooled condenser in a steam or combined cycle plant.
- Hot water from engine cooling or condenser in a steam or combined cycle plant.

Dryers may be indirect in which the material being dried contacts a heat-exchanger surface, such as hot pipe through which hot fluid is circulated, and direct in which the substance being dried contacts a hot fluid, such as hot synthesis gas from a gasifier. Three common configurations of biofuel dryers are rotary dryers (the most common type), flash dryers, and superheated steam dryers (Amos 1998). Rotary dryers are least sensitive to material size but present the greatest fire hazard and are not well suited to use with hot synthesis gas because their large volume poses an explosion danger. Flash dryers are more compact and easily controlled, but require a small particle size, which is not much of a disadvantage for biofuel gasifier feedstocks.

The steps involved in drying are (1) heating the material to the wet bulb temperature at which drying occurs; (2) generally rapid evaporation of surface moisture; (3) heating to drive water from inside particles of biomass to the surface where evaporation occurs, a process that slows as the material becomes dryer; and (4) heating of the dried material because there is insufficient water to evaporate and keep the material cool. At this last stage, which is generally avoided in most drying operations, under high-temperature conditions biofuel may begin to pyrolyze so that water chemically bound to the biofuel evaporates and with hot air drying fire is a major concern.

The treatment of process water from the production of ethanol by fermentation of grain, hydrolyzed wood wastes, and other biomass sources is a major consideration in the environmental acceptability of biomass-to-ethanol processes for the production of fuel ethanol (Anonymous 1998). Activated carbon is a key material in the removal of oxygen-demanding organic matter from water prior to its reuse or release. The production of low-cost powdered activated carbon from biomass near the source of the supply of biomass used in fermentation as proposed for this research could be very helpful to the ethanol industry.

Sugarcane is a natural source of ethanol either through sugarcane juice fermentation or through crystal sugar molasses fermentation. A surplus of these fermentation and distillation processes is the methane production from "vinasses", the liquid residues collected from the distillation columns, through anaerobic treatment, and the biomass production from the subsequent aerobic treatment, useful for fish feedlots. A final use of an-aerobic-aerobic treated "vinasses" as poor soils amendment is also an asset of the holistic use of sugarcane as a raw material (Durán-de-Bazúa et al. 1991; Durán-de-Bazúa et al. 1994; Chaux et al. 1997; Bautista-Zúñiga and Durán-de-Bazúa 1998; Pedroza-Islas et al. 1999; Bautista-Zúñiga et al. 2000a; Bautista-Zúñiga et al. 2000b).

8 Potential post applications

In the rural economy, both the demand for energy and the potential supply of energy from crop biomass byproduct are huge. If the process described in this proposal can be shown to be technically feasible and economically viable, the potential exists for business enterprises producing substantial amounts of energy and activated carbon product from renewable crop by-product sources in worldwide agricultural areas.

In addition to producing energy, gasification of crop biomass byproduct yields a synthesis gas mixture of carbon monoxide and hydrogen that can serve as a platform for synthesis of needed chemicals. Of particular significance in the agricultural economy is the potential for the production of elemental hydrogen, H_2, which can also be used to manufacture anhydrous ammonia, NH_3. Currently, methane from natural gas is used to generate hydrogen for ammonia synthesis. But demand for natural gas exceeds supply and the cost of this source of hydrogen, and, therefore, ammonia, are likely to increase. The potential of producing hydrogen for ammonia synthesis by gasification of crop biomass byproducts is a positive indicator for potential applications.

9 Final remarks

A strong argument may be made that the proposed process of producing energy and activated carbon from sugarcane biomass byproducts would have a very positive social impact by providing a means of enhancing the economies of rural communities and making such communities more self-sufficient, particularly in energy. Enhanced employment opportunities provided by crops cultivation and biofuel utilization facilities should have a positive effect upon communities in which these facilities are located.

Extreme drought in the US and record-setting high temperatures in Europe during the summer of 2003 have given impetus to considerations of reducing greenhouse gas carbon dioxide emissions. By using renewable biomass, the proposed process is greenhouse-gas-neutral and hence desirable from a societal viewpoint.

References

Amos WA (1998) Report on Biomass Drying Technology. NREL/TP-570-25885, National Renewable Energy Laboratory, Golden, CO, http://www.ott.doe.gov/biofuels/predownload.html#3953

Anonymous (1998) Wastewater Treatment Options for the Biomass-To-Ethanol Process. Merrick & Company. http://www.afdc.doe.gov

Anonymous (2003) Biopower in the US. US Department of Energy, January, http://www.bioproducts-bioenergy.gov/news

Bapat H, Manahan SE, Larsen DW (1999) An Activated Carbon Product Prepared from Milo (*Sorghum Vulgare*) Grain for Use in Hazardous Waste Gasification by ChemChar Coccurrent Flow Gasification. Chemosphere 39:23–32

Bautista-Zúñiga F, Durán-de-Bazúa C (1998) Análisis del beneficio y riesgo potenciales de la aplicación al suelo de vinazas crudas y tratadas biológicamente /Analysis of the benefits and potential risks derived from the application of crude and biologically treated vinasses to soil. Rev Int Contam Ambient 14(1):13–19

Bautista-Zúñiga F, Durán-de-Bazúa C, Lozano R (2000a) Cambios químicos en el suelo por aplicación de materia orgánica soluble tipo vinazas / Soil chemical changes by application of soluble organic matter present in vinasses (Spanish). Rev Int Contam Ambient 16(3):89–101

Bautista-Zúñiga F, Durán-de-Bazúa C, Reyna-Trujillo T, Villers-Ruiz L (2000b) Agroindustrial organic residues: Handling options in canesugar processing plants. Sugar y Azúcar 95(9):32–45

Brammer JG, Bridgewater AV (1999) Drying in a Biomass Gasification Plant for Power or Cogeneration. In: Proceedings of the Biomass Conference of the Americas, 4th. Vol 2, pp 281–287

Brammer JG, Bridgwater AV (2002) The Influence of Feedstock Drying on the Performance and Economics of Biomass Gasifier-Engine CHP System. Biomass and Bioenergy 22:271–281

Chaux D, Durán-de-Bazúa C, Cordovés M (1997) Towards a cleaner and more profitable sugar industry. An information package on waste minimization and pollution abatement in the cane sugar industry. Miranda-da-Cruz S, Luken R (eds) United Nations Industrial Development Organization. Vol 1, 82 p, Vol 2, 40 p, Vienna, Austria

Chum HL, Overend RP (2001) Biomass and Renewable Fuels. Fuel Processing Technology 71:187–195

Durán-de-Bazúa C, Noyola A, Poggi HM, Zedillo LE (1991) Biodegradation of process industry wastewater. Case problem: Sugarcane industry. In: Martin AM (ed) Biological Degradation of Wastes. Ch. 17, pp 363–388, Elsevier Science Pub Ltd, London

Durán-de-Bazúa C, Noyola A, Poggi H, Zedillo L (1994) Water and energy use in sugar mills and ethyl alcohol plants. In: Efficient Water Use. Eds. Garduño-Velasco H, Arreguín-Cortés F, Pub UNESCO-ROSTLAC. Montevideo, Uruguay, pp 361–37027

Haney P (2003) Gasification of Granular Cornstalk Biomass for Drying Biomass Fuel and Generation of Activated Carbon. Ph.D. Dissertation, University of Missouri, Department of Chemistry

Haq Z (2003) Biomass for Electricity Generation. http://www.eia.doe.gov/oiaf/analysispaper/biomass/

Kinner LL, McGowin A, Larsen DW, Manahan SE (1991) ChemChar Process for Carbon Re-activation. Chemosphere 22(12):1197–1209

Manahan SE (1990) Process for the Treatment of Hazardous Wastes by Reverse Burn Gasification. US Patent Number 4,978,477, ChemChar Research, Inc., Columbia, MO, March 30

Manahan SE (1992) Process for the Regeneration of Activated Carbon Product by Reverse Burn Gasification. US Patent Number 5,124,292, ChemChar Research, Inc., Columbia, MO, June 23

McAuley B, Kunkel J, Manahan SE (2001) A New Process for the Drying and Gasification of Sewage Sludge. Water Engineering and Management May:18–23.

Medcalf BD, Larsen DW, Manahan SE (1997) Fate of Nitrogen in Nitrogen-Containing Compounds during Cocurrent Flow Gasification (ChemChar Process). Environmental Science and Technology 31:194–197

Medcalf BD, Larsen DW, Manahan SE (1998) Gasification as an Alternative Method for the Destruction of Sulfur Containing Waste (ChemChar Process. Waste Management 18:197–201

National Research Council (1999), Biobased Industrial Products: Priorities for Research and Commercialization. Board on Biology, Commission on Life Sciences, National Academic Press, Washington, DC

Pedroza-Islas R, Vernon-Carter EJ, Durán-Domínguez C, Trejo-Martínez S (1999) Using biopolymer blends for shrimp feedstuff microencapsulation. I Microcapsule particle size, morphology and microstructure. Food Res Intl 32:367–374

Prasertsan S, Krukanont P (2003) Implications of Fuel Moisture Content and Distribution on the Fuel Purchasing Strategy of Biomass Cogeneration Power Plants. Biomass and Bioenergy 24:13–25

Raiko MO, Gronfors THA, Haukka P (2003) Development and Optimization of Power Plant Concepts for Local Wet Fuels. Biomass and Bioenergy 24:27–37

Robertson A, Fan Z, Froehlich R, Lu C (2002) Partial Gasification Tests with Subbituminous Coal. In: Proceedings Annual International Pittsburgh Coal Conference, 19th. University of Pittsburgh, Pittsburgh, PA, pp 1215–1226

SENER (2005) Secretaría de Energía. Balance nacional de energía. Poder Ejecutivo Federal, Mexico, http://www.energia.gob.mx

Turnbull JH, Hulkkonen S, Simons HA (1996) Options for Biomass Drying: What's Feasible? What's Practical? Bioenergy'96 – 7th National Bioenergy Conference. September 15–20. pp 557–564 van der Grift A, van Doorn J, Vermeulen JW (2001) Ten Residual Biomass Fuels for Circulating Fluidized Bed Gasification. Biomass and Energy 20:45–56

Wyman ChE (2003) Potential Synergies and Challenges in Refining Cellulosic Biomass to Fuels, Chemicals, and Power. Biotechnology Progress 19:254–262

Hydrogen: The Ecological Fuel for Mexican Future

Suilma Marisela Fernández-Valverde

Instituto Nacional de Investigaciones Nucleares, Apartado Postal 18-1027, México DF 11801, México. E-mail: smfv@nuclear.inin.mx

Abstract

Mexico, as many other countries, needs the substitution of fossil fuels to avoid the environmental and health problems produced by their burning. Hydrogen is an important alternative energy source for the growing energy demand and the answer to the present need of a clean, efficient, and environmentally-friendly fuel. While the proved hydrocarbon reserves in Mexico will last for just some more decades, hydrogen, in counterpart, is the most abundant element in the universe and the third one in the earth's surface. Burning hydrogen in a combustion engine produces water and a small amount of nitrogen oxides, which can be eliminated in proton exchange membrane fuel cells or in alkaline fuel cells. Fuel cells working with hydrogen are electrochemical devices where the recombination of hydrogen with oxygen produces electricity, heat and water in a silent manner. In the seventies, most countries started in hydrogen research and development due to the petroleum crisis; being Germany, Canada and Japan the most advanced countries in the area. USA announced in 2003 a $1.2 billion Hydrogen Fuel Initiative to develop technology for commercially viable hydrogen-powered fuel cells. Iceland, one of the countries with large contaminant emissions, is working to become the first hydrogen based economy. In the other hand, two hydrogen isotopes, deuterium and tritium, could produce energy for the future through thermonuclear reactions. Mexico needs to include in its energetic program research and development in hydrogen for its use as an energy carrier of the renewable and still non renewable energy sources. Mexico should also explore the possibility of becoming a member of the International Thermonuclear Experimental Reactor ITER project, like other countries (e.g. India), to develop a clean energy future.

1 Introduction

Mexico City is one of the most contaminated cities in the world. The Mexican Ministry for the Environment reports that the emissions to the atmosphere due to the transport are superior to 16 millions of tons per year. Among these emissions there are particles suspended in the air and gases that lead to respiratory and cardiovascular diseases, decay of the building materials and other surfaces. These particles interfere in the photosynthesis of plants which produce the oxygen needed for life. During the winter season, Mexico City is covered by a pollution layer that causes the green-house effect and health and visibility problems. The particles emitted by combustion engines lead to respiratory system diseases and even causing some types of cancer that include that of the lung, uterus and prostate (Zweig et al. 1998).

The gases generated in fossil fuel combustion are: CO_2 (carbon dioxide), CO (carbon monoxide), SO_2 (sulphur dioxide), and NO_X (nitrogen oxides) among others. O_3 (ozone) and NO_2 compounds come from non-burned fuel and polyaromatic hydrocarbons formed by the solar ultraviolet irradiation (SMA 2006). While CO_2 is considered the principal causative agent of the atmosphere warming in our planet, CH_4 (methane) is also very harmful with an atmosphere warming power five times superior to that of the CO_2. SO_2 along with ultraviolet photochemical reactions forms sulphur trioxide and sulphates. These compounds then react with atmosphere humidity forming sulphuric acid, a common agent of and acid rain that destroys forests, monuments and constructions. Other aggressive contaminants are caused by the NO_X, which by means of photochemical reactions produce nitrate peroxides, nitric acid and ozone. Recent investigation shows the cooling and contracting of the upper atmosphere as a result of rising greenhouse gas concentration.

The fuel outlined as most suitable today to avoid pollution is hydrogen (Das and Veziroglu 2001; Fernandez-Valverde 2002; DOE 2006a). In Fig. 1 appears the hydrogen cycle. It is observed that hydrogen obtained for any renewable or non renewable source can be used as fuel in almost all human activities: fuel in transportation either by air, sea or land; as fuel in the domestic area and as electricity in all levels: commercial, domestic, and manufacturing. If hydrogen is burnt in a combustion engine some amounts of nitrogen oxides are formed; if the hydrogen is recombined with oxygen in a hydrogen fuel cell only water and heat are obtained.

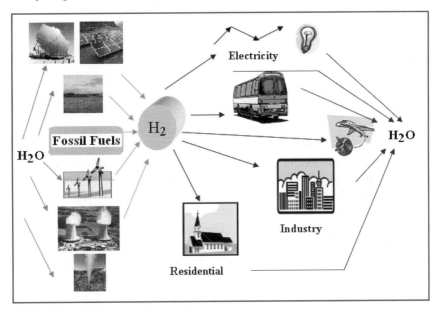

Fig. 1. Energy sources for hydrogen production and areas to be used as fuel

The price of this fuel is high; nevertheless there are research and development projects in many countries to bring down the costs to make it competitive with the other fuels. The International Partnership for the Hydrogen Economy has censed most of 300 projects worldwide (IPHE 2006). Most of the projects are in Asia, EU and USA; in Latin America or Africa no projects are censed so far. If the financial costs in health and the environmental damage are taken into account, it is clear why hydrogen is a candidate to solve the environmental problems. Table 1 shows the energy content of different fuels and hydrogen has the highest energy amount (Ni et al. 2006). If hydrogen is obtained from renewable energy sources like solar, wind, geothermal, etc. it can solve one of the other big problems of our times, the storage of the intermittent renewable energies to be used in any moment. Hydrogen can also be obtained from non renewable energies like nuclear or fossil fuels. Hydrogen obtained from fossil fuels has lower dioxide of carbon emissions for the same quantity of energy than fossil fuels.

Table 1. Energy contents of different fuels

Fuel	Energy content [MJ/kg]
Hydrogen	120
Liquefied natural gas	54.4
Propane	49.6
Aviation gasoline	46.8
Automotive gasoline	46.4
Automotive diesel	45.6
Ethanol	29.6
Methanol	19.7
Coke	27
Wood (dry)	16.2
Bagasse	9.6

The last information of the Mexican Energy Secretariat (SE 2006) reported the proved hydrocarbons reserves are 46,417.500 millions of barrels and the installed capacity of electric power generation in 2004 of 46,552 MW. Of this total wind power provides only 0.0004 %, geothermal 2 % and nuclear 2.9 % (SENER 2006). The solar power generation is not reported because it is limited to little and remote towns, in some park installations or for highways telephones. Dr. Huacuz from the IIE (Instituto de Investigaciones Electricas) reported 2.5 MW installed of wind energy in 2006. Solar energy capacity is less than 2.5 MW and municipal waste producing biogas is in the order of 7.4 MW.

2 Hydrogen production

Hydrogen can be obtained from renewable or not renewable energy sources; Fig. 2 shows the hydrogen classification production methods based in the sources from whom it is obtained. In the next section, different hydrogen production technologies are presented.

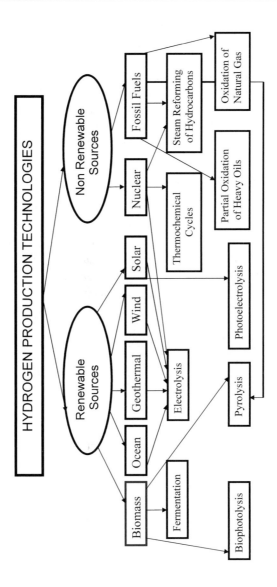

Fig. 2. Technologies of hydrogen production and energy sources

The methods for hydrogen production from fossil fuels are: steam reforming of hydrocarbons, catalytic oxidation of natural gas and partial oxidation of heavy oils; all of them are commercial techniques. Steam reforming of methane is presented as an example of a fossil fuel method. A zinc oxide catalyst is used for desulphurization. The reforming reaction takes place at temperatures between 1073 and 1273 K with ceramic catalysts.

$$CH_4 + H_2 \Leftrightarrow CO + 3H_2$$

The CO and the H_2 are separated and the shift reaction was carbon monoxide, which reacts with water over a catalyst of nickel to produce more hydrogen.

$$CO + H_2O \Leftrightarrow CO_2 + H_2$$

The warm gases are cooled and separated; the efficiency of the process is approximately 75 %. The moles of produced hydrogen per number of carbon in the hydrocarbon used is 4 for methane, liquefied petroleum gas (C_4H_{10}), kerosene ($C_{12}H_{26}$) give 3.25 and 3.08 moles of hydrogen and 2 to 2.5 moles of coal, respectively (INSC 2004).

The biological methods of hydrogen production could be performed in the presence or absence of light and in the absence of air. They could be photobiological or fermentative. In biological water splitting, certain photosynthetic microbes produce hydrogen from water in their metabolic activities using light energy. Photobiological technology holds great promise, but oxygen is produced along with the hydrogen. The technology must overcome the limitation of oxygen sensitivity of the hydrogen-evolving enzyme systems. Researchers are searching naturally occurring organisms that are more tolerant of oxygen, and creating new genetic forms of the organisms that can sustain hydrogen production in the presence of oxygen. A new system that uses a metabolic switch as sulphur deprivation to cycle algal cells between a photosynthetic growth phase and a hydrogen production phase is also being developed. Fermentations can be done in light or in the dark with anaerobic or facultative bacteria (Reith et al. 2003; Das and Veziroglu 2001). The hydrogen production yields are higher with fermentations than in photobiological process. In 2002 Woodward et al. reported that the enzymes of the oxidative pentose phosphate cycle can be coupled to hydrogenase purified from the bacterium *Pyrococcus furiosus* to generate 11.6 mol H_2 per mol of glucose-6-phosphate (Woodward et al. 2002). The biological processes can be summarized in the utilization of algae bacteria or enzymes that in certain conditions, as nitrogenasas in the absence of nitrogen, produces hydrogen. These processes are very interesting because they allow the degradation of organic compounds found in municipal waste. All the developed countries such as Japan, the EU, USA, and others carry out projects of investigation in this area; in Mexico an incipient research was undertaken.

The electrolysis of water (Kreuter and Hoffman 1998) is another technique for hydrogen production from renewable or non renewable energy sources as can be seen in Fig. 2.

Alkaline water electrolysis is based on the reactions:

$$2H_2O + 2e \Leftrightarrow H_2 + 2OH^-$$

$$2OH^- \Leftrightarrow O_2 + 2H_2O + 4e$$

The alkaline electrolysers use nickel as cathode due to his corrosion resistance. Recent investigations are related to the development of materials that allow covering big surfaces with metal transition compounds. The kinetic of the oxygen evolution reaction at the anode is the determining step for the water electrolysis; a 4 electron transfer is needed for the oxygen evolution reaction. The commercial anodes are of nickel, nevertheless research is performed to replace it by oxides and part of our research is orientated to this area.

Wind power can produce electricity to be used in an electrolyser. This system has been reported to generate huge quantities of hydrogen in the Patagonia with exporting purposes. In December 2005, the first wind power-electrolyser plant for the demonstration of hydrogen production in Palo Truncado in Argentina started its operation. It is the only electrolyser in Latin America working with wind power. Mexico has excellent wind conditions in almost all the country that could be used for hydrogen production.

In Fig. 3 a clean energy hydrogen system working with an electrolyser fed with solar energy transformed in solar cells is presented (it can function with any energy source). The hydrogen obtained can be stored and used in fuel cells. Burning hydrogen in an internal combustion engine produces a small quantity of nitrous oxides, which could be eliminated with catalytic converters. Mexico has also a huge solar energy potential. In the figure, a panel of solar cells produces the energy used for the electrolyser in which water is separated in its components.

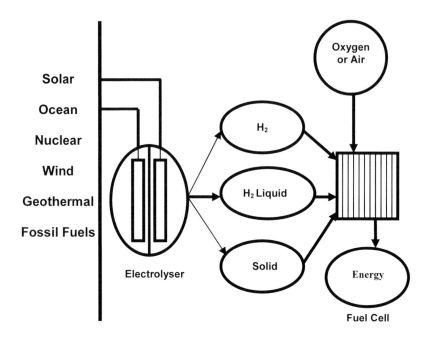

Fig. 3. Clean energy system. Photovoltaic hydrogen production; hydrogen storage – fuel cell

In the electrolysis of water a very interesting process is under research which consists in the use of semiconductors capable to decompose water directly using only solar energy. The process is called photoelectrolysis. Commercial electrolysers can be alkaline or acid. In the first case, hydroxide solutions are used during the electrolysis and the hydrogen produced must to be cleaned. In the second case, a proton exchange membrane is the electrolyte and the hydrogen can be used in a fuel cell without purification. The alkaline electrolyser is a mature technology and it can produce high-pressure hydrogen, up to 10,000 psi, needed for efficient hydrogen storage and distribution without the need to have a separate compressor (2006 H2Fc). The hydrogen storage and distribution forms are described in the next section.

The hydrogen thermochemical production has been investigated for more than four decades and more than 180 cycles have been studied (Funk 2001). There are pilot plants installed for hydrogen production using this technology. The hydrogen is obtained when the manganese (II) oxide reacts with the sodium hydroxide at 1173 K producing $NaMnO_2$, which is hydrolysed. Then, the products formed are: oxide of manganese (III)

Mn_2O_3, sodium hydroxide in solution and water. To close the cycle, the Mn_2O_3 is reduced with solar radiation to regenerate the MnO. It enters again to the cycle and oxygen is obtained a by-product. In a schematic way this is the principle of thermochemical cycles. In Japan, with the development of the new reactor of high temperature, research is done for the installation of a thermochemical cycle that could also use solar power directly. A project exists between Japan, the UE and the USA for the development of the thermochemical iodine-sulphur cycle (IS). This cycle is represented in Fig. 4.

The nuclear energy research for hydrogen production in a High Temperature Test Reactor (HTTR) is going at the Japan Atomic Energy Research Institute. The HTTR has power capacity of 30 MW, of which 10 MW will be coupled to a steam reforming natural gas plant for the production of 4200 m^3/h of hydrogen. Hydrogen production with this processes is expected to be ready for 2008 and in the future the thermochemical IS (Fig 4) process will be used for the hydrogen generation (JAERI 2004). A reactor of the same type with a generation power of 600 MW working at 50 % efficiency will produce almost 200 metric tons per day or 73,000 tons per year. That is equivalent to 200,000 gallons of gasoline per day or 3 millions of oil barrels per year. Preliminary calculations give an approximate cost of $1.15 US dollars per kg of hydrogen with methane steam reforming process and $1.42 US dollars per kg with the IS thermochemical cycle (Schultz et al. 2003; Francoise 2006).

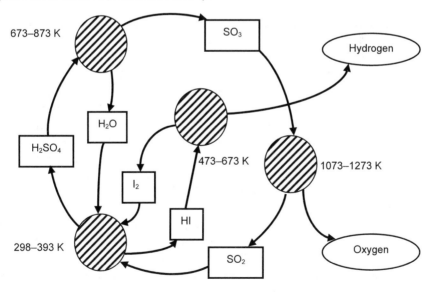

Fig. 4. Thermochemical iodine-sulphur cycle for hydrogen production

3 Hydrogen storage and distribution

Hydrogen can be stored in different forms: gas, liquid or solid. The storage technique depends of its later hydrogen utilization. The storage as gas-phase is a well known commercial technique and liquid hydrogen storage is now being developed. Both of them are in demonstration for fuelling stations in Japan, Germany and the USA. Some hydrides are commercial for hydrogen storage and research is being done for the storage of hydrogen in other materials (MH 2006; Hanneken 1999). In a transportation system, an ideal chemical hydrogen storage material would be inexpensive, would have rapid kinetics for absorbing and desorbing H2 in the 198–393 K temperature range and the reaction must be reversible. It would also have to store large quantities of hydrogen and the ideal material would have a low molar weight in order to decrease the reservoir/storage mass. At present, no single material fulfills all of these requirements although some researches take the physicochemical properties of hydrides and changes the composition by the addition of transition metals to improve the hydrogen storage. Carbon materials are being studied for physical sorption of hydrogen. The methods of hydrogen storage can be classified as: underground, liquid, compressed gas, metal hydrides and physisorption. The former two are for large quantities and long-term storage times; the third is for small quantities and short-term storage times and the last two are for small quantities and long-term storage times. The hydrogen transportation can be done by pipeline, liquid, compressed gas or metal hydrides (Dunn 2001).

Fuel cells

There are several types of fuel cells that are being developed for a wide range of applications: from small as a cellular phone of only 0.5W energy demand, to as large as a small power plant for an industrial facility or a small town with a capacity of 10 Megawatts. Fuel cells are classified by their electrolyte material.

Proton Exchange Membrane Fuel Cells (PEMFC) is believed to be the best type of fuel cell for vehicular-power use to eventually replace the gasoline and diesel internal combustion engines. PEMFC use a solid polymer membrane as the electrolyte. This polymer is permeable to protons when it is saturated with water, but it does not conduct electrons. The fuel for the PEMFC is hydrogen and the charge carrier is the proton. At the anode, the hydrogen molecule splits into protons and electrons. The protons permeate across the electrolyte to the cathode while the electrons flow through an external circuit and produce electric power. Oxygen, usually in the form of air, is supplied to the cathode and combines with the proton to produce water. A lot of research is done to understand the materials of the PEMFC to improve its kinetics (Kreuer et al. 2004).

Direct Methanol Fuel Cells (DMFC) technology is still in the early stages of development. It has been successfully demonstrated powering mobile phones and laptop computers. DMFC is similar to the PEMFC in the way that the electrolyte is a polymer and the charge carrier is the hydrogen ion (proton). However, the liquid methanol (CH_3OH) is oxidized in the presence of water at the anode generating CO_2, hydrogen ions and electrons. The electrons travel through the external circuit as the electric output of the fuel cell; the hydrogen ions travel through the electrolyte and react with oxygen from the air to form water at the anode.

The Alkaline fuel cells (AFC) are one of the most developed technologies and have been used since the mid-1960's by NASA in the Apollo and Space Shuttle programs. The fuel cells on board provide electrical power for on-board systems, as well as drinking water. AFCs operate at relatively low temperatures and are among the most efficient in generating electricity at nearly 70 %. The concentration of KOH can be varied with the fuel cell operating temperature, which ranges from 273–493K. The AFCs are very sensitive to CO_2 that could be present in the fuel or air. The CO_2 reacts with the electrolyte, poisoning it rapidly, and severely degrading the fuel cell performance. Electrodes are relatively inexpensive compared to the catalysts required for other types of fuel cells. Therefore, AFCs are limited to closed environments.

Phosphoric Acid Fuel Cells (PAFC) were the first fuel cells to be commercialized. Developed in the mid-1960's and field-tested since the 1970's, they have improved significantly in stability, performance, and cost. Such characteristics have made the PAFC a good candidate for early stationary applications. They use an electrolyte that is phosphoric acid at almost 100 % concentration. The ionic conductivity of phosphoric acid is low at low temperatures, and then the PAFCs are operated at the upper end of the range 423–493K.

Molten Carbonate Fuel Cells (MCFC) are in the class of high-temperature fuel cells. The higher operating temperature allows them to use natural gas directly without the need for a fuel processor and have also been used with low-Btu fuel gas from industrial processes and other sources and fuels. MCFCs work quite differently from other fuel cells. These cells use an electrolyte composed of a molten mixture of carbonate salts. Two mixtures are currently used: lithium carbonate and potassium carbonate, or lithium carbonate and sodium carbonate. To melt the carbonate salts and achieve high ion mobility through the electrolyte, MCFCs operate at high temperatures: 923K.

The Solid Oxide Fuel Cell (SOFC) is currently the highest-temperature fuel cell in development and can be operated over a wide temperature range from 873–1273K allowing a number of fuels to be used. To operate at such high temperatures, the electrolyte is a thin, solid ceramic material (solid oxide) that is conductive to oxygen ions (O^{2-}). The SOFC has been in development since the late 1950's and has two configurations planar and tubular. The extremely high temperatures for SOFCs operation results in a significant time required to reach operating temperature and slowly response to changes in electricity demand. It is therefore considered to be a leading candidate for industrial and large scale central electricity generating stations.

Zinc-Air Fuel Cells (ZAFC) share characteristics with a number of the other types of fuel cells as well as some characteristics of batteries. The electrolyte for a ZAFC is a ceramic. It operates at 973K. The anode is composed of zinc and is supplied with hydrogen or even hydrocarbons. The cathode is separated from the air supply with a gas diffusion electrode (GDE), a permeable membrane that allows atmospheric oxygen to pass through. The high operating temperature of the ZAFC enables internal reforming of hydrocarbons, eliminating the need for an external reformer to generate hydrogen.

The Regenerative Fuel Cell is a system that can operate in a closed loop and could serve as the basis of a hydrogen economy operating on renewable energy. Hydrogen would be generated from the electrolysis of water. The electrodes would be reversible and the cell can act as electrolyser for hydrogen production when the renewable energy sources such as wind, solar, or geothermal are in place or as a fuel cell to generate electricity from the hydrogen stored.

Standards for hydrogen and fuel cells

The hydrogen economy needs standards for hydrogen, fuel cells and energy converters running with hydrogen. Most countries along with different organizations are working in the harmonization of International Standards, like the International Organization for Standardization (ISO) with the ISO Technical Committee ISO/TC 197 for hydrogen technologies and the International Electrotechnical Commission (IEC). ISO is the world's largest developer of technical standards, with a network of the national standards institutes of 146 countries. The ISO Technical Committee

ISO/TC 197 for hydrogen technologies was created in 1990. It has 15 participating countries, 15 observers and collaborates with 15 other ISO/IEC committees. Four standards have been published to date in Liquid hydrogen – Land vehicle fuelling system interface; Hydrogen fuel – Product specification and Basic considerations for the safety of hydrogen. The United Nations' World Forum for Harmonization of Vehicle Regulations works closely with the Committee and also with the IEC, the leading global organization that prepares and publishes international standards for all electrical, electronic and related technologies.

Fusion

This article will be incomplete without showing the utilization of hydrogen in a project that is in stage of research and development and could be the best way to obtain energy in great quantities (Aymar 2001). The transformation of matter into energy is the most efficient way to generate energy. The difference between the energy that produces the hydrogen burned or recombined with oxygen is very small corresponding to electronic binding energy. Only some eV, compared with that produced by the nuclear reaction of two hydrogen isotopes: deuterium and tritium. The reaction is made at very high temperatures; the gas highly ionized known as plasma is placed in a magnetic confinement and the fusion of deuterium en tritium generates 12.85 MeV.

Caderache in Provance Alpes-Côte d'Azur, in France, has been chosen as site for the construction of the first experimental fusion reactor, to show the feasibility of the utilization of this type of energy. The project is known as ITER (International Thermonuclear Experimental Reactor), there are seven national and supranational parties participating in the ITER program: China, the European Union, India, Japan, Russia, South Korea, and the USA (ITER 2006). The project is expected to cost about €10 billion (US$12.1 billion) over its thirty year life. Construction of the ITER complex is planned to begin in 2008, while assembly of the tokamak itself is scheduled to begin in the year 2011. Unforeseen political, financial, or even social issues could alter these estimated dates substantially (Nuclear 2006). However the Kyoto protocol has been ratified at the annual United Nations conference held in Nairobi. (Stone and Bohannon 2006) and USA, Australia, Canada, European Union and Japan has already developed Roadmaps as guides to their transition to a hydrogen economy (DOE 2006b).

Conclusions

Most of the developed countries have the vision of hydrogen and fuel cells as the energy future, between them Canada and the USA. Mexico, as a commercial partner of these two countries, needs to be in the same way. Hydrogen, at the moment, is not a primary energy source like coal and gas. It is an energy carrier. Initially, it will be produced using different conventional primary energy sources, like carbon and fossil fuels. These systems with capture and safe storage of CO emissions are almost completely carbon-free energy pathways. In the longer term, renewable energy sources will become the most important source for the hydrogen production. Hydrogen produced from nuclear sources do not generate greenhouse gases and with fossil-based energy conversion the can produce hydrogen in the large quantities necessary for the transport and stationary energy power. Markets could become a barrier to progress beyond the initial demonstration phase, starting now in Mexico. If cost and security of supply are dominant considerations, then coal gasification and CO_2 sequestration may be of interest for Mexico, only if politics move towards renewable energies as biomass, solar, wind and ocean energy, more or less viable according to regional geographic and climatic conditions. For example, concentrated solar thermal energy is a potentially affordable and secure option for large-scale hydrogen production, especially in the north of Mexico, so it is wind energy in the Istmo of Tehuantepec. The Ocean and geothermal energy resources can become significant; Mexico has 10,600 Km of costs and a volcanic axe is situated along the pacific cost.

A Mexican hydrogen program should include the wide range of options for hydrogen production and also the different kind of fuel cells that could be used in a wide range of fields: from mobile applications like cars, vehicles, buses and ships, to heat and power generators in stationary applications in the domestic and industrial sector, all the way through very small fuel cells in portable devices such, as laptops and mobile phones. The energy systems will also include conventional energy converters running on hydrogen as internal combustion engines, Stirling engines, and turbines. The hydrogen program can improve the electric generation from alternative energy sources and avoid the environmental and healthy problems and other damages produced from the burning of fossil fuels. It would include all the hydrogen areas to introduce the hydrogen economy in Mexican life: Hydrogen production, storage, delivery, fuel cells, technology validation for portable mobile and stationary fuel cells and energy converters running on hydrogen, safety codes and standards and also education to promote the hydrogen use. Mexico should also explore the possibility of becoming a member of the International Thermonuclear Experimental Reactor ITER project like other countries (e.g. India) to develop a clean energy future.

References

Aymar R (2001) Overview of ITER-FEAT – The future international burning plasma experiment. Nuclear Fusion 41:301–310

DOE (2006a) http://www.hydrogen.energy.gov/annual_review06

DOE (2006b) http://www.hydrogen.energy.gov/roadmaps.html

Das D, Veziroglu TN (2001) Hydrogen Production by Biological Process: a Survey of Literature. Int J of Hydrogen Energy 26:13–28

Dunn S (2001) Hydrogen Futures: Toward a Sustainable Energy System In: Peterson JA (ed) World Watch Paper 157. Washington DC

Funk JE (2001) Thermochemical Hydrogen Production: past and present. Int J of Hydrogen Energy 26:185–190

Fernandez-Valverde SM (2002) Hydrogen as an energy source to avoid environmental pollution. Geofisica International 4:223–228

Francoise J (2006) Private communication

Hanneken JW (1999) Hydrogen in metals and other materials: a comprehensive reference to books, bibliographies, workshops and conferences. Int J of Hydrogen Energy 24:1005–1026

INSC (2004) Nuclear Production of hydrogen. Technologies and perspectives for Global Development. International Nuclear Society Council. Illinois, USA

IPHE (2006) http://www.iphe.net/newatlas/atlas.htm

ITER (2006) http ://www.spacewar.com/news/nuclear-civil-05zzzx.html. ITER Members.

JAERI (2004) "JAERI, High-temp Engineering Test Reactor (HTTR) Used for R&D on Diversified Application of Nuclear Energy". http://www.jaeri.go.jp/english/ff/ff45/tech01.html

Kreuer KD, Padisson SJ, Spohr E, Schuster M (2004) Transport in proton conductors for fuel-cell applications: simulation, elementary reactions, and phenomenology. Chem Rev: 104:4637–4678

Kreuter W, Hoffman H (1998) Electrolysis: the important energy transformer in a world of sustainable energy. Int J of Hydrogen Energy 23:661–666

MH (2006) International symposium in metal hydrogen systems. 2–6 October, Hawai, USA

Ni M, Leung MKH, Sumathy K, Leung DYC (2006) Int J of Hydrogen Energy 31: 1401–1412

Nuclear (2006) Nuclear technology review 2006. Ed. International Atomic Energy Agency. Vienna

Reith JH, Wijffels RH, Barten H (2003) Bio-methane & Bio-hydrogen. Status and perspectives of biological methane and hydrogen production. Dutch biological hydrogen foundation. The Netherlands

SE (2006) http://sie.energia.gob.mx/sie/bdiController?action=login

SENER (2006) Secretaría de Energía. Prospectivas del sector eléctrico 2004–2014. México

SMA   (2006)   http://www.sma.df.gob.mx/sma/modules.php?name=Biblioteca&d_op=viewdownload&cid=2

Stone R, Bohannon J (2006) Puts Spotlight on Reducing Impact of Climate Change. Science 314:1224–1225

Woodward J, Orr M, Cordray K, Greenbaum E (2000) Biotechnological hydrogen production Nature 405:1014–1015

Zweig RM, Chair MD, Provenzano J (1998) Pollution solution evolution new opportunity for hydrogen. In: Bolcich JC, Veziroglu TN (eds) Proceeding of the 12th World Hydrogen Energy Conference. Buenos Aires, Vol. I pp 169–175

Nuclear Fusion as an Energy Option for the 21ˢᵗ Century

Julio E Herrera-Velázquez

Instituto de Ciencias Nucleares, Universidad Nacional Autónoma de México, A.P. 70-543, Ciudad Universitaria, Del. Coyoacán, 04511 México, D.F. México. E-mail: herrera@nucleares.unam.mx

1 Introduction

For more than 50 years, controlled nuclear fusion has been promised as a safe, clean and environmentally acceptable energy alternative for the future. Fusion was actually known from particle accelerator experiments well before nuclear fission, and by the time the latter was being discovered, there were already well developed theories of how fusion is the source of energy in the stars, including our Sun, and how stars work as the element factories in the Universe (Bethe 1939). Yet, after several decades of work by researchers in several countries, producing more energy than is invested in fusion devices has been elusive. This has led to the common joke that the date in which fusion reactors will become available is a new constant in physics; always 30 years away. Actually, the international controlled fusion research programme is sound and healthy, and has achieved significant progress (International Fusion Research Council 2005), but the road to the reactor is more difficult than originally envisioned. In the process, it has influenced the development of plasma science as an interdisciplinary endeavour which requires the collaboration of physicists and engineers, and has led to important spin-offs in other applications. The limitations of fusion reactors will depend as much on physics issues as on engineering and materials design, and its competitiveness will depend on the results of further research and development in these areas.

November 21st, 2006 was an important date, as ministers from the seven parties of the International Thermonuclear Experimental Reactor (ITER, www.iter.org), namely China, the European Union, India, Japan, the Russian Federation, South Korea, and the United States of America, signed the agreement to establish the ITER Organization, and to proceed to build this long awaited machine in Cadarache, France. This certainly meant good news, since ITER will be the first burning plasma experiment to be built, and will provide necessary information for the future of the programme. It will also be an important test bed for crucial engineering components. Unfortunately, the road to design this machine and the decision to build it took too long. It was first conceived in 1985, and will not produce its first plasma until 2016. Its building cost is estimated at €5 billion, and roughly the same amount will be necessary to operate it in the next 20 years. Other burning plasma machines have been proposed, such as IGNITOR, in the USA and Italy, as well as the Compact Ignition Torus (CIT) and FIRE in the USA. The first two could already be in operation if the decision to build them had been reached when they were first proposed, in the late 1980s. Only IGNITOR is still in progress in Italy, at a very slow pace set by the funding available. This is the result of the premise that it is better to invest in a single more ambitious and flexible design, but it has led to the loss of a link between two generations of tokamak experiments, which would have provided important information on the behaviour of burning plasmas.

Opposite to other alternative energy sources, fusion is a clear example of extreme engineering, which stands on advanced technology and requires highly trained scientists. The purpose of this essay is to highlight the major hurdles which have been negotiated and which will still need to be negotiated in the future, in order to reach the final goal. A word on how Mexico could participate in the fusion energy effort will be advanced, with some recommendations on the steps to follow.

2 The challenge of controlled nuclear fusion

Nuclear fusion reactions were first studied in the particle accelerators of the 1930s. Since nuclear forces act at a short range, which means distances close to the size of the nuclei, it is necessary to overcome the Coulomb force between them. Several ways of inducing fusion reactions in the laboratory can be devised, and contrary to common impression, that is not an issue. The real challenge is to device a way in which the energy output is much greater than the energy input. It is generally recognized that in order to do so, a massive burning of the fuel is needed. For this purpose, it must be confined and heated at high temperatures, such that the kinetic energy between the nuclei is large enough to overcome the Coulomb forces. At such temperatures, the fuel is in a state of plasma, which is an electrically quasi-neutral gas consisting in positive ions and electrons. When the temperature reaches more than 10 keV (1 keV = 1.6×10^7 Kelvin), nuclear reactions ensue, and the plasma is said to reach thermonuclear conditions. In the case of the stars, this is achieved by means of the gravitational collapse. On Earth, the necessary conditions can be obtained in thermonuclear weapons compressing the fuel by means of a fission device. However, it has proven to be more challenging to design a way of producing the reactions in a controlled way in the laboratory.

In 1955 J.D. Lawson established the necessary criteria for plasmas to compensate the energy losses by energy gain from the fusion reactions (Lawson 1957). The triple product of plasma density, energy confinement time and temperature should reach a minimum value, which depends on the particular reaction being considered.

Taking into account the necessary values of temperature, density and energy confinement time, as well as fuel availability, it is generally agreed that the most suitable reaction for a first generation of fusion reactors would be

$$^2\text{H} + {}^3\text{H} \rightarrow {}^4\text{He}(3.52 \text{ MeV}) + \text{n}(14.06 \text{ MeV}).$$

Deuterium (^{2}H) would be obtained from sea water, while tritium (^{3}H), being a short lived radioactive isotope, would be bred in the reactor using the fusion neutrons (n) in exothermic and endothermic reactions with lithium (Li) in a mantle surrounding the plasma:

$$\text{n} + {}^6\text{Li} \rightarrow {}^4\text{He} + {}^3\text{H} + 4.3 \text{ MeV}$$

$$\text{n} + {}^7\text{Li} \rightarrow {}^4\text{He} + {}^3\text{H} + \text{n} - 2.5 \text{ MeV}.$$

While the first reaction would be more desirable, ^{7}Li is more abundant in nature. The energy balance would still be favourable.

The energetic alpha particles (^{4}He) would help to heat the plasma, and keep it burning. Once they give up their energy, they would be removed. For the deuterium-tritium reaction, the triple product is 5×10^{21} keVm^{-3}s for a full compensation of energy losses. Since no additional heating would be needed from external sources, this is known as the ignition condition. Defining Q as the ratio of fusion power and input heating power, this would mean $Q = \infty$. However, lower values of Q might still be acceptable for a working reactor.

The idea of developing a nuclear fusion reactor dates back to 1947, when G.P. Thomson and M. Blackman patented in Britain the idea of producing thermonuclear plasma in a toroidal pinch discharge (Haines 1996). The essential idea looks simple enough at first glance; to induce a current through a plasma discharge, magnetically confined, so that it can be heated by Joule effect up to thermonuclear conditions. Similar ideas arose independently in the Soviet Union, where I. Tamm and A. Sakharov proposed the tokamak machine in 1950, and in the USA, where L. Spitzer proposed the Stellerator in 1951 (Bromberg 1982). Although the details vary considerably, they are all based essentially on the same idea of producing a toroidal plasma with a symmetry (toroidal in the cases of the Z-pinch and the tokamak, and helical in the one of the stellerator) which yields a quasi-two-dimensional magnetic configuration, necessary for the existence of robust closed magnetic field surfaces. The magnetic confinement programmes remained classified until 1956, when I.V. Kurchatov gave a lecture at the Harwell laboratory in Britain, in which he voluntarily described the Soviet advancements on the subject. This led to an era of naïve optimism, when the main physics and engineering hurdles were still unknown. Detection of fusion produced neutrons lead the ZETA team in Britain to claim success in 1958 (Thoneman et al. 1958). This originated undue enthusiasm, and it was suggested in the *Daily Telegraph* that there might be nuclear fusion reactors in 20 years. Unfortunately, it was soon realized that the neutrons were not originated by thermonuclear temperatures, but by beam-target effect, due to accelerated ions induced by plasma instabilities (Rose 1958).

It may be interesting to quote, from the sceptic's point of view, the recollection John D. Lawson gave in a recent interview, about his early work in controlled fusion:

"I never was really in fusion. I spent most of my working life working on particle accelerators. My main original achievement here was to show that the parameters suggested for a strong focusing machine were not realistic, although it's still a very strong and powerful principle. Sharing an office with Peter Thonemann I saw what the fusion problem was. I produced the criterion, produced the report, and then I got involved with lots of other discussions and wrote the other report, a survey of different methods. And that was it. Then I was back to accelerators."

"I wrote one or two other papers surveying the other ideas that had been suggested and showing that most of them wouldn't work. I also knew that I wouldn't see fusion power in my own lifetime, although most people were talking about it coming in 20 years or so. They still are. My work was always negative and was tending to be showing things that wouldn't work, or surveying an area to see whether it might possibly be feasible."

It has been a major success of the past fifty years of fusion research to tame a wide variety of instabilities. Understanding such problems, and the nature of other hurdles, such as the transport of energy in the plasma, has led to the development of plasma physics as a very active branch in the process. In terms of Lawson's criterion, the advancement in the performance of fusion devices is comparable to that of the well known Moore's law for computing devices (Manheimer 2004). However, controlling the plasma has proven to be more difficult than originally envisioned. In terms of physics and engineering, the most advanced and better understood device is the Tokamak (Wesson 2004). The contribution from the Joint European Tokamak (JET) (Pamela 2005), as well as from a large fleet of devices, mainly in Europe, Japan, Russia and the USA, have been crucial in gathering a wide database, which summarises a deep understanding of the physics and operating scenarios for the experiments built so far. It is possible nowadays to produce plasma shots of several seconds in tokamaks, with a ratio of fusion power to input heating power greater than 0.6. Equilibrium and global stability of the plasma is satisfactorily controlled. Energy transport still needs to be fully understood, but recent codes give a fairly good description of it.

As a result of this advancements (ITER Physics Basis Editors 1999), a next generation tokamak has been designed; the International Thermonuclear Experimental Reactor (ITER). The main issue to be understood in this new development is that, since in the existing experiments have only investigated $Q < 1$ regimes, no plasmas for which there is a significant effect of the α particles from the fusion reactions have been studied so far.

3 Burning plasma experiments: the uncharted territory

From the physics point of view, the next generation of fusion experiments are to address the question of how the burning plasma behaves. There will be a significant population of energetic 3.5 MeV alpha particles from the deuterium-tritium reaction, which will be confined and will heat the plasma.

Since in a deuterium-tritium reaction the fusion energy is approximately five times the energy in the alpha particles, we can write $P_{fusion} = 5P_\alpha$. Thus, $Q = P_{fusion}/P_{heat} = P_\alpha/P_{heat}$. Defining the alpha heating power as $f_\alpha = P_\alpha/(P_\alpha + P_{heat})$, we get $f_\alpha = Q/(Q+5)$.

Thus we have:

$$
\begin{array}{ll}
Q = 1 & f_\alpha = 17\,\% \\
Q = 5 & f_\alpha = 50\,\% \\
Q = 10 & f_\alpha = 60\,\% \\
Q = 20 & f_\alpha = 80\,\% \\
Q = \infty & f_\alpha = 100\,\%
\end{array}
$$

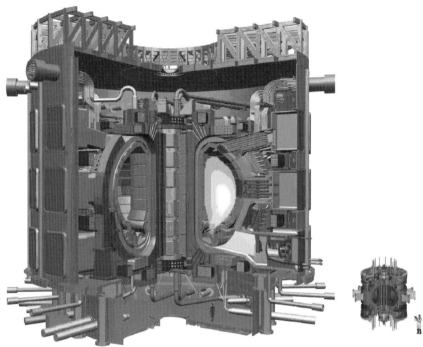

Fig. 1. Comparison of the sizes of the International Tokamak Experimental Reactor (ITER) (left) and IGNITOR (right)

The nature of the problem can be appreciated if we realise that, in present day experiments, the break-even regime ($Q = 1$) has not been reached yet (reaching an equivalent deuterium-tritium regime in pure deuterium plasmas is not valid, since there is no alpha heating in such cases). ITER, on the other hand, is expected to reach $Q = 5$ and possibly $Q = 10$. The only device designed to reach ignition ($Q = \infty$) is IGNITOR (Coppi et al. 2001; Horton et al. 2002; Bombarda 2004), but although considerable work has been done in building components and systems for it, it remains uncertain whether it will eventually be commissioned, due to lack of sufficient financial support. It must be stressed, however, that the goals and missions of both machines are quite different, and should not exclude each other. Quite on the contrary, since they would both produce important information for the next step, precisely because of the difference in their design and goals, they are complementary. While ITER is expected to be flexible enough to test engineering components needed for a reactor, in long plasma pulses, IGNITOR is expected to provide information on the behaviour of burning plasmas and the path to ignition, in shorter pulses.

Table 1. Main parameters of JET, the only deuterium-tritium experiment in operation, and the largest one ever built, IGNITOR and ITER

Parameter	JET	IGNITOR	ITER
R (Major radius)	3 m	1.32 m	6.2 m
a (minor semi-axis)	1.25 m	0.47 m	2 m
δ (elongation)	1.8	0.4	1.7
Volume	100 m^3	10 m^3	840 m^3
Pulse length	60s (plateau current)	4+4 s	400 s (plateau current)
Plasma Current I$_P$	7 MA	11 MA	15 MA
Toroidal Field B$_T$	4 T	13 T	5.3 T
Central density	2×10^{20} m^{-3}	10^{21} m^{-3} (at ignition)	10^{20} m^{-3}
Electron Temperature	20 keV	11.5 keV (at ignition)	21 keV
Ion Temperature	40keV	10.5 keV (at ignition)	18 keV
Q	0.6	∞	10

At $Q > 5$, alpha particle effects on stability and turbulence are expected. At $Q > 10$, strong non-linear coupling between alphas, pressure driven current, turbulent transport, and magnetohydrodynamic stability would be affected. That is how far ITER would go. At $Q > 20$ it would be necessary to control the fuel input, as propagation of fusion burn and fusion ignition transient phenomena would occur. Only IGNITOR would be able to test this regime. In a few words, a whole new uncharted territory opens up for fusion research, as the experiments move to the burning plasma regime, and as often happens in uncharted territory, it is difficult to predict what surprises; new hurdles and solutions will arise. It is to be noted that theory has seldom been able to predict the major advancements in fusion research, and these can only be achieved by conducting the necessary experiments.

4 The engineering challenge

There have been a number of engineering objections to fusion reactors which can be summarised as follows:

4.1 Complexity as Compared to other Alternative Energy Sources

A fusion energy reactor is in fact an authentic example of extreme engineering. While the core confines a plasma with record temperatures on Earth, higher than those in the centre of the Sun (albeit at a much smaller densities), the confinement is achieved with magnetic fields produced by superconducting coils at about 4K. The energy of the neutrons will be absorbed by a lithium mantle surrounding the device, which would heat and breed tritium. This mantle, from which the energy is absorbed by a heat exchanger, should also protect the more costly equipment, including the coils, from neutron radiation. The system should be engineered in such a way that tritium leaks and lithium fires should never happen. On the other hand, since the first wall in the vacuum chamber will be activated, any repair will be remotely handled. High precision robotics, already being tested at JET, is thus necessary.

All this means that a very precise, well planned and disciplined engineering will be needed to run a fusion reactor, in a reliable way. However, there are good examples in which this has been presently achieved, not only in the nuclear energy industry, but also in airlines, which are able to keep flying reliably complex jets for long periods of time on schedule, and in most cases cost effectively.

It is also to be recognised that technology will improve with time. Using the achievements of the aerospace industry as an example again, present day jet planes, with all their complexity and friendliness would have not been imagined just a few decades ago.

4.2 High energy flux on the first wall

Since the first wall will be subject to high radiation and neutron fluxes, its heat transfer capacity will play an important role in determining the size of the reactor. The materials currently under development will be able to stand 6 times as much heat as those being used today, which will mean that the size will be mostly determined by the physics, and not by the engineering (Baluc 2006).

4.3 Radiation damage on the reactor components

The first wall will be activated by the neutron flux, so the frequency at which it will be replaced and the rate of production of radioactive waste will depend on the material used. This, as well as the issue mentioned in the past section, makes material research a crucial activity for a successful fusion reactor design.

It is estimated that, using reduced activation ferritic steel, the radioactivity produced 10 years after shutdown will be above 10^{-2} Curies/Watt of thermal power produced in the life time of the reactor. Although this would be only slightly smaller than for a conventional light water fission reactor, the wastes are short lived, so 100 years after the shutdown, the radioactivity of the fusion reactor would be reduced to 10^{-6} Curies/Watt, while that of a fission reactor would have decayed only an order of magnitude. Yet, if vanadium alloys were used instead, the initial radioactivity would be 10^{-6} and 10^{-8} Curies/Watt, 10 years and 100 years after shutdown, respectively. Even if the first wall needs to be removed frequently, it will stand the radiation, and shield the more costly equipment, such as the vacuum chamber and the superconducting coils, which can be designed to last the lifetime of the reactor.

4.4 Economic competitivity

With the high uncertainties that prevail, since there is still much research to be done, it is not easy to estimate in a reliable way the cost of electricity. However, extensive work on future reactors has been done (see for instance the ARIES programme, Najmabai et al. (1997) and the ARIES web site), which gives a first approximation. As research advances, the costs will probably be reduced. According to present day projections, the cost of electricity in a nuclear fusion reactor would be 25 % greater than for a coal fired plant, 50 % greater than for a boiling water fission reactor, and 100 % greater than for a natural gas plant. Yet, these estimates do not take into account the impact produced by CO_2 or the conventional disposal of radioactive waste.

4.5 Divertor design

This is presently recognised as the major problem to be solved (Rebut 2006), since there is no ideal material. The poloidal divertor is regarded as an essential component in conventional tokamaks. It rests in a separate chamber connected to the main plasma chamber, and its main role is to produce an open magnetic field which allows the collection of alpha particles, once they have given up their energy, and divert higher atomic number impurities, which may result from the sputtering of low temperature plasma on the walls. In the case of ITER, operating at $Q = 10$, 4 m^2 of divertor plates will be subject to 30 MW when edge localized modes occur; enough to melt or evaporate droplets off the plate, if they are made of copper, for instance. These droplets would capture tritium, and redeposit as dust on the plates, creating a radioactive inventory problem.

Interesting solutions are being proposed to solve this problem, such as the use of liquid lithium limiters, with an external divertor, i.e.: not connected to the plasma chamber. Proof of principle of liquid lithium limiters have been made in a simulator at Argonne (IMPACT facility) and in the CDX-U tokamak at Princeton (Nieto et al. 2006). An active limiter has been tested in the T-11M tokamak (Mirnov et al. 2006). More recently, work in this direction has been started in the Frascatti Tokamak (FT) in Italy. Although the results are promising it remains to be seen if either ITER or IGNITOR will test the concept in a burning plasma machine.

5 The hybrid fission-fusion option

As mentioned earlier, ITER is expected to operate in the $Q = 10$ region, and even if IGNITOR is built and successfully operated, it would still be short from what would be required for a reactor, since it is designed on a science first basis, and it is not its mission to address many of the important engineering problems. Its goal is mainly to understand burning plasmas, and the path to ignition. On the other hand, burning plasmas are so far an uncharted territory, which means there are issues we shall ignore until the experiments are built.

For a pure fusion reactor, working on deuterium-tritium, at least $Q = 50$ would be necessary, since part of the energy would be lost when converting the neutron energy into thermal energy, and there would be further losses in the conversion to electrical energy (Rebut 2006). Thus, as it stands, it is uncertain if a pure fusion reactor will be possible within a foreseeable future.

Yet, fission-fusion scenarios look very promising (Manheimer 2001, 2004, 2006; Rebut 2006), although they must be approached with care. In this scheme, fusion could help fission in two different ways: as a breeder for fissile fuel and as a source of fast neutrons for nuclear waste transmutation.

As a breeder, neutrons from a low gain fusion reactor can multiply in the mantle, in order to turn non-fissile into fissile material, using either ^{238}U in the form

$$n + {}^{238}U \longrightarrow {}^{239}U \longrightarrow {}^{239}Np + e-$$

$$^{239}Np \longrightarrow {}^{239}Pu + e-,$$

or ^{232}Th in the form

$$n + {}^{232}Th \longrightarrow {}^{233}Th \longrightarrow {}^{233}Pa + e-$$

$$^{233}Pa \longrightarrow {}^{233}U + e- \,.$$

While there may be serious objections to the former, due to the possibility of using ^{239}Pu for weapons' proliferation, the latter might be more acceptable.

Transmutation is the process in which neutrons are used to degrade long-lived nuclear waste ($\sim 10^4$ years) into short lived nuclear waste (~ 100 years). There are four possible neutron sources for transmuting nuclear waste: thermal neutrons form LWRs (Light Water Reactors, 6×10^{-3} n/MeV), fast neutrons from ALMRs (Advanced Liquid Metal Reactors, 7.5×10^{-3} n/MeV), accelerators (1.5×10^{-2} n/MeV) and fusion reactors (4.5×10^{-2} n/MeV). Furthermore, a fusion reactor can provide either fast or thermal neutrons, as required. In the latter case a moderator could be added in the blanket, including a neutron breeder, such as lead or beryllium. However, the fusion reactors should be used for transmuting long lived species rather than for burning plutonium, which would be extremely dangerous in a fusion reactor, in the event of plasma disruptions. The idea would be to burn plutonium in conventional reactors, not mixed with uranium, but with lithium, so the neutrons can contribute to breeding tritium for the fusion reactors.

There is a wide spectrum of fission-fusion scenarios, which range from a hybrid reactor, and would lead to very complex designs, down to ITER technology, in which fission and fusion reactors could work separately in a symbiotic way. One can even envisage a nuclear park in which no enriched fuel would be brought in, and the nuclear waste could be transmuted and treated within the park.

6 A strategy for a Mexican fusion programme

Up to date, there has been no participation from Mexico in the world fusion programme, since serious fusion research requires an investment which is beyond the scale of science and technology spending in the country. On the other hand the uncertainty in the future availability of fusion reactors makes it difficult to justify a greater spending in fusion, in a country which needs to catch up fast with other more "down to Earth" technologies, including fourth generation fission reactors.

However, it is necessary for Mexico to recognise the importance of this alternative energy source, just as China, India, South Korea, all of which have brand new, state of the art, superconducting tokamaks, have already done, if only to benefit from the spin-offs of the emerging technology. To a lesser degree, the only Latin American country with significant contributions in the field is Brazil.

The high energy physics Mexican community has shown that it is possible to get involved in state of the art science with the scale of spending of Mexican science. A good example may be collaboration in data processing. Just as the Large Hadron Collider (LHC), ITER will produce a considerable amount of data which will need to be processed in a short timescale. The Mexican high energy physics community is setting the path and the infrastructure for this kind of collaborations for LHC, which will start operating in 2007. Since the first plasma in ITER is not expected until 2016, let alone the first burning plasma, there is time to build up a reasonable fusion community which should learn to collaborate with researchers in the main laboratories. A good example of this kind of research has been put forward by Portuguese researchers, who have developed diagnostics for JET and follow the experiment, sited in Britain, in real time. This has been an important test bed for future collaborative activities (Varandas 2006).

From the theoretical point of view, the magnetically confined plasma is seen as a complex system in which coupled phenomena at different space and time scales need to be taken into account. Therefore, better computing techniques and infrastructure need to be developed (Batchelor 2005). Mexico is certainly in a good position to contribute along these lines, with its computational infrastructure, and should undertake it as soon as possible.

Finally, although it is difficult to do significant fusion research with small experiments, if a group is to be formed which can be in a position to collaborate with the groups working in larger machines, it is important to have some experimental work in Mexico, either in fusion oriented or basic plasma science research, which can be innovative enough to produce publishable and applicable results. Such experimental facilities will be necessary in any case, in order to train people who can join experimental campaigns in the larger laboratories. The International Atomic Energy Agency (IAEA) has been particularly helpful to establish such links, but it is a necessary condition that the participants have hands-on experimental experience.

Whether it is data interpretation, theoretical-computational, or experimental work, it is clear that if a Mexican fusion programme is to be meaningful, it will have to be done in an international collaboration context, and the Mexican group must be able to have something to offer if its participation is to be welcomed.

7 Conclusion

Human technology has been sufficient to send manned missions to the moon and explore the Solar System by means of robots. Man has been able to explore the depths of the seas, build supersonic transports, long bridges and dig long tunnels, but being able to build a workable fusion reactor requires technology which goes beyond that available up to date. It is only honest to admit that fusion is beyond the present state of development. However, disciplined research in plasma and fusion science has produced significant achievements in the past 50 years, and it can be predicted that following this tradition will bring fusion as an alternative source of energy. However, although as noted by John D. Lawson, there have always been people predicting fusion reactors are just a few years away, just as fusion scientists were ignorant of their own ignorance in the late 1950s, in spite of the achievements reached, we are still ignorant about the behaviour of burning plasmas, and only building the necessary experiments will quench our ignorance.

From the engineering point of view, there is still a long way to go before a competitive fusion reactor can be built. However, for every hurdle in the way different ways of negotiating it are being devised, and the future looks bright, as long as there is budget to support the work.

Fusion can provide a real contribution for supplying the necessary energy for development, in a sustained way, in mid term, if it works in symbiosis with fission, both providing the fuel and transmuting nuclear waste. However, if this approach is to be developed safely and in an economically competitive way, it needs to be developed intelligently.

As seen in the past sections, fusion research requires determination, discipline, and is not for the light hearted. While other sources of energy, particularly the renewable ones must be harnessed in Mexico, fusion is the most challenging, as it requires state of the art technological and scientific resources. In a way, along with fission technology, it sets a milestone which differentiates the *developed* from the *developing* status for countries. Brazil, China, India and South Korea, recognising the need for energy sources in a scale necessary to sustain development, have already taken necessary steps for moving into the developed world. It is up to Mexico to take the challenge, or to remain in a comfortable *developing* status forever.

References

ARIES (2006) http://aries.ucsd.edu

Baluc N, (2006) Materials for fusion power reactors. Plasma Phys and Controlled Fusion 48:B165–B177

Batchelor DB (2005) Integrated simulation of fusion plasmas. Phys Today (Feb. 2005):35–40

Bethe H (1939) Energy production in stars. Phys Rev 55:434–456

Bombarda et al. (2004) Ignitor: Physics and progress towards ignition. Brazilian J Phys 34 (4B):1786–1791

Bromberg JL (1982) Fusion: Science, Politics and the Invention of a New Energy Source. MIT Press, Cambridge, Mass

Coppi B et al. (2001) Optimal regimes for ignition and the Ignitor experiment. Nuclear Fusion 41:1253–1257

Haines M, (1996) Fifty years of controlled fusion research. Plasma Phys. and Controlled Fusion 38:643–656

Horton W et al. (2002) Ignitor physics assessment and confinement projections. Nuclear Fusion 42:169–179

International Fusion Research Council (2005) Status report on fusion research. Nuclear Fusion 45:A1–A28

ITER Physics Basis Editors (1999) ITER Physics Expert Group Chairs and Co-Chairs and ITER Joint Central Team and Physics Integration Unit ITER Physics Basis Editors, ITER Physics Expert Group Chairs and Co-Chairs and ITER Joint Central Team and Physics Integration Unit. ITER Physics Basis. Nuclear Fusion 39:2137–2638

ITER (2006) http://www.iter.org

Lawson JD (1957) Some criteria for a power producing thermonuclear reactor. Poc Phys Soc B 70:6–10

Manheimer W, (2001) An alternate development path for magnetic fusion J. Fusion Energy 20:131–134

Manheimer W (2004) The fusion hybrid as a key to sustainable development J. Fusion Energy 23:223–235

Manheimer W (2006) Can fusion and fission breeding help civilization survive? J Fusion Energy 25:121–139

Mirnov SV et al. (2006) Perspectives of the lithium capillary-pore system application to fusion: Experiments with lithium limiter on T-11M tokamak In Plasma and Fusion Science: Proc. of the 16th IAEA Technical Meeting on Research using Small Fusion Devices, AIP Conference Proceedings Series 875:83–88

Najmabai F et al. (1997) Overview of the ARIES-RS reversed-shear tokamak power plant study. Fusion Eng Design 38:3–25

Nieto M et al. (2006) Plasma material interaction studies on lithium and lithiated substrates during compact tokamak operation. In: Plasma and Fusion Science: Proceedings of the 16th IAEA Technical Meeting on Research using Small Fusion Devices, AIP Conference Proceedings Series 875:78–82

Pamela J et al. (2005) Overview of JET results. Nuclear Fusion 45:63–85

Rebut P (2006) From JET to the reactor. Plasma Phys and Controlled Fusion 48:B1–B13

Rose B (1958) Meaurement of the neutron spectrum from ZETA. Nature 181:1630–1632

Thonemann PC et al. (1958) Production of High Temperature and Nuclear Reactions in a Gas Discharge. Nature 181:217

Varandas C et al. (2006) Real-time plasma control tools for advanced tokamak operation. In: Plasma and Fusion Science: Proceedings of the XI Latin American Workshop on Plasma Physics, AIP Conference Proceedings Series 875:385–390

Wesson J (2004) Tokamaks. Oxford University Press